Geodynamics of Iceland and the North Atlantic Area

NATO ADVANCED STUDY INSTITUTES SERIES

Proceedings of the Advanced Study Institute Programme, which aims
at the dissemination of advanced knowledge and
the formation of contacts among scientists from different countries

The series is published by an international board of publishers in conjunction
with NATO Scientific Affairs Division

A	Life Sciences	Plenum Publishing Corporation
B	Physics	London and New York
C	Mathematical and	D. Reidel Publishing Company
	Physical Sciences	Dordrecht and Boston
D	Behavioral and	Sijthoff International Publishing Company
	Social Sciences	Leiden
E	Applied Sciences	Noordhoff International Publishing
		Leiden

Series C – Mathematical and Physical Sciences

Volume 11 – Geodynamics of Iceland and the North Atlantic Area

Geodynamics of Iceland and the North Atlantic Area

Proceedings of the NATO Advanced Study Institute
held in Reykjavik, Iceland, 1–7 July, 1974

edited by

L. KRISTJANSSON
Science Institute, Reykjavik, Iceland

D. Reidel Publishing Company
Dordrecht-Holland / Boston-U.S.A.

Published in cooperation with NATO Scientific Affairs Division

Library of Congress Catalog Card Number 74–27848

ISBN-13:978-94-010-2273-6   e-ISBN-13:978-94-010-2271-2
DOI: 10.1007/978-94-010-2271-2

Published by D. Reidel Publishing Company
P.O. Box 17, Dordrecht, Holland

Sold and distributed in the U.S.A., Canada, and Mexico
by D. Reidel Publishing Company, Inc.
306 Dartmouth Street, Boston, Mass. 02116, U.S.A.

CONTENTS

PREFACE

During the revolution in earth science that has taken place in recent years, studies of the North Atlantic ocean floor and of Iceland have played an increasingly significant role. Icelandic geoscientists have followed, and taken part in, these studies with a keen interest; one of the first tasks of the Geoscience Society of Iceland was to organize an Icelandic symposium on "Iceland and Mid-Ocean Ridges" in 1967.

At the suggestion of Dr. G. Pálmason, the Society and various local research institutions formed in 1972 an Organizing Committee for an international meeting on earth science. It was decided that it should be devoted to examining the various expressions of geodynamical forces in the North Atlantic area, in particular at the ocean ridges passing through Iceland. Apart from the scientific content of such a meeting, the organizers also felt it was highly important for scientists from both sides of the Atlantic to meet in Iceland, to become acquainted with recent progress in earth science research there and to coordinate their future research projects in the area.

The meeting was held in Reykjavik 1 - 7 July, 1974, and was followed by field trips in Iceland. Generous financial support from the NATO Scientific Affairs Division, the Inter-Union Commission on Geodynamics, and many other sources, is gratefully acknowledged.

In addition to the papers included in this volume, invited review lectures on respectively the evolution of the Norwegian Sea and on rare earth geochemistry were given at the meeting. Manuscripts, as received in July-September, have been retyped by Mrs. Unnur Halldorsdottir on a standard format, but minor variations in spelling and notation between these have been left unchanged.

Various short research papers and progress reports presented at this meeting will be published elsewhere. No resolutions or recommendations were passed, but the reader is referred to the extensive recommendations for research on continental and oceanic rifts published in Report no. 5 of the Inter-Union Commission on Geodynamics (Paris, 1974).

Reykjavik, 5 October 1974.   Leo Kristjansson

Official organizers: Geoscience Society of Iceland

Organizing Committee members:

    S. Björnsson   Science Institute
    L. Kristjansson   University of Iceland
    S. Steinthorsson
    S. Thorarinsson

    G. Palmason   National Energy Authority
    K. Saemundsson

    R. Hugason   National Research Council
    V. Ludviksson

    S. Jakobsson   Museum of Natural History

    K. Grönvold   Nordic Volcanological Institute

This volume is Scientific Report No. 12 of the Geodynamics
Project.

The Geodynamics Project is an international programme of research on the dynamics
and dynamic history of the Earth with emphasis
on deep-seated foundations of geological phenomena.
This includes investigations related to movements
and deformations, past and present, of the
lithosphere, and all relevant properties of the
Earth's interior and especially any evidence for
motions at depth. The programme is an inter-
disciplinary one, coordinated by the Inter-Union
Commission on Geodynamics (I.C.G.) established
by the I.C.S.U. at the request of I.U.C.G. and
I.U.G.S., with rules providing for the active
participation of all interested I.C.S.U. Unions
and Committees.

CONTINENTAL DRIFT IN THE ATLANTIC AND THE ARCTIC

W.C. Pitman III and E.M. Herron

Lamont-Doherty Geological Observatory
Palisades, New York 10964, U.S.A.

This report summarizes investigations of the history of drift in the Atlantic by Pitman and Talwani [1] and in the Arctic by Herron, Dewey and Pitman [2].

The magnetic lineation pattern in the Atlantic is symptomatic of sea floor spreading and hence of continental drift. Crustal accretion takes place at the axis of mid oceanic ridge in a very narrow zone. The plates separate along this line – new material that wells into the void divides lengthwise down its middle and roughly equal strips of this material became attached to the trailing edge of each of the two plates.

The magnetic anomalies reflect the fact that the basaltic material which is intruded at the ridge axis acquires a permanent magnetization in the direction of the main field at the time the newly intruded material cooled through its Curie point. Because the earth's magnetic field reverses polarity in a non-periodic manner the polarity history of the earth's magnetic field forms a unique stratigraphic sequence. Since the crustal accretion process is nearly continuous, the pattern of magnetic anomalies, bilaterally symmetric about the ridge axis forms a record of the earth's polarity history. The spatial sequence of magnetic anomalies on the ocean floor may be correlated with the temporal sequence of the reversals of earth's magnetic polarity and thus the age of the oceanic crust may be determined. Each lineation can be regarded as an isochron and as the lineations are bilaterally symmetric, the pattern of magnetic lineations may be viewed also as a pattern of isochrons bilaterally symmetric with respect to the ridge axis and becoming systematically older away from the ridge axis.

Kristjansson (ed.), Geodynamics of Iceland and the North Atlantic Area. 1-15. *All Rights Reserved.*
Copyright © 1974 *by D. Reidel Publishing Company, Dordrecht-Holland.*

Since the lithospheric plates are internally torsionally
rigid, each magnetic lineation within an oceanic plate should
duplicate the shape of the ridge axis at the time of its formation.
Its bilaterally symmetric counterpart should similarly duplicate
the shape of the ridge axis at the time of formation. In the
Atlantic the process of crustal accretion described above has been
the cause of continental drift. The continents that surround the
Atlantic have been passively rafted away from each other as separation and crustal accretion occurred at the mid Atlantic Ridge
axis. Then following the arguments presented above in order to
obtain the relative position of, say, Europe and North America
for the -65 m.y., one needs only to superimpose magnetic lineations
of this age from opposite sides of the intervening ridge axis.
Reconstructions using this technique have been done in a number
of oceans and in particular for the North Atlantic by Pitman and
Talwani [1].

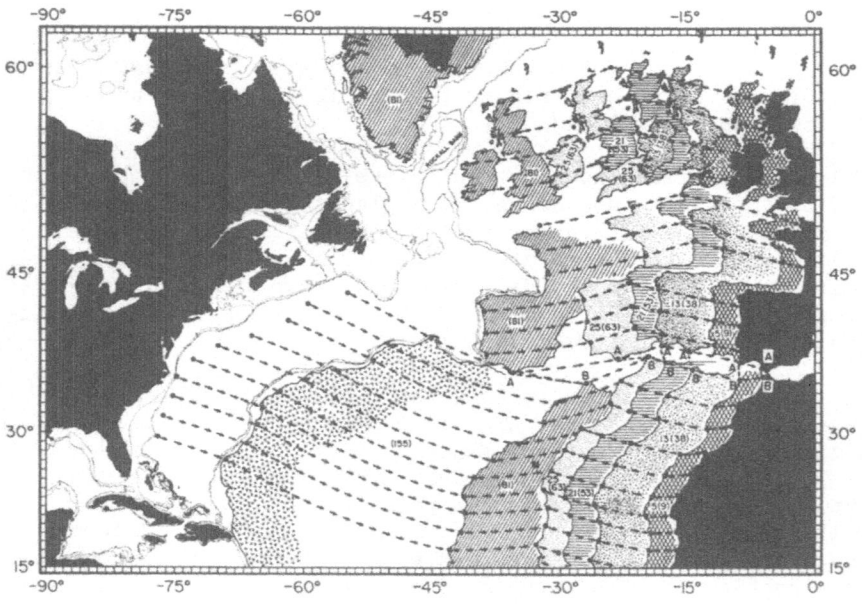

Fig. 1. (Adapted from ref.1). The relative positions of Europe and
Africa with respect to North America for specific times are shown.
Blacked-in continents represent present positions. Dates for
earlier positions are indicated in the map. Greenland is shown
at -81 m.y. and in its present position. At -81 m.y., Greenland is
assumed to have been attached to Europe according to the reconstruction of Bullard and others [3]. The arrows show the path of
drift of Africa and Europe away from North America.

The results of their analysis are shown in Fig. 1 giving
the relative positions of Africa, Europe and North America for
several consecutive instants in time beginning at -180 m.y.
The magnetic anomalies that lie within that portion of the Atlantic
between the Azores and about 57°N (where the Reykjanes Ridge bends
off to the east) are symptomatic of drift between Europe and North
America: north of 57°N, the magnetic lineation pattern reflects
not only the pattern of drift between North America and Eurasia
but also the movement of Greenland relative to these two. Pitman
and Talwani did not attempt to reconstruct a drift history in the
Labrador and Greenland-Norwegian Sea as the data for such an analysis
were unavailable. But sufficient information was known to at least
provide constraints on the movement of Greenland. The first
constraint was that the oldest anomaly identified in the Norwegian-
Greenland Sea which lay adjacent to the continental margin on the
Greenland side and the Norwegian side was the negative anomaly
between 24 and 25 (-63 m.y. in age). Continental drift in the
Norwegian-Greenland Sea area thus did not begin until -63 m.y.
A second constraint was that, from the magnetic anomaly pattern
in the Labrador Sea, it could be inferred that spreading at least
slowed drastically at about -50 m.y. It is now known (Kristoffersen,
personal communication) that spreading in the Labrador Sea finally
ceased at about -38 m.y. However, on the basis of anomalies identified to the south between Europe and North America [1,4] drift
between North America and Europe began about 81 m.y. ago. From
these data the following stage by stage interpretation could be
made for the evolution of the North Atlantic.

At about -81 m.y., Europe and North America began drifting
apart along a line that extended from between the Iberian peninsula and Newfoundland northward through the Labrador Sea to the
Arctic region. At this time, Greenland drifted away from North
America as part of Eurasia. Opening continued exclusively along
this Atlantic-Labrador rift system until about -63 m.y. when
rifting began in the Norwegian Sea. From -63 m.y. to about -50 m.y.
opening occurred simultaneously in both the Greenland-Norwegian
Sea and in the Labrador Sea. At about -50 m.y., spreading slowed
considerably in the Labrador Sea but continued in a very subdued
fashion until -38 m.y. Thus Greenland drifted as a part of Europe
from -81 to -63 m.y.; as a separate plate between -63 m.y. and
-38 m.y.; and as a portion of the North American plate since.

Pitman and Talwani [1] further considered the sequence
of paleogeographic reconstruction in terms of poles and angles
of relative motion. For example, in a reference frame held fixed
with respect to North American they computed a pole and angle
necessary to translate Eurasia from one paleoposition to the next.
They made similar computations for a reference frame fixed with
respect to Eurasia.

TABLE 1. Finite Difference Rotations (from [1]).

Between anomaly numbers (age in parentheses)	Time span m.y.	Coordinates of pole of rotation				Finite rotation (degrees)	Rate of rotation assuming pole remains fixed 10^{-7} dg/yr
		Reference frame fixed to North America		Reference frame fixed to Eurasia			
		Lat	Long	Lat	Long		
A. North of Azores							
Axis (0) and 5 (9)	9	68.0 N.	137.0 E.	68.0 N.	137.0 E.	2.50	2.78
5 (9) and 13(38)	29	63.5 N.	131.1 E.	63.5 N.	131.6 E.	5.11	1.76
13(38) and 21(53)	15	27.6 N.	155.7 E.	29.0 N.	161.1 E.	2.79	1.84
21(53) and 25(63)	10	67.3 N.	152.0 W.	72.3 N.	150.1 W.	4.55	4.55
25(63) and 31(72)	9	76.0 N.	49.8 W.	76.8 N.	11.1 W.	7.70	8.56
31(72) and (81)	9	76.0 N.	49.8 W.	76.8 N.	11.1 W.	7.70	8.56
(81) and		70.9 N.	20.4 W.	70.4 N.	15.1 E.	11.19	

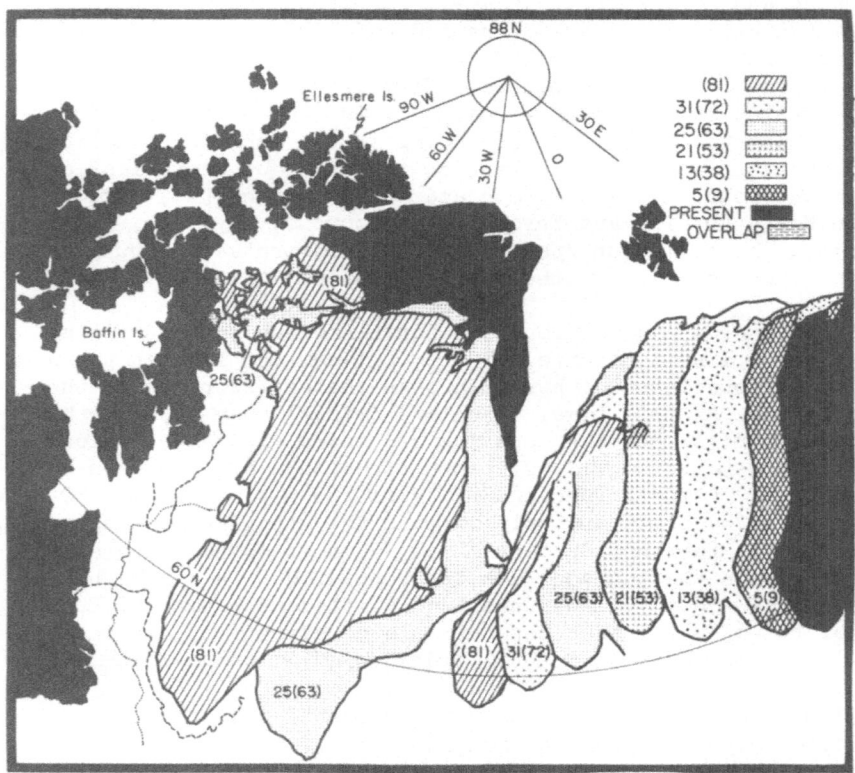

Fig. 2. The same paleographic arrangements as given in Fig. 1 are shown. Here the projection is azimuthal equidistant.

The poles and angular rates are given in Table 1. The connecting series of arcs drawn using the tabulated sequence of poles and angles compared well with the shape of well surveyed fracture zones. As a consequence they concluded that the poles and angular rates thus derived could be regarded as approximating the path and rate of drift of the plates as they separated and that position of the pole is often approximately stationary for some millions of years. This may be so for the following reasons. As is well known transform faults follow small circles about the instantaneous pole of relative motion [5]. If a ridge system has lengthy transform offsets that are both right and left lateral, then if the instantaneous pole changes position it will be necessary to form some new transforms by shearing through old rigid oceanic crust [6]. For this reason, this type of ridge-transform fault system may tend to stabilize, with the instantaneous pole remaining within a small region for long periods of time or drifting slowly with time so as to minimize the consequent alteration of the ridge-transform fault pattern. (This is similar to arguments used by Le Pichon and Hayes [7] in discussing the pattern of early rifting in the South Atlantic). The general concept of the stability of poles of relative motion is supported by the fact that very long segments of fracture zones found in the Atlantic, Pacific and Indian Ocean may be approximated by arcs drawn about a single pole [5,8]. Thus the finite difference poles computed by Pitman and Talwani* may in fact approximate the real relative motion that has occurred between plates. Although poles may migrate with time we suggest that drastic or sudden large changes in the poles' position probably occur only when there is a corresponding reorganization of the plates involved. Pitman and Talwani computed numerically precise poles of relative motion. They recognized that due to: 1) the incomplete nature of the data, 2) the fact that large time intervals were considered and 3) the probability that poles migrate with time, the numerical precision meant little but rather each of these positions defined a region within which the pole may have moved about during the time interval. With these generalities in mind we may now consider the evolution of the North Atlantic and its northern extension the Arctic Ocean in terms of three different sequential arrangements of plates and different patterns of relative motion.

The first occured when Europe began to separate from North America during the period from -81 m.y. to -63 m.y. and Greenland drifted as a part of the European plate. During this interval the pole of relative motion of Eurasia with respect to North America lay in northern Greenland (Table 1 and Figs. 2 and 3). The significance of this pattern of early motion is shown in Fig. 3. It means that as the Eurasian plate was rotating away from the North American plate in the Atlantic, it must have been rotating away from the North American plate in the Arctic (and of course toward the Pacific). As the pie shaped wedge of new ocean was created by accretion in the North Atlantic a corresponding amount of compression

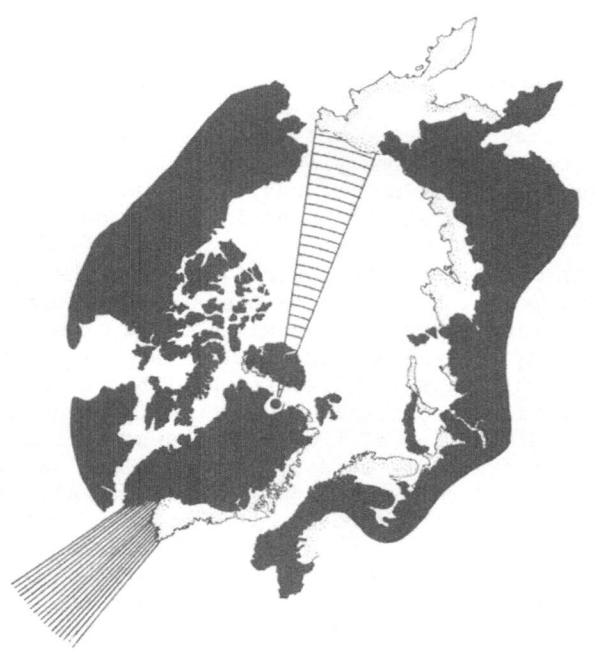

Fig. 3. (Azimuthal equidistant projection). The black gives the
relative positions of Europe, Greenland and North America at
-81 m.y. The dotted border gives the position of Eurasia and
Greenland relative to North America at -63 m.y. The locus of
the rotational pole is given by the black dot in Northern Greenland.

must have occurred on the Arctic-Pacific side of the pole (Fig. 3).

      The second stage occurred between -63 m.y. and -38 m.y.
At -63 m.y. there was a major reorganization of plate boundaries
as spreading was initated in the Norwegian-Greenland Sea but
continued in the Labrador. Also at this time the locus of poles
of relative motion moved from northern Greenland to the north-
eastern Pacific. Thus the relative motion of Eurasia with respect
to North America was extensional not only in the North Atlantic
but all across the Arctic Ocean. An interesting observation made
by Pitman and Talwani is that the reconstruction of North Atlantic
at anomaly 25 (-63 m.y.) results in the complete closure of the
Eurasian basin between the Lomonosov Ridge and the Eurasian coast
(see Fig. 4). That is, if the Lomonosov Ridge is considered a part
of the North America to its relative position at -63 m.y. it is

found that the Lomonosov Ridge fits nicely against the Eurasian continental shelf edge (Fig. 4), Pitman and Talwani concluded that all of the spreading due to drift in the Arctic that has occurred in the past 63 m.y. has occurred in the Nansen basin causing the separation of a ridge of the continental crust, Lomonosov Ridge [9] from the Eurasian margin.

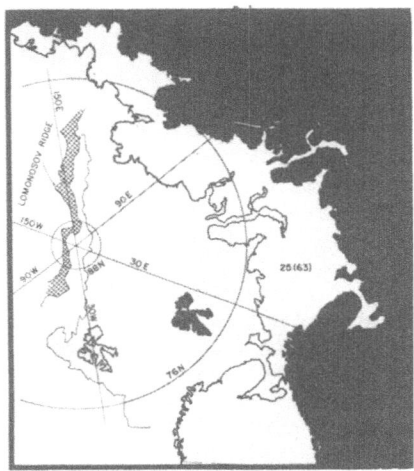

Fig. 4. (Azimuthal equidistant projection). The relative position between the Lomonosov Ridge and the Barents Shelf (as defined by the 2,000 m isobath) is shown for 63 m.y. ago. The present positions of Eurasia and Spitsbergen are shown in black. The positions of Eurasia, the Barents Shelf, and Spitsbergen for 63 m.y. are shown by stippling.

The third stage has occurred between -38 m.y. and the present. Spreading ceased entirely in the Labrador Sea at -38 m.y. and Greenland became a rigidly attached part of the North America plate. Also from that time until the present the locus of the poles has lain in the western North Pacific. And as in the previous stage, the style of relative motion of Eurasia with respect to North America has continued to be extensional in the North Atlantic and across the Arctic into Siberia. As stated previously this spreading in the Arctic occurred only in the Eurasian Basin between -63 m.y. to the present. The youth of the Eurasian basin has recently been confirmed by identification of the magnetic anomaly lineations over the Nansen basin (Fleming et al., 1974).

But the important unanswered question now is where could the implied compression have been accommodated in the Arctic during

the interval from -81 to -63 m.y. The amount of crustal contraction
necessary ranges from zero near northern Greenland to approximately
800 km on the Siberian side of the Arctic. We suggest that this
compression was accommodated by subduction beneath the Alpha-Mendeleyev Ridge system. We propose that this ridge system is in fact
a fossil island arc.

Vogt and Ostenso [10] proposed that the Alpha Cordillera
had been a spreading ridge and that the spreading which had
begun perhaps in the Upper Cretaceous ceased about 40 m.y.b.p.
Several lines of evidence may be used to argue against these
hypothesis. The first is that the available magnetic data do not
show the bilaterally symmetric pattern associated with sea floor
spreading ridges. Attempts have been made to find such a pattern
in the data and correlate various profiles with the magnetic polarity time scale of Heirtzler et al. [11]. The results of these
efforts have been unconvincing.

A second serious problem relates to the elevation of the
Alpha Cordillera. Sclater et al. [12] have shown that the
general morphology of all sea floor spreading ridges in terms of
age vs. depth may be described by a single age vs. depth curve
[13,14]. Within narrow limits very nearly all ridges regardless
of their spreading history follow this curve (shown in Fig. 5).
The two notable exceptions are the southeast Indian Ocean Ridge
which in one area is uniformly 500 meters deeper than expected.
The second area is Iceland and vicinity. Where the anomalously
high topography has been explained as the result of a hot spot
located under Iceland. In fact the entire region of the mid-
oceanic ridge system north of the Charlie-Gibbs fracture zone
to the latitude of Spitzbergen is anomalously shallow with the
anomalous apex being Iceland itself. But on the average this
region is elevated about 500 meters too high and not 2,000-
3,000 meters too high as is the axis of the Alpha-Mendeleyev
Ridge complex. Fig. 7 shows a geophysical profile across the
extinct Labrador Sea ridge. If the spreading had ceased at the
Alpha Ridge at -40 m.y. as Vogt and Ostenso suggested the axis
of this ridge should have subsided to a depth of about 4,700 meters
(Fig. 5). But as can be seen the axis is at about 2,500 meters.
For comparison we can see that the axis of the Labrador Sea ridge
which ceased spreading at -38 m.y. has subsided to the expected
depth. Furthermore, whatever the origin of the Alpha Ridge, the
presence of Upper Cretaceous silicoflagellates [15] in a sediment
core taken from the crest of the Alpha Cordillera suggest a pre-
Cenozoic age. In this case if sea floor spreading were the origin
the axis should have subsided to greater than 5,500 meters. Again
Iceland itself is 3,000 meters above the predicted depth but this
is but one localized spot in the entire mid oceanic ridge system
whereas the Alpha-Mendeleyev Ridge system extends over 2,000 km
in length. Although we can not conclusively reject the idea that

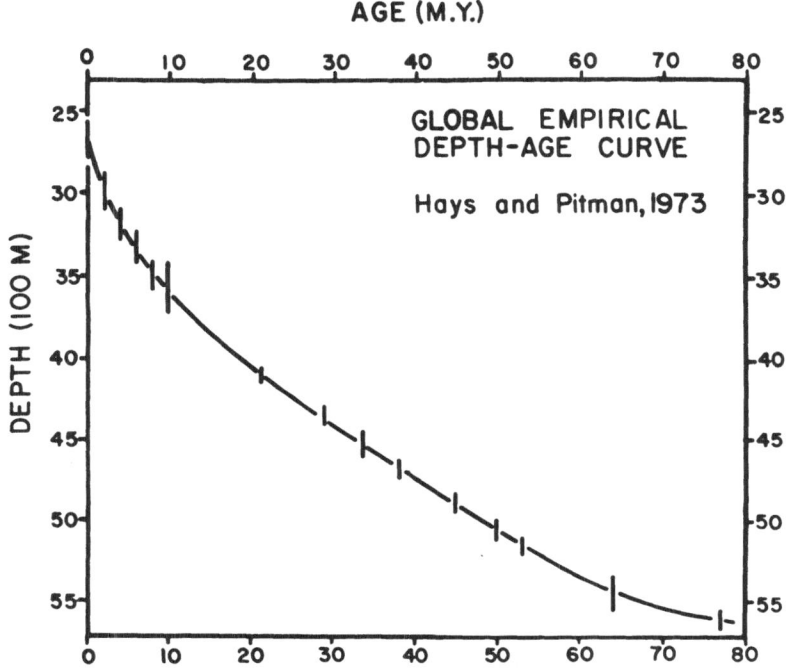

Fig. 5. (Adapted from [16]). Age vs depth for oceanic crust, constructed from the data of [12], is given. The data were taken from various regions along the Mid-Oceanic Ridge System and have been averaged. The vertical bar indicates the standard deviations of the averages.

this feature is an extinct spreading ridge we prefer the alternative suggestion that this is a fossil islands arc. The silico-flagelles mentioned above were associated with a yellow tuffaceous sediment which Clark [17] interpreted as supporting an island arc origin. It is our suggestion that the compressive motion that occurred between North America and Eurasia in the Upper Cretaceous (-81 to -63 m.y.) was accommodated by subduction. The Alpha-Mendeleyev Ridge is an island arc behind this subduction zone. Island arcs may be approximately regarded as zones where crustal thickness is doubled that of the adjacent ocean basin. Because there is no associated thermally maintained low density crust-mantle reservoir, they would not tend to subside very rapidly with time. In contrast the elevation of mid ocean ridges is mostly thermal in origin [18] and hence the ridge subsides rapidly as

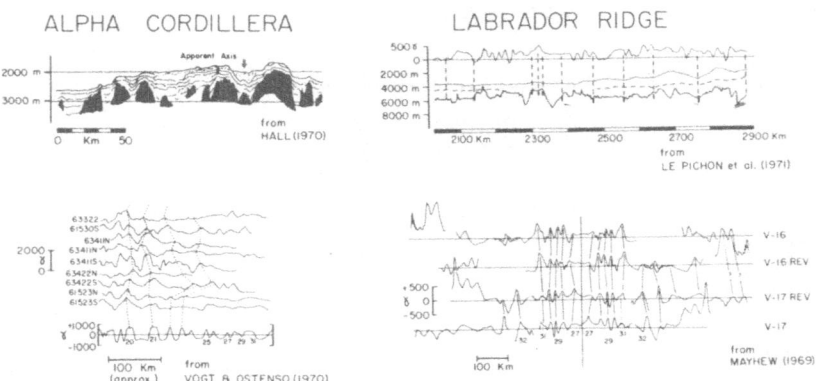

Fig. 6. Bathymetric and magnetic profiles across the Alpha Ridge
and the Labrador Sea Ridge; both features have been interpreted as
fossil spreading centers which were active during the Early Cenozoic. We agree with this interpretation of the Labrador Sea Ridge
but suggest that the Alpha Ridge is more probably a fossil island
arc.

the crust cools.

In this study of the North Atlantic and Arctic there still
remains the question of the origin of the Amerasian Basin. If
it was not created during the Meso-Cenozoic by spreading at the
Alpha-Mendeleyev Ridge system how then was it generated? Before
presenting our model we wish to review briefly several geologic
observations which are particularly relevant to the problem (see
Fig. 7). First, North America consists basically of a central
craton surrounded at its borders by orogenic belts of Paleozoic or
younger age. The most recent of these, the Laramide, reached a
climax in Upper Cretaceous to Eocene times. Each of these orogenic
belts represents a fossil (or still active) compressional plate
boundary. In several instances (for example the Caledonide orogeny)
actual plate collision occurred. Eurasia is also bordered by orogenic
belts but in contrast to North America it also has two such features
which traverse the continent internally. The Ural Mountains mark
a former compressional boundary between the Siberian and Russian
cratons. These two cratons were sutured together along the Urals
during the Permian orogeny [14].

The Verkhoyansk Mountains separate the Kolymski Massif from
the Siberian craton [20] and this belt also represents a former

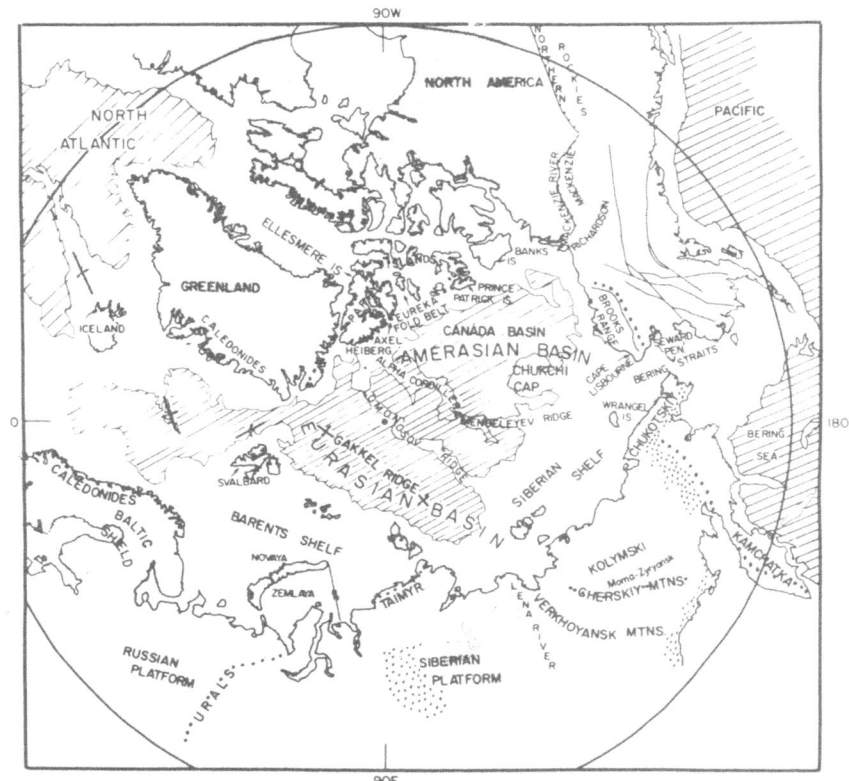

Fig. 7. (Adapted from [2]). Location map showing the major geographic and geologic features of the Arctic. Shield areas on the continents are stippled; major basalt fields are indicated by the V pattern; ophiolite localities are shown by the large dotted pattern. In the ocean, areas underlain by oceanic crust have been hachured and axes of presently active spreading centers are shown by the heavy solid lines with arrows leading away from the line.

compressional boundary and is in fact a suture indicative of a continent-continent collision. The major phase of deformation of the Verkhoyansk Mountains occurred in the Upper Jurassic-Lower Cretaceous [21,20]. We will propose that a collision occurred at that time between the Kolymski Plate and the Siberian Plate.

Our Hypothesis begins in the Lower Paleozoic when what we

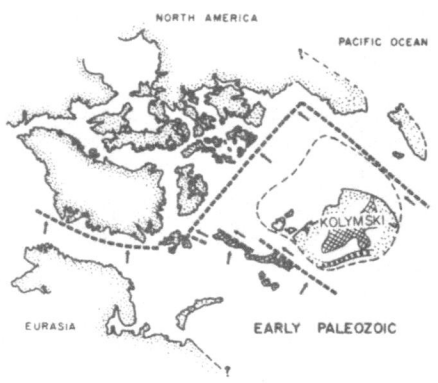

Fig. 8. A schematic model for the early history of the Amerasian
Basin. Major plate boundaries and the direction of movement relative
to North America are shown by the heavy dashed lines and arrows.
Fig. 8a illustrates the closure of the proto-Amerasian Basin in
the Early Paleozoic. This motion culminated with the Mid-Paleozoic
folding of the Franklinian geoclinal sediments and the shedding of
clastics onto North America and the Chukotskiy Peninsula. Fig. 8b
outlines our model for the opening of the Amerasian Basin during the
Jurassic magnetic quiet period as Kolymski broke away from North
America.

shall call the Kolymski Plate (see Figs. 7 and 8a) drifted into
North America generating the Parry Island belt during the Upper
Devonian to Lower Mississippian [22]. Highlands generated by this
orogeny provided the source for the clastic quartzose sediments
that were shed southward onto the Canadian Arctic Islands, the area
of Brooks range in Alaska as well as onto the Chukotskiy Peninsula [20] (Fig. 5a). In the Lower Jurassic rifting occurred and
the Kolymski Plate separated from North America drifting across
the Arctic to collide gently with Eurasia along the line of the
Verkhoyansk Mountains (Fig. 8b). Rifting occurred about a triple
junction. One arm failed and evidence for this failed arm is found
today in the graben that lies beneath the Mackenzie delta and River.
We have assigned most of this latter phase of drift to the Lower
Jurassic because during this interval that the earth's magnetic
field appears to have reversed polarity infrequently if at all, and
this constant polarity explains the absence of magnetic anomaly
lineations in the Canada basin [14]. Collision with Siberia and
folding in the Verkhoyansk occurred in the Upper Jurassic-Lower
Cretaceous. The absence of volcanics in the Verkhoyansk indicates,
however that Kolymski gave the Siberian shelf structures only a
glancing blow.

The next event of importance occurred when the North Atlantic
began to open at -81 m.y. From -81 to -63 m.y. the style of motion
in the Arctic was compressional and this compression was taken up
by subduction which formed the Alpha-Mendeleyev Ridge system as an
island arc. We further propose that there may have been an additional plate which we call the Laramide Plate (Fig. 9). Thrusting
between the Laramide Plate and the North American caused the major
uplift and crustal shortening of the Canadian or Northern Rockies
[23] and right lateral shear in the Richardson Mountains [24].
At -63 m.y. the locus of pole of relative motion between Eurasia
and North America shifted across the Arctic to the Pacific side
where it has remained to the present. Because of this shift in the
pole position spreading again occurred in the Arctic but this time
forming the Eurasian Basin.

Also at -63 m.y. underthrusting at the Alpha-Mendeleyev Ridge
system ceased. Subsequently erosion and subsidence have caused the
Alpha-Mendeleyev complex to sink to its present depths. The large
nonlinear magnetic anomalies that do occur over this ridge are
probably related to the volcanics emplaced there when it was an
active island arc.

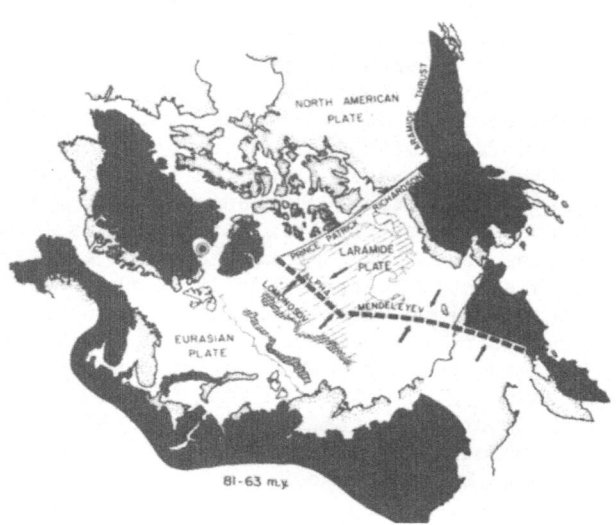

Fig. 9. Schematic model for the evolution of the Alpha-Mendeleyev
Ridge complex as an incipient subduction zone - island arc, produced
by the initial rotation of Eurasia away from North America. The
pole of opening of the North Atlantic relative to North America
as determined by Pitman and Talwani [1] for the period between 81
and 63 m.y. ago (Laramide time) is shown by the bull's eye on
northernmost Greenland. The position of Greenland and Eurasia
relative to North America at -81 m.y. is shown by the solid black,
and the position at -63 m.y. has been stippled. We have also
suggested relative motion within what is now the North American
continent, along the line of the Laramide thrust, and have further
assumed that the "Laramide" Plate included Chukotskiy and Koryak-
Kamchatka as well as most of what is now the Amerasian Basin. The
boundary between the North American and Laramide Plates is continued north of the Laramide thrusts into the Arctic via the
Richardson Mountains and the Prince Patrick uplift, which we have
interpreted as strike-slip fault zones, analogous to the San
Andreas. The Alpha Cordillera-Mendeleyev Ridge form a compressive
plate boundary between the Eurasian and Laramide Plates. Again,
areas underlain by oceanic crust in the Arctic have been hachured,
and the position of the Lomonosov Ridge at -81 and -63 m.y. is
shown by the cross-hatching.

REFERENCES

1. W. C. Pitman and M. Talwani, Geol. Soc. Am. Bull., 83, 619, 1972.
2. E. M. Herron, J. F. Dewey and W. C. Pitman III, Geology, 2, 377, 1974.
3. E. C. Bullard, J. E. Everett and A. G. Smith, Phil. Trans. R. Soc. London, A 258, 41, 1965.
4. C. Williams and D. McKenzie, Nature, 232, 168, 1971.
5. W. J. Morgan, J. Geophys. Res., 73, 1959, 1968.
6. J. W. Menard and T. A. Atwater, Nature, 219, 463, 1968.
7. X. LePichon and D. E. Hayes, J. Geophys. Res., 76, 6283, 1971.
8. X. LePichon, J. Geophys. Res., 73, 3661, 1968.
9. J. T. Wilson, Nature, 198, 925, 1963.
10. P. R. Vogt and N. A. Ostenso, Nature, 215, 810, 1967.
11. J. R. Heirtzler, G. O. Dickson, E. M. Herron, W. C. Pitman III and X. LePichon, J. Geophys. Res., 73, 2119, 1968.
12. J. G. Sclater, R. N. Anderson and M. Lee Bell, J. Geophys. Res., 76, 7888, 1971.
13. H. W. Menard, Science, 157, 923, 1967.
14. P. R. Vogt and N. A. Ostenso, J. Geophys. Res., 75, 4925, 1970.
15. H. Y. Ling, L. M. McPherson and D. L. Clark, Science, 180, 1360, 1972.
16. J. D. Hays and W. C. Pitman III, Nature, 246, 18, 1973.
17. D. L. Clark, Geology, 2, 41, 1973.
18. D. McKenzie and J. G. Sclater, Bull. Volcanol., 33, 101, 1969.
19. W. Hamilton, Geol. Soc. Am. Bull., 81, 2533, 1970.
20. M. Churkin Jr., Geol. Soc. Am. Bull., 83, 1027, 1972.
21. D. V. Nalivkin, The geology of the U.S.S.R. A short outline; Pergamon Press, N.Y., 170 p., 1960.
22. H. P. Trettin, Early Paleozoic evolution of the Canadian Arctic Archipelago, in: Arctic Geology, Am. Assoc. Petr. Geol., Memoir 19, 57, 1973.
23. D. H. Roeder, J. Geophys. Res., 78, 5005, 1973.
24. R. J. W. Douglas, Geology and economic minerals of Canada; Canadian Geol. Survey Econ. Geol. Rept., 1, 838 p., 1970.

Lamont-Doherty Geological Observatory
Contribution Number 2142

THE REYKJANES RIDGE - A SUMMARY OF GEOPHYSICAL DATA

U. Fleischer

Deutsches Hydrographisches Institut
Hamburg, F.R.G.

ABSTRACT. A review of the morphology, the deep structure, and
the history of the Reykjanes Ridge is given, using all available
geophysical data. The topographic features, the sediment distribution, and the magnetic anomaly pattern are discussed in their
relation to the Icelandic mantle plume. Then the crustal structure from different seismic and earthquake studies is evaluated
and compared with gravity models. Finally, the chronological
development is briefly outlined.

1. MORPHOLOGY

On the one hand, the Reykjanes Ridge is an arbitrary section of
the mid-ocean rift system; on the other hand, it has an excep-
tional position within it, forming a transition link between a
hot spot and a deep-sea ridge. Although there is a natural termination by topographic features, i.e. the Icelandic shelf to the
north and the Gibbs *) Fracture Zone to the south, it does not
represent a homogeneous matter. There are a main part, about
1000 km long, extending from the Reykjanes Peninsula to about
57°N latitude showing a uniform strike of 35°, and a north-south
striking southern element.

The ridge as a whole becomes successively smaller and
decreases gradually as far as the Gibbs Fracture Zone, i.e. from
the sea level down to 1500 m water depth for the crest summits,

---

*) also the "Charlie" or "Charlie Gibbs" Fracture Zone.

Kristjansson (ed.), Geodynamics of Iceland and the North Atlantic Area. 17-31. *All Rights Reserved.*
Copyright © 1974 *by D. Reidel Publishing Company, Dordrecht-Holland.*

or more than 3500 m in the Gibbs fracture valleys [1,2]. This decreasing southwards and northwards from Iceland which is due to the hot spot beneath Iceland [3,4], is a common tendency of the North Atlantic sea floor.

Connected with a common descent of the ridge to the south its topography becomes rougher and rougher. Finally there appears a pronounced rift valley like in other parts of the Mid-Atlantic Ridge [5]. Therefore, in contrast to several authors, the Reykjanes Ridge as a whole does not lack a rift valley and is not fundamentally different from other parts of the Mid-Atlantic Ridge.

The northern part of the Reykjanes Ridge, however, is relatively smooth and flat. The crestal zone and the inner flanks region both have a nearly constant width, up to 100 km off the crest [6]. Generally, we find a linearity and an axial symmetry of the mid-ocean ridge topography [7].

The southern third of the Reykjanes Ridge, i.e. the area between the bending at 57°N and the Gibbs Fracture Zone is relatively heterogeneous. It is split into about 10 crest segments, interrupted by east-west trending fracture valleys, as indicated on a seismic reflection profile [8] (Fig. 3). These could be pure transform faults so that Johnson et al. [9] assumed a constant strike of the different segments. Magnetic data of Fleischer et al. [10], however, indicate changes of the strike direction at five of these interruptions some of which amount to several tens of degrees. Bathymetric valleys found at the same positions go down to a depth of about 2500 m - to be compared with the adjoining crests near 1500 m depths.

2. MAGNETIC ANOMALIES

Magnetic anomalies over and beside the Reykjanes Ridge as far as published, are shown in Fig. 1. Most of the anomalies, in particular those from number 1 to 5, could very well be correlated. This is true also for the southern ridge part in an area where a rift valley is present [11] and is in contradiction to the supposition of Pitman and Talwani [12] that a wide-spread dike injection in those areas may obscure linear anomalies. In addition, the observed anomalies fit very well the theoretical anomalies, calculated for an alternating magnetization of 0.015 e.m.u. down to 3.5 km, according to the Heirtzler time scale with a spreading rate of 1.10 cm/yr parallel to the Gibbs Fracture Zone [10].

As commonly known, it were the pronounced linear magnetic anomalies on the northern Reykjanes Ridge [13] which confirmed the hypothesis of Vine and Matthews [14] and enabled Heirtzler et al. [15] to deduce their geomagnetic time scale. In the mean-

time such magnetic strips were detected in connection with all
mid-ocean ridges. In particular, they were verified in a very
large portion of the North Atlantic Ocean, as summarized by
Vogt and Avery [16].

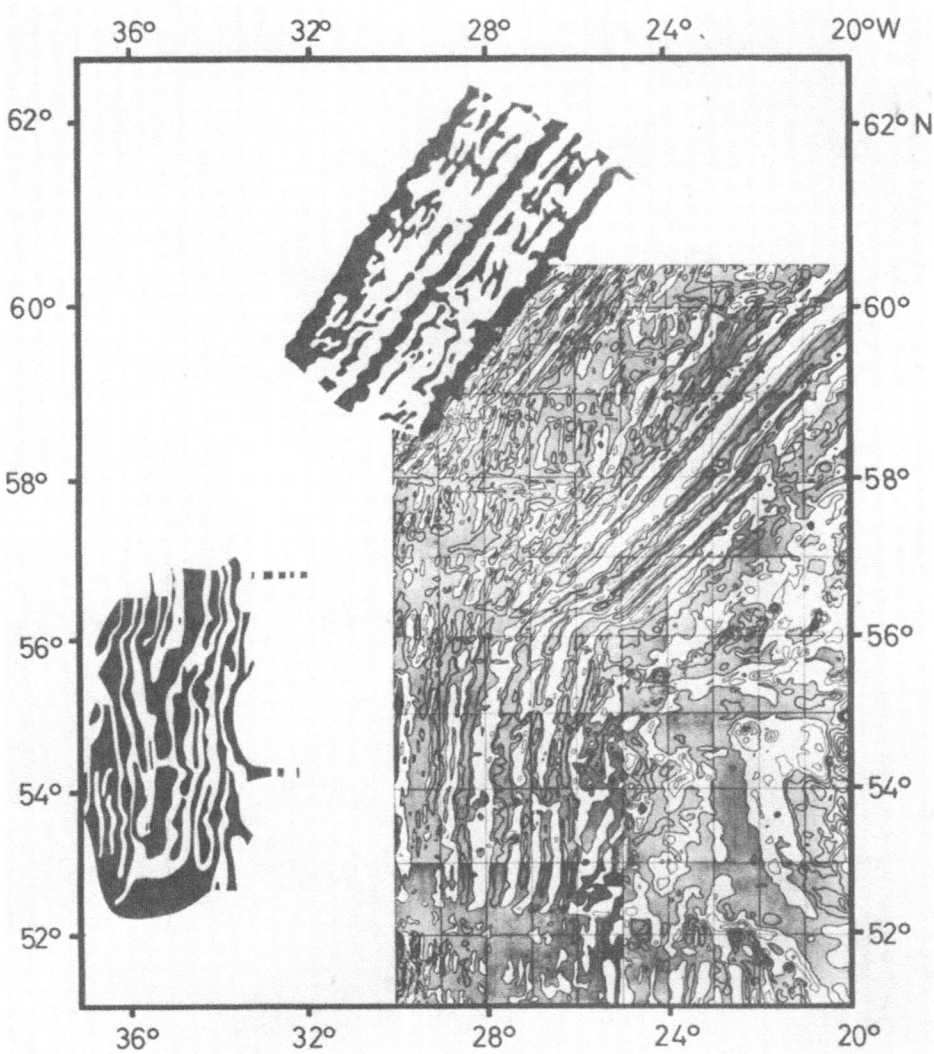

Fig. 1. Magnetic anomaly map [10,15,16] (see Fig. 8). Contour
interval 250 or 500 gammas, resp., positive areas are shaded.

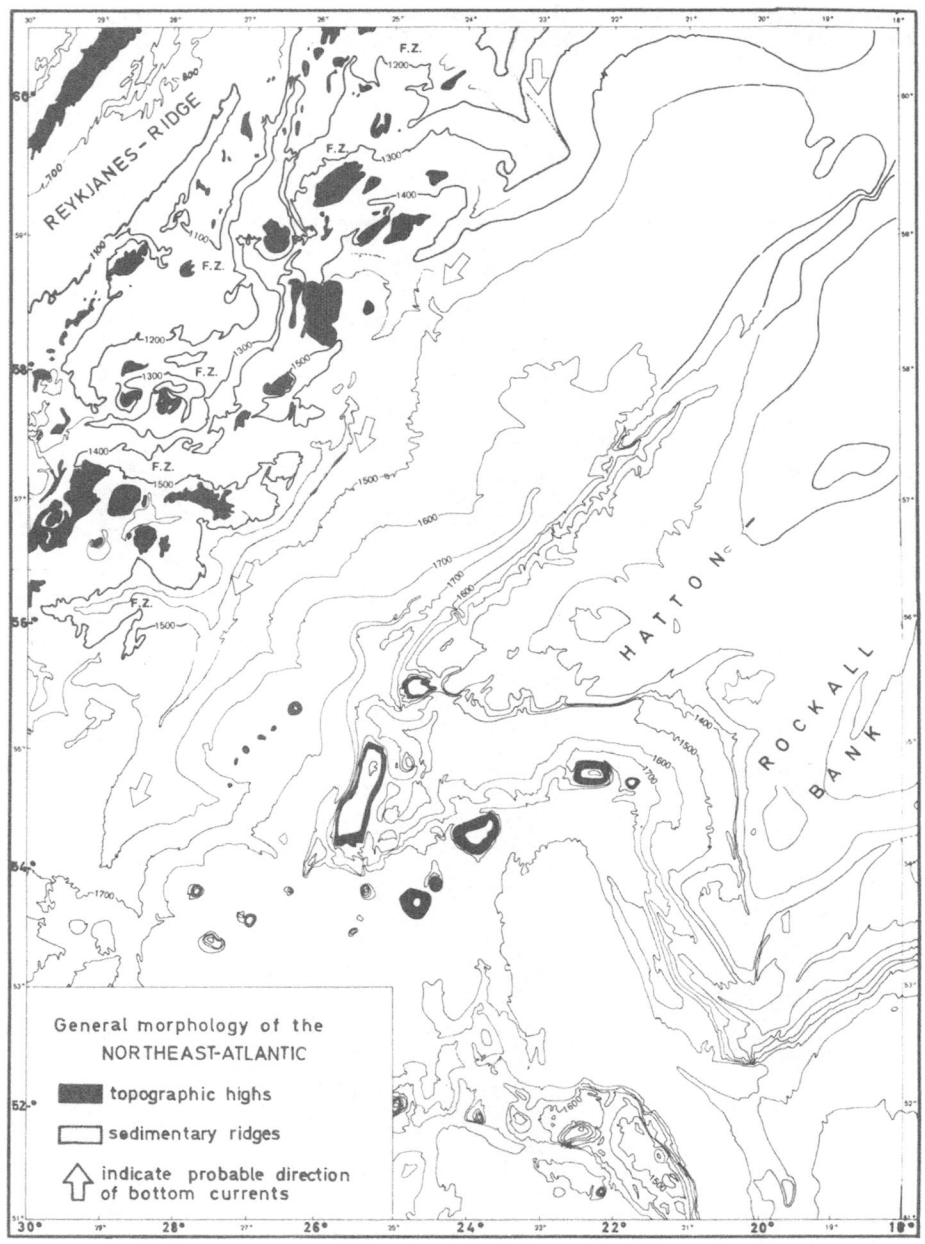

Fig. 2. Bathymetric chart by Johnson et al. [9].

Independent of a median valley the rift axis is associated with an extreme positive magnetic anomaly with a maximum near 1500 gammas. This central strip has a continuous straight course north of 57° latitude, but in the southernmost part of the ridge it is split into several small sections and a branching at 55°15'N [10]. Between 55° and 56°N both main directions, i.e. the north-south orientation of the southern part and the 35° orientation of the northern ridge appear simultaneously. The main branch trends northeast, but before it achieves this final orientation at 56°40' again a north-south running section exists. A similar controversy between both these trends also occurs farther away from the axis, between anomalies 10 and 19 [16]. Up to anomaly 5 (10 m.y.B.P.) a good parallelity to the present axis is evident, and the same is true for anomalies 19 to 24 (about 42 and 60 m.y.B.P., resp.). The area in between, north of 56°N, is characterized by anomalies of non-linear pattern indicating a number of fractures and a frequent change from the north-south to the 35° strike.

3. SEDIMENT DISTRIBUTION

This segmentation is given also by the bathymetric chart derived from the same survey [9] (Fig. 2), i.e. a relatively smooth sea-floor with isobaths roughly parallel to both the ridge axis and the Hatton-Rockall Bank margin and a differentiated structure in a zone extending from 100 km to 350 km off the crest, as far as surveyed.

In addition, there is a smooth ridge bordering the fracture zone to the east, named Gardar Ridge (Fig. 2). This has proved to be a sedimentary ridge [9] being constructed along a fragmented basement ridge roughly 42 m.y. in age.

East-west seismic reflection lines obtained at 8 n.m. intervals in this area reveal a pattern of sediment cover controlled by the reaction of bottom water movement to crustal topography [17]. It indicates prevalent erosion and redisposition of bottom sediments. The thickest sediment cover observed within the fractured area is 1.75 km. Most of this fracture zone filling with an increased smoothness and reflectivity suggests localized turbid downslope flow at the ridge flank. The crestal zone is nearly devoid of sediment. At the flanks the sediment thickness is highly variable but over all increasing with distance from the axis. After a lack of sediments at the outer scarp of the inner flanks there are thick sediment deposits. For the most part, the thickest deposits lie over local basement depressions. Their stratification is more nearly conformable with the sea-floor surface than with the basement topography [6].

The basement topography shows deviations from parallelity, i.e. an oblique trend of basement highs and lows (see also Fig. 9) relative to both the ridge axis and the crustal isochrons as predicted by Wilson [18]. In detail, these secondary ridges are split into a great number of segments which are oriented parallel to the main axis, as shown by Vogt and Johnson [19] for a "ridge B" segment (area 3, Fig. 4).

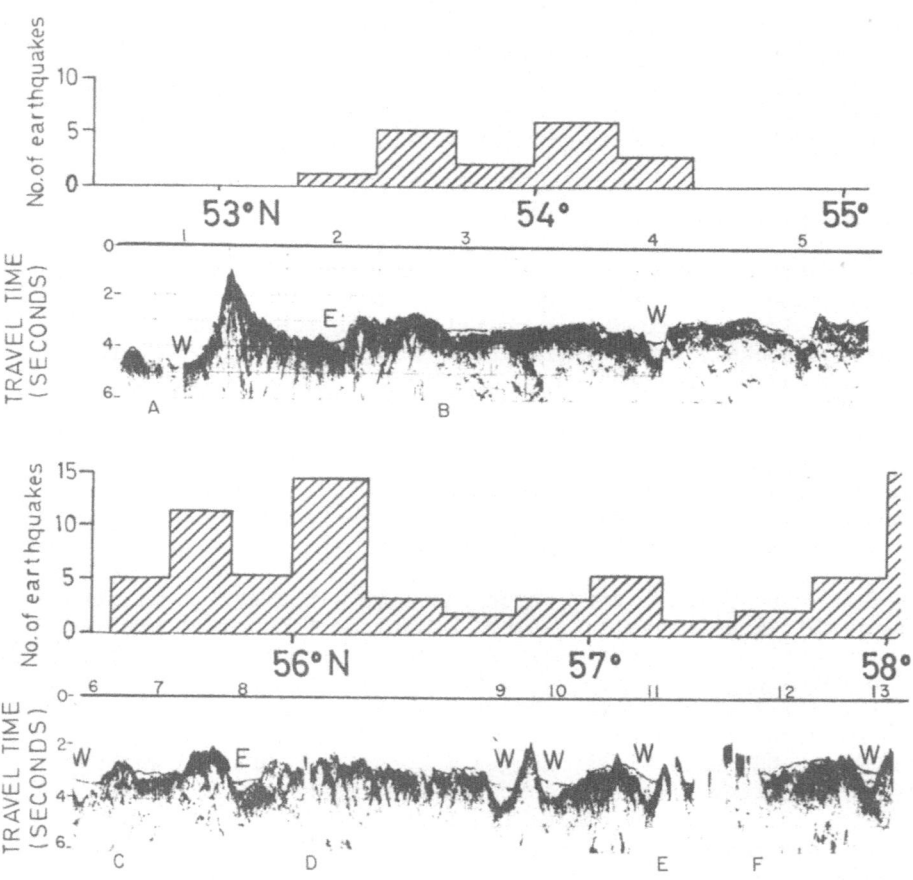

Fig. 3. Seismic reflection profile B (Fig. 4) [8] and number of earthquakes 1963-71 per 15 n.m. [21] along Reykjanes Ridge. A to F are turning points, 1 to 13 fracture valleys and 'E' or 'W' inferred directions of bottom currents.

From the angle of oblique trend, Vogt [20] estimated a southward asthenosphere flow near the Icelandic mantle plume of about 20 cm/yr.

Other striking elements are the linear seamount chains with an azimuth of about N 50°E. Lying in an area of nearly north-south (7°) striking magnetic anomalies, these lines of seamounts intersect the crustal isochrons with an angle of 45°. If these seamounts are understood as mantle plume discharges puncturing the moving lithosphere this angle indicates a southward astheno-sphere flow of about 1 cm/yr in 1200 km distance from the centre (see Fig. 9).

A longitudinal seismic reflection profile near the east of the ridge axis (B in Fig. 4) shows a number of hidden fracture valleys, nearly all of them lying between 52°N and 57°N (Fig. 3). From the slope of the sediment filling, visible in this figure, Vogt and Johnson [8] could determine an alternating sequence of easterly (E) and westerly (W) directed bottom currents. A comparison with the graphically displayed distribution of ridge earthquakes, occurring in swarms [21], shows only moderate agreement between local centres of seismic activity and individual fractures. Generally, however, they are well correlated and in particular they occur only scarceley up to a distance of 1000 km from the Iceland centre.

4. DEEP SEISMIC INVESTIGATIONS

Another interesting feature discovered during the same reflection survey [8] on the northern Reykjanes Ridge are ultra-flat opaque reflectors, abbreviated as UFOR (Fig. 4). These are basement areas with an inclination of less than 1:1000, found at about 1500 m water depth beneath 300 m to 500 m of poorly stratified sediments. Within slightly tilted terraces they seem to be separated by basement scarps or valleys. Vogt and Johnson [22] explain this phenomenon by a discharge of atypically low-viscosity lavas and thus see a relationship to the smooth surface topography near the hot spot. In their opinion the fractured topography farther away is due to discontinuous injection of moderate-temperature mantle material into a slow-spreading lithosphere.

In Fig. 4, localities of specific geophysical investigations on the Reykjanes Ridge are plotted as far as they had come to the author's knowledge [2,6,8,9,10,13,16,17,19,23,24,25,26,31,32 and others].

At locations E3-E5, Ewing and Ewing [26] found a basement of about 4 km thickness and a 5.7 km/sec P-velocity and from 6 km depth downwards a 7.4 km/sec-layer. Tryggvason [27] and Francis [28]

found an extension of this anomalous mantle to a depth of about
250 km from P-arrivals of Reykjanes Ridge earthquakes. But there
is some doubt about such interpretation [29].

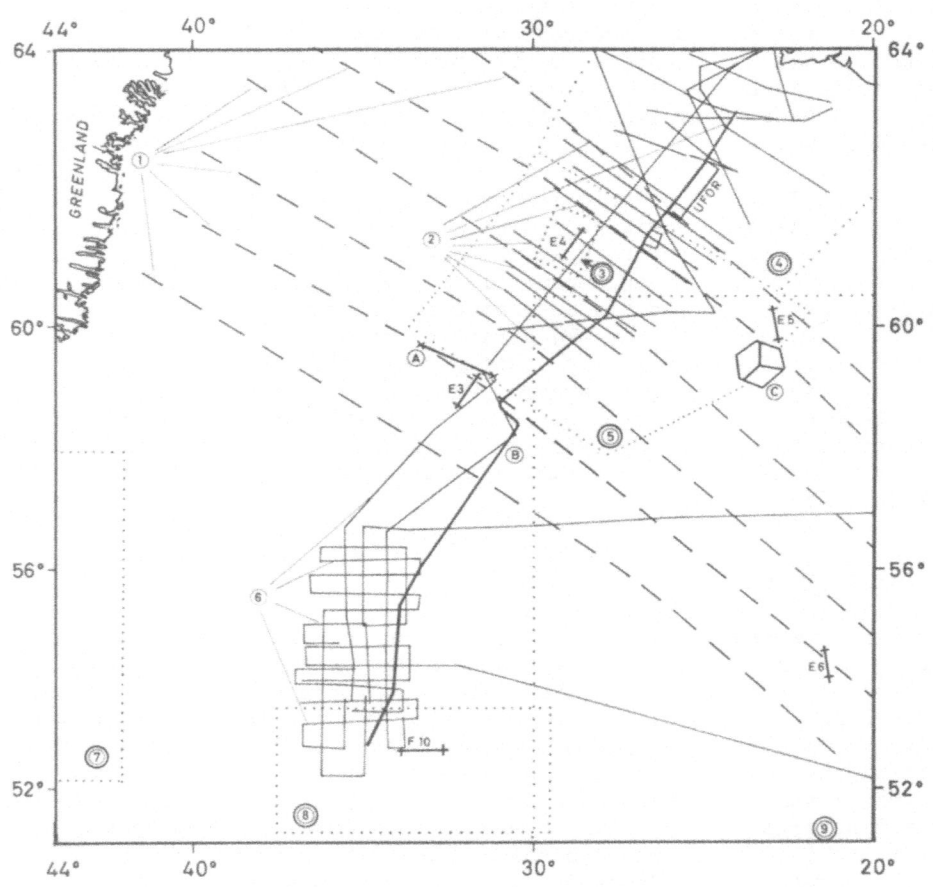

Fig. 4. Profiles and areas (dotted) of geophysical investigations.
1,4,5: aeromagnetic surveys [23,24,13]
2,6 : mainly magnetic and gravity surveys [6,10]
3   : detailed bathymetric and seismic survey [19]
7,8 : bathymetric and magnetic surveys [37,2]
9   : detailed bathymetric, magnetic, and seismic
        reflection surveys [9,16,17]
A,B : seismic reflection lines [25,8]
C,E,F: seismic refraction lines [31,26]

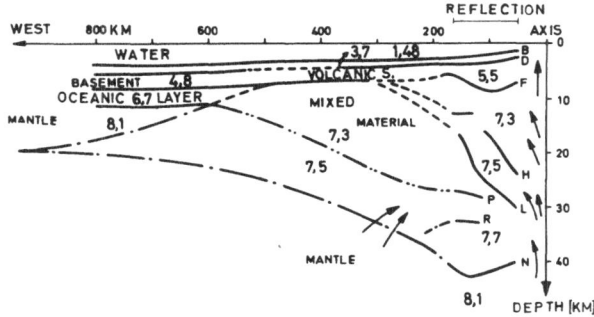

Fig. 5. Deep structure of the eastern ridge flank by Ariç [25].

This material, probably a crust-mantle mixture, disappears at
lateral distances of 300 to 500 km from the ridge axis. Correspondingly, the normal oceanic layer of about 6.3 km/sec which has
a mean thickness of 5 km, thins towards the crestal zone. A
normal layering was found, for instance, at location C (Fig. 4)
300 km southeast of the Reykjanes Ridge crest and, in contrast to
other data [30] there is no symptom of an anisotropy of the uppermost mantle [31].

Underneath the crest there is no indication of a mantle boundary, neither from refraction seismics nor from surface wave dispersion [32] except at 33°N in 7 km depth [33]. Ariç [25], however,
could verify a Moho depth of about 40 km at the central Reykjanes
Ridge ('A', Fig. 4) by seismic reflections (Fig. 5).

The reflection results belonging to the crestal zone give a
multiple stratification within the anomalous mantle zone. There
is, of course, an uncertainty in the assumed velocity and therefore
in real depths, but no contradiction to assume a limitation of the
anomalous mantle at relatively shallow depths.

A summary of all available data of the North Atlantic area is
given in Fig. 6 [25-35]. Obvious are the general facts: a down-
warping of the Moho and an upwelling of the 7.3 km/sec-layer towards
the crest. Merely beneath Iceland we find an intervention of material of 6.3 km/sec - the P-velocity of the oceanic layer.

5.   GRAVITY EFFECTS

From combined refraction results of the Mid-Atlantic Ridge near
32°N [33] assuming a Moho depth of 35 km, Talwani et al. [36]

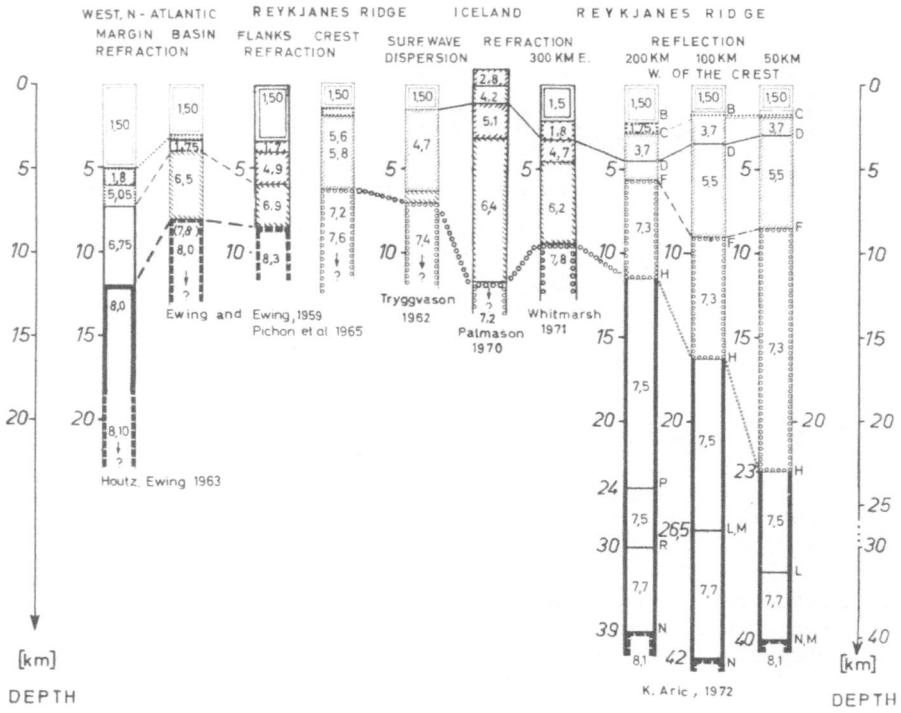

Fig. 6. A summary of seismic results in the North Atlantic area updated after Ariç [25,26,31-35].

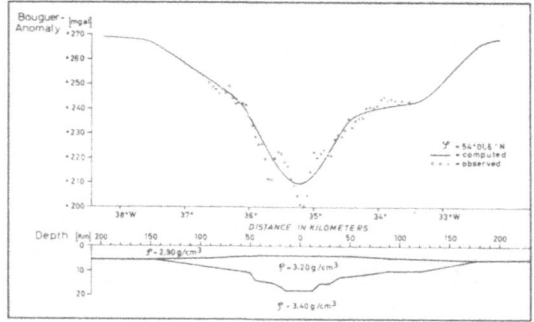

Fig. 7. Observed data (a) and two-dimensional model (b) of Bouguer gravity anomalies for a profile across the southern Reykjanes Ridge [10].

calculated a gravity model which is in good agreement with the
observed gravity anomalies. For the southern Reykjanes Ridge an
equivalent model was derived by Fleischer et al. [10], which shows
a smaller horizontal and vertical extent of the anomalous layer
(Fig. 7). This is due to the facts that here the Mid-Atlantic
Ridge has its least width south of Iceland and also a lesser
elevation ($\sim$ 1800 m) above the surrounding deep sea floor than
farther south ($\sim$ 2800 m). The central minima of Bouguer anomalies
are the same (+ 180 mgal) in both cases and therefore the prediction of Talwani et al. [6] for a compensation by a thicker
anomalous layer underneath the (northern) Reykjanes Ridge is not
valid here.

On the other hand, the level of free-air gravity anomalies
near +40 mgal (Fig. 8) points to an incomplete isostatic equilibrium. Negative values were reached only over the great depths of
the Gibbs Fracture Zone. A correlation with the topography is
evident. Because of a central rift valley in the southern ridge
section, here the rift axis is indicated by a trough in the free-
air anomalies, interrupted by transverse faults. Over the northern
Reykjanes Ridge, however, the axis coincides with the ridge summit
and therefore with a high of free-air anomalies. The same effect is
evident for the secondary ridges so that the V-shaped trend appears
again in the adjacent gravity highs (Fig. 8).

Possible V-shaped structures in the southern area may be
detected within the Bouguer anomaly pattern ([10], insert Fig. 8).
This would indicate deep asthenosphere flow scars and reveal the
same velocity near 1 cm/yr, as was derived from the seamount chains
to the east. A synopsis of all these results is given in Fig. 9
showing a slowing down of the asthenosphere flow inversely proportional to the distance from Iceland.

6. HISTORY OF THE AREA

The development of this part of the North Atlantic area mainly
derived from magnetic anomalies (Fig. 9) has been presented by
several authors [7,12,37,38,39 and others], especially by Vogt and
Avery [16,40]. Their results as far as they concern the Reykjanes
Ridge will be briefly reviewed:

60 m.y. ago, when the Mid-Labrador Sea Ridge was already
spreading, anomaly 24 developed in a rift along the Hatton-
Rockall Bank, and Greenland began to split from Europe.

Up to anomaly 19 (approx. 42 m.y.B.P.) the spreading rate
decreased from about 1.7 cm/yr to 0.7 cm/yr and linear
anomalies were created. Spreading was probably perpendicular to the northeast striking branch. To the south of

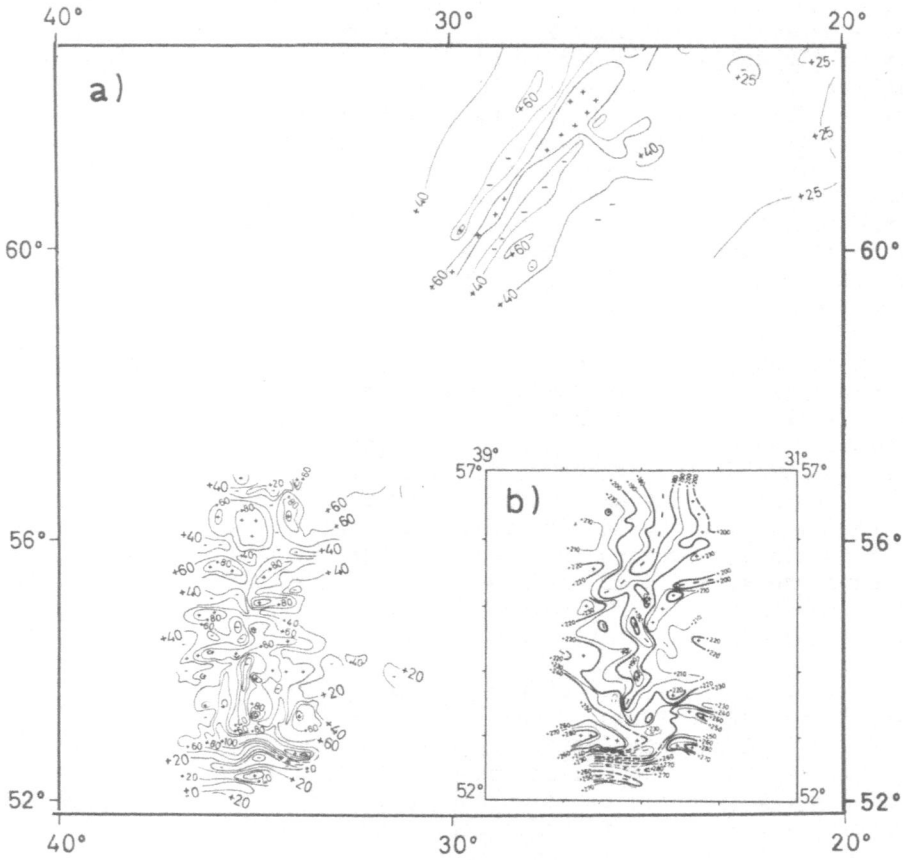

Fig. 8. a) Free-air anomaly map [6,11], b) Bouguer anomalies [10] for parts of the Reykjanes Ridge.

the triple junction to Labrador, however, till present north-south (7°) striking anomalies were generated.

After 40 m.y.B.P. when widening of Labrador rift ceased, the America-Greenland and the Europe-Rockall plates separated in east-west direction (90°-100°) which is given by the fracture zones north and south of Iceland (Fig. 9). In addition, a hypothetical "Reykjanes Fracture Zone" is proposed [41] indicated by the offsets of magnetic lineations near Reykjanes Peninsula and to possible identified strips on the Iceland-Faeroe Ridge [42]. This assumption implies a fit of the Faeroes to an East Greenland bight

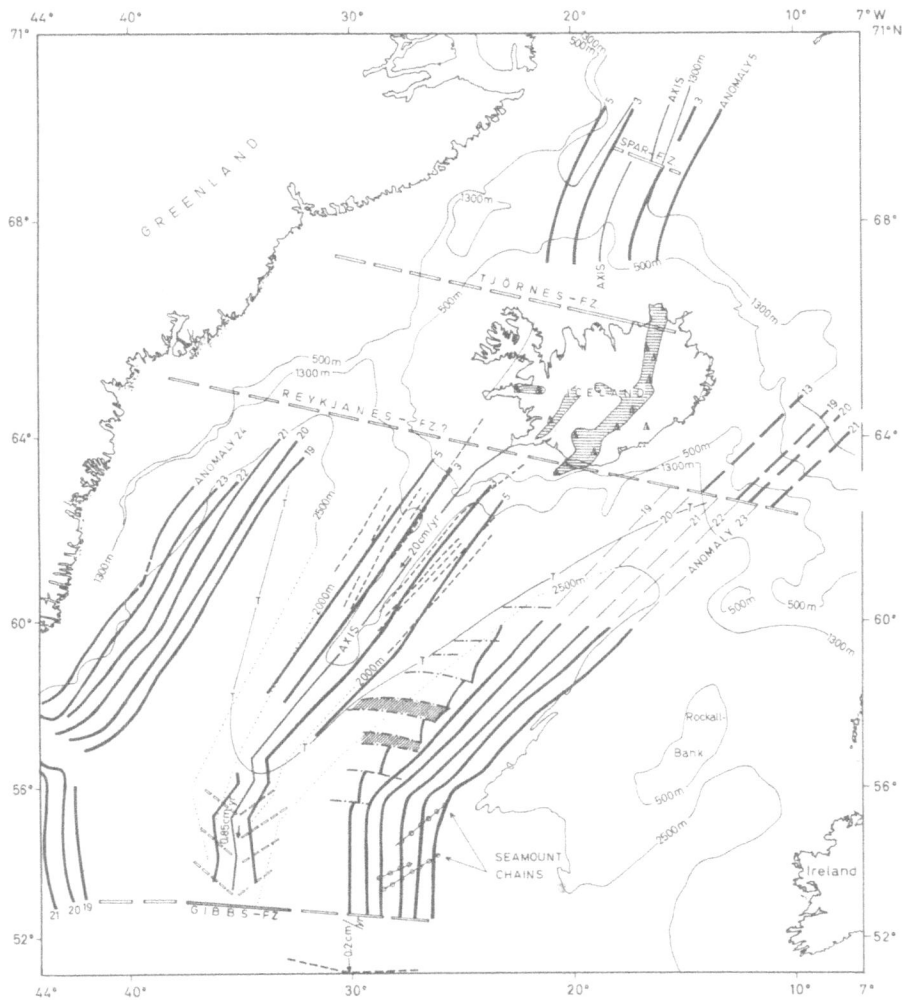

Fig. 9. Median rift, crustal isochrons, fracture zones, V-shaped features and inferred asthenosphere flow south of Iceland (see text).

350 km farther south than suggested by Brooks [43].

North of 56°, lineations perpendicular to this plate movement are restricted to small pieces. The lengths of offsetting fractures increase southwards and are followed

by 35° striking magnetic strips. This pattern
indicates a reorientation event 'T' (Fig. 9)
running at a rate of about 2 cm/yr away from the
plume centre where it started some 35 m.y. ago [16,20].

Since 10 m.y. the Reykjanes Ridge was completed in its
present figuration which is coupled with a unique
spreading behaviour: the rift axis is partly perpendicular, partly nearly 45° oblique to the spreading
direction.

7. PROSPECTS

Several questions remain: How did the plates move prior to anomaly
19 time? Is there an exact mirror image of the "reorientation area"
to the west of the ridge and how is the pattern farther north? How
is the detailed pattern of ridge strike and fracture zones in the
southern branch and how is the transition from the continuous feature to both the southern element and Iceland? How is the deep
structure, esp. the Moho depth and the anomalous mantle, and their
relation to the driving mechanism?

Further seismic and gravity investigations are to be made and
a complete knowledge of the magnetic pattern and topographic details
is necessary in order to understand all the complex processes of
ocean-floor generation.

ACKNOWLEDGEMENTS

The author is much indebted to Dr. M. Talwani and Dr. P. R. Vogt
for providing him with original drawings, and to D. Voppel for
helpful discussions and review of the manuscript.

REFERENCES

1. G. L. Johnson, Earth Planet. Sci. Lett., 2, 445, 1967.
2. H. S. Fleming, N. Z. Cherkis and J. R. Heirtzler, Marine
    Geophys. Res., 1, 37, 1971.
3. B. I. Haigh, Geophys. J. R. Astr. Soc., 33, 405, 1973.
4. E. E. Davis and C. R. B. Lister, Earth Planet. Sci. Lett.,
    21. 405, 1974.
5. U. Fleischer, Deut. Hydrogr. Z., 22, 205, 1969.
6. M. Talwani, C. C. Windisch and M. G. Langseth Jr.,
    J. Geophys. Res., 76, 473, 1971.
7. G. L. Johnson and B. C. Heezen, Deep-Sea Res., 14, 755, 1967.
8. P. R. Vogt and G. L. Johnson, Earth Planet. Sci. Lett., 18,
    45, 1973.

9. G. L. Johnson, P. R. Vogt and E. D. Schneider, Deut. Hydrogr. Z., 24, 49, 1971.

10. U. Fleischer, A. Korschunow, G. Schulz and P. R. Vogt, "Meteor" Forschungs-Ergebnisse, Reihe C, No. 13, 64, 1973.

11. U. Fleischer, Marine Geophys. Res., 1, 314, 1971.

12. W. C. Pitman III and M. Talwani, Geol. Soc. Am. Bull., 83, 619, 1972.

13. J. R. Heirtzler, X. LePichon and J. G. Baron, Deep-Sea Res., 13, 427, 1966.

14. F. J. Vine and D. H. Matthews, Nature, 199, 947, 1963.

15. J. R. Heirtzler et al., J. Geophys. Res., 73, 2119, 1968.

16. P. R. Vogt and O. E. Avery, J. Geophys. Res., 79, 363, 1974.

17. W. F. Ruddiman, Geol. Soc. Am. Bull., 83, 2039, 1972.

18. J. T. Wilson, Nature, 207, 907, 1965.

19. P. R. Vogt and G. L. Johnson, Earth Planet. Sci. Lett., 15, 248, 1972.

20. P. R. Vogt, Earth Planet. Sci. Lett., 13, 153, 1971.

21. T. J. G. Francis, Earth Planet. Sci. Lett., 18, 119, 1973.

22. P. R. Vogt and G. L. Johnson, Earth Planet. Sci. Lett., 18, 49, 1973.

23. E. A. Godby, P. J. Hood and M. E. Bower, J. Geophys. Res., 73, 7637, 1968.

24. T. Sigurgeirsson, Soc. Sci. Islandica, 38, 91, 1967.

25. K. Ariç, Deut. Hydrogr. Z. Ergänzungsh. 11, 1972 (Abstract in Z. Geophys., 36, 229, 1970).

26. J. Ewing and M. Ewing, Geol. Soc. Am. Bull., 70, 291, 1959.

27. E. Tryggvason, Bull. Seism. Soc. Am., 54, 727, 1964.

28. T. J. G. Francis, Geophys. J. R. Astr. Soc., 17, 507, 1969.

29. R. Stefánsson, Soc. Sci. Islandica, 38, 80, 1967.

30. R. B. Whitmarsh, Nature, 218, 558, 1968.

31. R. B. Whitmarsh, Bull. Seism. Soc. Am., 61, 1351, 1973.

32. E. Tryggvason, Bull. Seism. Soc. Am., 52, 359, 1962.

33. X. LePichon, R. E. Houtz, C. C. Drake and J. E. Nafe, J. Geophys. Res., 70, 319, 1965.

34. R. E. Houtz and J. I. Ewing, J. Geophys. Res., 68, 5233, 1963.

35. G. Pálmason, Science Inst. Univ. Iceland, 239 pp, 1970.

36. M. Talwani, X. LePichon and M. Ewing, J. Geophys. Res., 70, 341, 1965.

37. A. S. Laughton, Nature, 232, 612, 1971.

38. X. LePichon, R. D. Hyndman and G. Pautot, J. Geophys. Res., 76, 4724, 1971.

39. W. C. Pitman III, M. Talwani and J. R. Heirtzler, Earth Planet. Sci. Lett., 11, 195, 1971.

40. P. R. Vogt, G. L. Johnson, T. L. Holcombe, J. G. Gilg and O. E. Avery, Tectonophysics, 12, 211, 1971.

41. P. L. Ward, Geol. Soc. Am. Bull., 82, 2991, 1971.

42. U. Fleischer, F. Holzkamm, K. Vollbrecht and D. Voppel, Deut. Hydrogr. Z., 27, 97, 1974.

43. C. K. Brooks, Nature Phys. Sci., 244, 23, 1973.

DEEP STRUCTURE, EVOLUTION AND ORIGIN OF THE ICELANDIC TRANSVERSE
RIDGE

M.H.P. Bott

Department of Geological Sciences, University of Durham,
Durham DH1 3LE, England.

ABSTRACT. The elevated Icelandic transverse ridge extends between
the shelves of East Greenland and the Faeroe Islands. It is underlain by a three to four layered crust of oceanic affinity which is
much thicker and more variable than normal oceanic crust but does
not resemble normal continental crust. The thick crust accounts
for the elevation relative to a normal oceanic cross section.
The transverse ridge is terminated at its south-eastern end by an
unusual type of margin between the Iceland-Faeroe Ridge and the
Faeroe Block. The Mid-Atlantic Ridge crosses the otherwise aseismic
transverse ridge at Iceland, beneath which the anomalous low
density mantle appears to be more pronounced than beneath normal
ocean ridge regions.

    The transverse ridge probably originated in two successive
stages by vigorous differentiation of crustal material from the
hot underlying mantle during evolution of the North Atlantic.
During stage 1 (about 60-42 My BP) the Iceland-Faeroe Ridge
formed; the Iceland Block formed during stage 2 (about 42 My BP
to present) which started with a major westward migration of the
spreading centre. The Icelandic transverse ridge forms the abrupt
culmination of a gradual northward shallowing of the Atlantic
from the Azores towards Iceland, which can be interpreted in terms
of raised asthenosphere temperature and slightly thinned
lithosphere towards a high temperature focus beneath Iceland. The
high upper mantle temperatures are tentatively attributed to a
convective overturn 60-65 My BP rather than to a narrow plume
rising from the lower mantle.

Kristjansson (ed.), Geodynamics of Iceland and the North Atlantic Area. 33-47. *All Rights Reserved.*
Copyright © 1974 by D. Reidel Publishing Company, Dordrecht-Holland.

1. INTRODUCTION

Iceland lies at the intersection of the Mid-Atlantic Ridge with a
seismically inactive transverse ridge crossing most of the north-
eastern branch of the North Atlantic between East Greenland and
the Faeroe Islands (Fig. 1). The Icelandic transverse ridge is
subdivisible into two main structural regions, the broad Iceland
Block comprising Iceland and its insular shelf in the west, and
the narrower Iceland-Faeroe Ridge in the east. The Iceland Block
is separated from the Greenland shelf by a relatively narrow
bathymetric channel beneath Denmark Strait with maximum depths
in excess of 600 m. The smooth crest of the Iceland-Faeroe Ridge
is about 400 m deep along its length and is separated from the
Iceland and Faeroe Blocks by steep bathymetric scarps.

Datable Raff-Mason magnetic lineations have not been
recognised over Iceland [2] or the Iceland-Faeroe Ridge [3-5]
so that the spreading history cannot be simply evaluated as in
most oceanic regions. Nevertheless, if other parts of the north-
eastern North Atlantic formed by the sea-floor spreading mechanism
as we assume, then it follows that this anomalous transverse ridge
must have formed contemporaneously by the same mechanism despite
lack of datable magnetic lineations [6].

The Icelandic transverse ridge is one of the most substantial
aseismic ridges of the oceans. It also marks the boundary between
a region of relatively symmetrical spreading and bathymetry to
the south-west [7], and a region typified by asymmetrical
bathymetry and migrating spreading axes to the north-east [8].
The continental regions adjacent to the ends of the transverse
ridge suffered unusually vigorous basaltic volcanism some 65-45
My BP starting just before or at the time of the split between
Greenland and North Europe [9,10]. The transverse ridge is of
interest because of its role in the evolution of the North
Atlantic and as an example of anomalous differentiation from the
mantle. Study of the ridge and its associated features is also
relevant to discussion of the continental splitting and plate
driving mechanisms.

2. CRUSTAL STRUCTURE

2.1 Iceland Block

The crustal structure of Iceland has been determined by refraction
seismology [11-13]. A three to four layered crust overlies an
anomalously low velocity upper mantle as follows [13]:

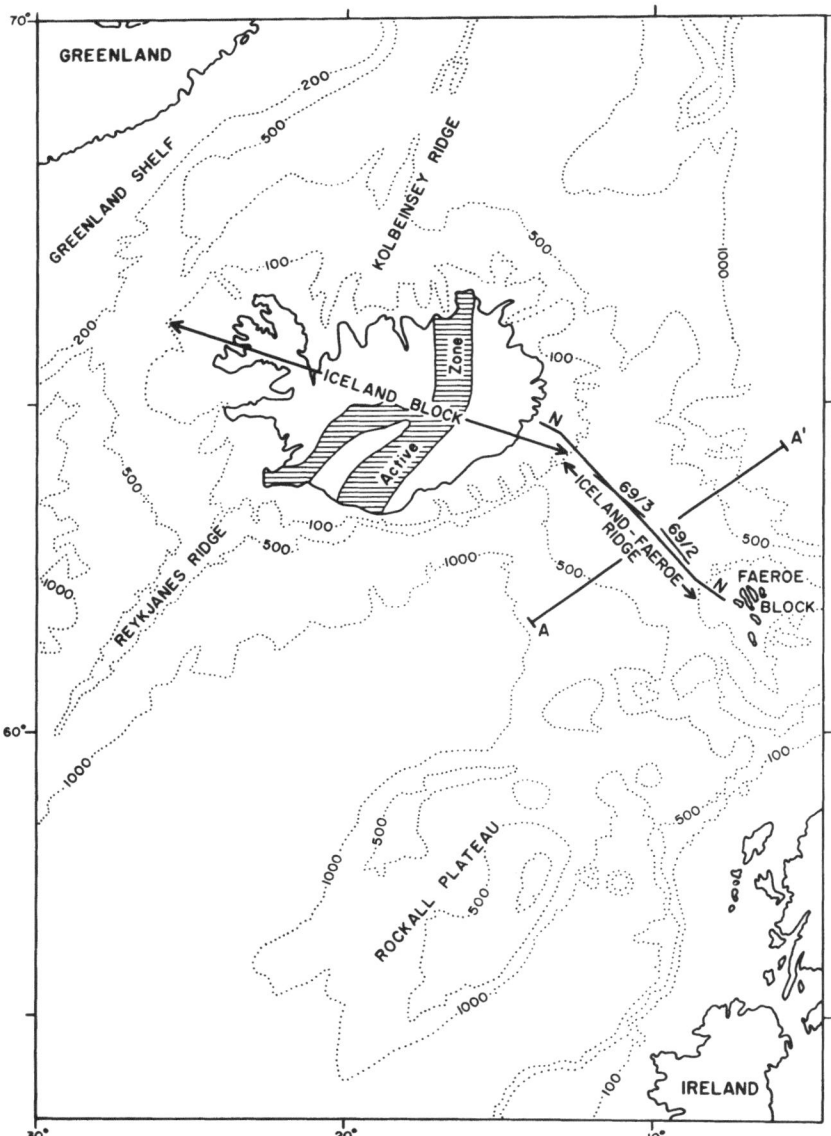

Fig. 1. Map of the Icelandic transverse ridge and neighbouring features, showing bathymetric depth contours at 100, 500 and 1000 fathoms, and 200 fathoms at the edge of the Greenland shelf. Also shown are seismic refraction lines 69/2 and 69/3, the North Atlantic Seismic Experiment line of shots NN, and the line of gravity profile AA' of Fig. 2. Part of bathymetry after Johnson et al. [1].

Layer	P-velocity (km s^{-1})	Thickness (km)
0	2.75 ± 0.37	0-1
1	4.14 ± 0.24	1.0 (average)
2	5.08 ± 0.24	2.2 (average)
3	6.35 ± 0.21	4-10
4	7.2	

Layer 0 is formed by post-Tertiary volcanic rocks and Layers 2 and 3 represent Tertiary lavas. Layer 3 is the main crustal layer for which a velocity of 6.5 km s^{-1} may be more representative than the value in the table above [13]. Small scale gravity anomalies correlate with the depth to the layer 2/layer 3 interface and suggest a density difference of 0.18-0.19 g cm^{-3} between these layers.

The velocity determined for layer 4 might normally be assigned to the lower crust, but surface wave studies [14] and teleseismic P-delay times [15,16] indicate that this "layer" extends down to at least 150-200 km depth. It is thus identified as the top of an anomalously low velocity upper mantle which underlies Iceland. The Moho varies between about 8 and 18 km depth, being 8-9 km deep beneath SW Iceland and 14-15 km deep beneath south-eastern and northern Iceland. Regional gravity survey [17] indicates that Iceland is in approximate isostatic equilibrium.

Sediments are effectively absent from Iceland and are generally thin on the Icelandic insular shelf [18]. They may thicken towards the shelf edge, particularly at the south end of the western shelf.

2.2 Iceland-Faeroe Ridge

Sediment is thin or absent over most of the smooth crestal region of the Iceland-Faeroe Ridge [3,5], probably because of strong bottom currents which carry Norwegian Sea water across it [19,20]. Troughs containing about 1 km thicknesses of sediments lie beneath the marginal bathymetric scarps which separate the Ridge from the Iceland and Faeroe Blocks [3,6]. Sediments are also thin on the south-western flank of the Ridge, but are about 1 km thick on the north-eastern flank.

Two refraction lines (Fig. 1) along the crest of the Ridge yielded the following layered structure [6]:

Layer	P-velocity (km s^{-1})	Thickness (km)
1	3.2–4.6	0–3
2	5.4–5.8	5+
3	6.8	line 69/3 only
4	7.8 (unreversed)	line 69/2 only

The layer 2/layer 3 interface is horizontal beneath line 69/3 and occurs at 7.5 km depth. Identification of the 7.8 km s^{-1} layer is dubious as it was only detected along a short segment shooting towards the south-east along line 69/2. Preliminary results from the North Atlantic Seismic Project [21] indicate that the Moho is slightly over 25 km deep beneath the south-eastern end of the Ridge and that layer 3 also probably occurs here despite its apparent lack of detection by line 69/2.

Gravity measurements [5,6,22] indicate that the Ridge is in approximate isostatic equilibrium. A gravity profile across the centre of the Ridge (Fig. 2) suggests that equilibrium occurs according to the Airy hypothesis with an underlying root of thickened crust relative to the Norwegian Sea [6]. The crust beneath the Ridge is estimated from the gravity profile to be about 22 km thick assuming a crust/mantle density contrast of 0.4 g cm^{-3}, or thicker if the density contrast is smaller.

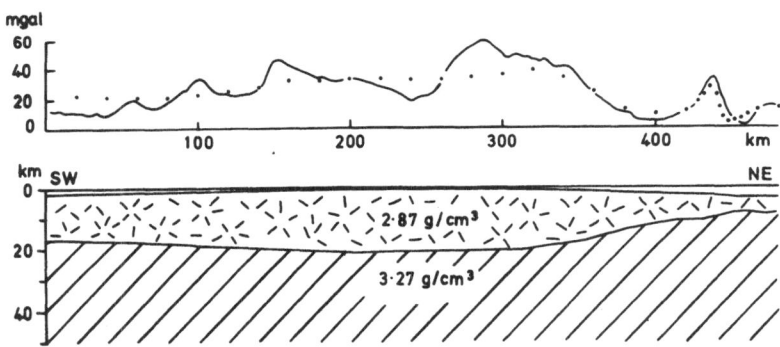

Fig. 2. Observed free air gravity anomalies along line AA' (Fig. 1) compared with a satisfactory theoretical model based on Airy isostatic equilibrium with a thickened crust beneath the Iceland-Faeroe Ridge, after Bott et al. [6].

Large-amplitude magnetic anomalies of short wavelength place
the magnetic basement at the seabed or not far beneath it [3-5].
Thus layers 1 and 2 are probably igneous in origin. Substantial
local gravity anomalies also occur over the Ridge crest [5,6]
and these show some correlation with medium wavelength magnetic
anomalies [4] and with variation in layer 1 thickness [6].
These may be partly explicable by varying depth to layer 3 (or 4)
but the steep gradients require a shallow density contrast of at
least -0.4 g cm^{-3} and are probably caused by variation in the
thickness of layer 1 rocks and overlying sediments. Layer 1 may
thus be essentially formed by pyroclastic rocks penetrated by
hypabyssal intrusions but may include some sediments [5]. Layer 2
is formed of denser lavas and associated intrusives including some
ring intrusions [23]. Layer 3 is the main crustal layer which
does not reach the surface and it may be equivalent to the layer
3 beneath Iceland.

2.3 Icelandic type crust

The crustal structure beneath the Iceland-Faeroe Ridge appears to
be similar to but not identical with that beneath Iceland. Both
have variable volcanic upper layers underlain by a well-defined
main crustal layer of velocity 6.4 to 6.8 km s^{-1}. The main
differences are as follows: corresponding layers have somewhat
higher velocities beneath the Ridge; layer 1 beneath the Ridge
representing deep local basins filled by pyroclastics and possibly
sediments is not present on Iceland; crustal thickness is
substantially greater beneath the Ridge than beneath Iceland;
fairly normal sub-Moho velocities occur beneath the Ridge in
contrast to the anomalously low values beneath Iceland. These two
varieties of crust resemble each other more closely than other
types of crust. We shall refer to the crust beneath both regions
as Icelandic type.

The individual layers are generally much thicker and more
variable than those of normal oceanic crust. However, layers 0 to
2 of Iceland and the Ridge could be equated to the layer 2 of the
normal oceanic crust, and our layer 3 could readily be identified
as the normal oceanic layer 3. Icelandic type crust differs from
most continental crust in the presence of the 6.4-6.8 km s^{-1}
layer 3 at shallow depth yielding first arrivals, and in the
apparent igneous origin of the overlying layers. How does
Icelandic crust originate? The answer depends on whether or not
one accepts the theory of ocean floor spreading and plate
tectonics for the evolution of the North Atlantic. If so,
Icelandic type crust must be an anomalous type of oceanic crust
formed by unusually vigorous differentiation, a viewpoint which
is taken in this paper.

Normal oceanic magnetic polarity zones are sharply defined because the top of the newly formed crust at the spreading centre is rapidly quenched by sea-water circulation [24] and does not suffer later remagnetization. In contrast, Icelandic type crust, which probably formed under sub-aerial conditions, lacks datable magnetic lineations for one or more of the following reasons: wide lateral extent of some lavas [25], multiple dyke intrusion, remagnetization by later intrusive bodies [2], migration of spreading centre, multiple spreading centres, etc.

3. THE ENDS OF THE ICELANDIC TRANSVERSE RIDGE

At its two ends, the Icelandic transverse ridge would be expected to abut against older crust of different origin. The position of the north-western end at the East Greenland shelf has not been defined but is unlikely to raise problems. The position of the south-eastern end has been more problematical because of uncertainty whether the Faeroe Block has affinities with the Icelandic transverse ridge or with Rockall Plateau which is known to be continental [26,27].

Two early, but inconclusive, indications that the Faeroe Block is continental and that the bathymetric scarp between the Block and the Iceland-Faeroe Ridge marks the end of the transverse ridge are as follows: (1) the continental fit between Rockall Bank and Greenland implies that the Faeroe Block itself must also be continental [28]; and (2) a decrease of about 80 mgal in the Bouguer anomaly between the Iceland-Faeroe Ridge and the Faeroe Block (Fig. 3) implies a change in crust beneath the intervening bathymetric scarp [6].

The most convincing new evidence comes from the North Atlantic Seismic Experiment of 1972 [29]. Time-term analysis for shots on the Faeroe Block recorded on the Faeroe and Shetland Islands indicates a sub-Moho velocity of 8.24 ± 0.35 km s^{-1} and a mean time-term of 3.60 s for the Faeroe Block. This yields a mean crustal thickness of 33 km assuming a mean crustal velocity of 6.1 km s^{-1}, or a larger thickness if the mean crustal velocity is higher. An upper crustal velocity of 6.0-6.2 km s^{-1} was determined for the north-western shelf region of the Faeroe Block, and there was no indication for presence of a 6.5-6.8 km s^{-1} layer. These factors enable the crust beneath the Faeroe Block to be identified as continental rather than Icelandic type.

Furthermore, arrivals from NASP shots along the Iceland-Faeroe Ridge [21] recorded at stations on the Faeroe Islands appear to travel beneath the Ridge as headwaves in layers 3 and 4 with apparent velocities of about 7.0 and 8.0 km s^{-1} respectively, but they travel through the Faeroe Block as a crustal P_g phase with velocity of about 6.0 km s^{-1}. Conversion to P_g appears to take place at the boundary between the Icelandic type crust and the continental crust beneath the vicinity of the bathymetric scarp. This is particularly convincing evidence that we have a fundamental crustal transition here.

Thus we conclude that the Faeroe Block is underlain by continental crust and that an anomalous continental margin beneath the bathymetric scarp separates it from the Icelandic type crust beneath the Iceland-Faeroe Ridge.

4.   UPPER MANTLE STRUCTURE

The upper mantle structure beneath Iceland itself is much better known than beneath less anomalous parts of the ocean ridge system because of the ease of making land observations. Elsewhere beneath the Icelandic transverse ridge, the only information on the mantle comes from gravity and crustal seismic observations on the Iceland-Faeroe Ridge.

A surface wave dispersion study [14] first indicated that the 7.4 km s^{-1} layer (now revised to 7.2 km s^{-1}) beneath Iceland belongs to an anomalous upper mantle rather than the lower crust. Recognition of a 1.3 second teleseismic P-wave delay relative to the J & B tables [15] and its subsequent confirmation [16] showed that the 7.2-7.4 km s^{-1} layer extends down to a depth of about 240 km, confirming that a highly anomalous upper mantle underlies Iceland. The apparent velocity across Iceland of body waves from Mid-Atlantic Ridge earthquakes confirms that low P-wave velocities extend down to at least 250 km depth and shows that the anomalous belt is about 300 km wide [30]. The ratio of S to P-velocity in the upper part of the zone is found to be anomalously low [30].

The geothermal gradient of Iceland at shallow depth outside thermal areas is typically within the range 35-$165°C$ km^{-1} indicating high heat flow of 1.7-7.4 μcal cm^{-2} s^{-1} [31]. Magnetotelluric measurements in south-western Iceland give a temperature estimate of $1000 \pm 200°C$ at the crust/mantle interface at 10-15 km depth indicating that the geothermal gradient can be approximately extrapolated to this depth [32]. The measurements indicate a much shallower temperature gradient between 0 and $2°C$ km^{-1} between the base of the crust and a depth of about 100 km. South-western Iceland thus has high temperatures at shallow mantle depths and a low geothermal gradient in the upper mantle, in agreement with the plate tectonic model of the spreading and cooling oceanic lithosphere.

The saucer-shaped Bouguer anomaly contours of Iceland [17] mirror image the broad topography. A minimum of -35 mgal is found

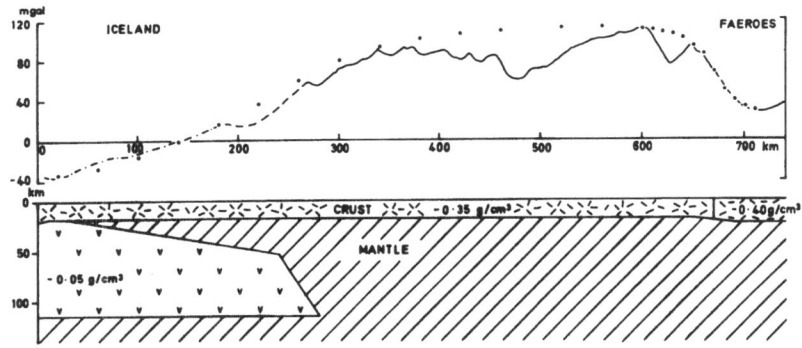

Fig. 3. A composite Bouguer anomaly profile from the centre of Iceland along the Iceland-Faeroe Ridge to the Faeroe Islands, showing a simplified two-dimensional interpretation of the underlying changes in deep structure, after Bott et al. [6].

near the centre and the average value around the coastline is about 60 mgal. The average Bouguer anomaly level is typical of a normal continental region, but the crust is 10-15 km thick and is unlikely to have an exceptionally low density. Thus an anomalously low density upper mantle must be present beneath Iceland [33]. Assuming an 0.2 g cm^{-3} reduction in density, this would need to have a vertical extent of at least 40 km, and smaller density reductions would require correspondingly larger thicknesses of anomalous material. The regional gravity anomalies of Iceland are not mainly caused by crustal thickness variations [13], and a more detailed study of the mantle-derived gravity anomaly will become possible when the variation in depth to the Moho has been more fully mapped.

More normal sub-Moho P-velocities occur beneath the southeastern end of the Iceland-Faeroe Ridge. The Bouguer anomaly also rises by about 140 mgal between the centre of Iceland and the Iceland-Faeroe Ridge [6]. This cannot be attributed to crustal thinning towards the Ridge or to higher crustal densities there, as the crust appears to thicken by at least 10 km between Iceland and the Ridge. The gravity profile (Fig. 3) thus indicates that the anomalously low density upper mantle beneath Iceland gives way to more normal densities beneath the Iceland-Faeroe Ridge. The true lateral density variation may be larger than shown in Fig. 3 because the thicker crust beneath the Ridge was neglected in the model and because of the imperfection of the two-dimensional assumption. Whatever the detail, the model indicates that Iceland bears much the same relationship to the Iceland-Faeroe Ridge as a normal ocean ridge does to its adjacent ocean basins. The Icelandic transverse ridge differs from a normal oceanic region mainly because of the thick Icelandic type crust which causes the relatively high elevation of both Iceland and the other parts of the transverse ridge.

Is the anomalous upper mantle beneath Iceland more pronounced than beneath normal ocean ridges? The deep seismic structure beneath ocean ridges is insufficiently well known for meaningful comparison, but isostatic considerations are significant. Central Iceland stands some 3 km higher in elevation than a normal oceanic ridge crest. The Moho is about 15 km deep in contrast to about 8 km beneath a normal oceanic ridge. Taking a density contrast of 0.4 g cm^{-3} at the Moho and an upper crustal density of 2.73 g cm^{-3}, the excess elevation of Iceland would require the underlying Moho to be over 20 km deep if the upper mantle density-depth distribution conformed to the normal pattern beneath ridges. This comparison suggests that the low density region in the mantle beneath Iceland is more pronounced than beneath a normal ridge.

If normal empirical relationships between P-wave velocity and density for rocks are assumed, then it is found that the P-wave delay times of Iceland are several times larger than would be expected for the underlying low density distribution [33]. This disagreement can be overcome if the seismic delays are mainly attributed to partial fusion or proximity to it [33] when the empirical relationships break down. The low densities may also be attributed partly to partial fusion, but the main contributors are probably thermal expansion and the presence of a region of low density plagioclase pyrolite.

5. HISTORY OF DEVELOPMENT

The Icelandic transverse ridge lacks datable magnetic lineations, and there are too great variations in crustal thickness along the ridge to apply the topographic method of dating [34] with any conviction. Nevertheless, the juxtaposition of the relatively narrow Iceland-Faeroe Ridge of north-westerly trend against the broader Iceland Block of more east-westerly trend suggests that these structural units have formed successively rather than contemporaneously. Furthermore, the magnetic anomaly pattern and the fracture zones indicate that the north-eastern North Atlantic has evolved in two such stages, respectively preceding and following the date of about 42 My ago when spreading ceased in the Labrador Sea [35], and involving a change in the pole of rotation which accounts for the contrasting trends of the two parts of the Icelandic transverse ridge. Thus two stages of formation are inferred as follows:

Stage 1: Between about 60 and 42 My ago, during the last stage of evolution of the Labrador Sea, a triple point junction was present at the south-western end of the Reykjanes Ridge and spreading occurred perpendicular to the Reykjanes Ridge. At this stage the spreading centre was centrally placed throughout the north-eastern North Atlantic. It is suggested that the Iceland-Faeroe Ridge formed during this stage. This is consistent with the north-westerly trend of the Ridge parallel to the spreading direction. It also explains the symmetrical bathymetry along the crest of the Ridge, with slightly shallower depths characterising the supposed youngest crust occurring in the middle of it.

Stage 2: At about 42 My ago the Labrador Sea ceased to open significantly. However, the pole of rotation of the North Atlantic remained nearly the same as before as indicated by the line of the Charlie Gibbs Fracture Zone [35,36]. Consequently, there was an abrupt change in the pole of rotation of the north-eastern branch, and the spreading direction and fracture zones became co-polar with the Charlie Gibbs Fracture Zone and oblique to the Reykjanes Ridge [35]. The west-north-westerly trend of the Iceland Block is in agreement with the new spreading direction. At the start of this stage, the spreading centre beneath the Icelandic transverse ridge migrated westwards to the Greenland margin or near it, in conformity with the region further north but in distinction to the Reykjanes Ridge which remained centrally placed. In contrast to the northern region where further westward migration of spreading centre occurred, the spreading centre beneath the transverse ridge migrated back towards the east to its present asymmetrical location, probably in stages. Thus transform faults developed first to the south and then to the north of the Iceland Block, these now being represented by the Reykjanes and Tjörnes Fracture Zones [37].

During stage 1, the underlying hot region in the upper mantle was of relatively short extent along the ridge crest but was probably hotter than now to give rise to the thick crust beneath the Iceland-Faeroe Ridge. After the spreading centre had migrated westwards at the start of stage 2, the underlying hot spot was broader and less intense, accounting for the wider Iceland Block with its thinner crust. According to this interpretation, the bathymetric step separating the Iceland Block from the Iceland-Faeroe Ridge marks a break in age of about 20 My between the older and cooler lithosphere with thicker crust beneath the Ridge and the younger lithosphere beneath the Block.

The 320 km long Iceland-Faeroe Ridge would take about 13 My
to form at the appropriate spreading rate of 1.2 cm/yr [8,35],
so that if the 80 km wide channel between the Iceland Block and
the Greenland shelf formed at the start of stage 1, then the
stage would finish about 42 My ago. On the other hand, if the
channel formed after the main westward migration of spreading
centre at the start of stage 2, then stage 1 would terminate
somewhat before 42 My ago. It should be possible to test the
hypothesis advanced for the development of the Iceland-Faeroe
Ridge by drilling near its ends and in the middle. Similarly the
dating of the channel west of the Iceland Block must await
drilling.

The submerged parts of the Icelandic transverse ridge were
presumably underlain by anomalously low density upper mantle,
such as now occurs beneath Iceland, at the time of their formation.
Assuming isostatic equilibrium prevailed, they would have stood
up to 2 km above their present elevation at the time of their
formation and have subsequently subsided as the underlying
lithosphere cooled [6]. Thus a continuous land bridge probably
existed between Greenland and the Faeroe Block until 30 to 40 My
ago, probably accounting for the absence of strong bottom currents
in the North Atlantic during the early Tertiary [38].

6. ORIGIN OF THE TRANSVERSE RIDGE

The Icelandic transverse ridge forms the abrupt culmination of a
gradual shallowing of the whole ridge and basin structure of the
North Atlantic from the latitude of Azores towards Iceland. This
is accompanied by a corresponding northward reduction in the
cross-sectional area of the Mid-Atlantic Ridge and a decrease in
the ridge to basin elevation difference [39]. This trend is
reversed north of Iceland. Haigh [39] accounted for these trends
by a simple thermal model involving cooling of the spreading
lithosphere above an asthenosphere which increased in
temperature northwards by about 150°C, towards a high temperature
"focus" beneath Iceland (Fig. 4). The temperature variation caused
the thickness of the lithosphere to decrease from 95 to 75 km
towards the north.

The Icelandic transverse ridge does not itself fall within
the scope of Haigh's lithospheric model, as the elevation is
mainly attributable to thick Icelandic type crust. However,
Haigh's hypothesis provides the setting of exceptionally high
upper mantle temperature beneath the transverse ridge required
for the vigorous differentiation of crustal material. Haigh [39]
also showed that the removal of additional crustal material from
the upper mantle during formation of the transverse ridge, while
depleting the upper mantle in silicon and aluminium, is unlikely

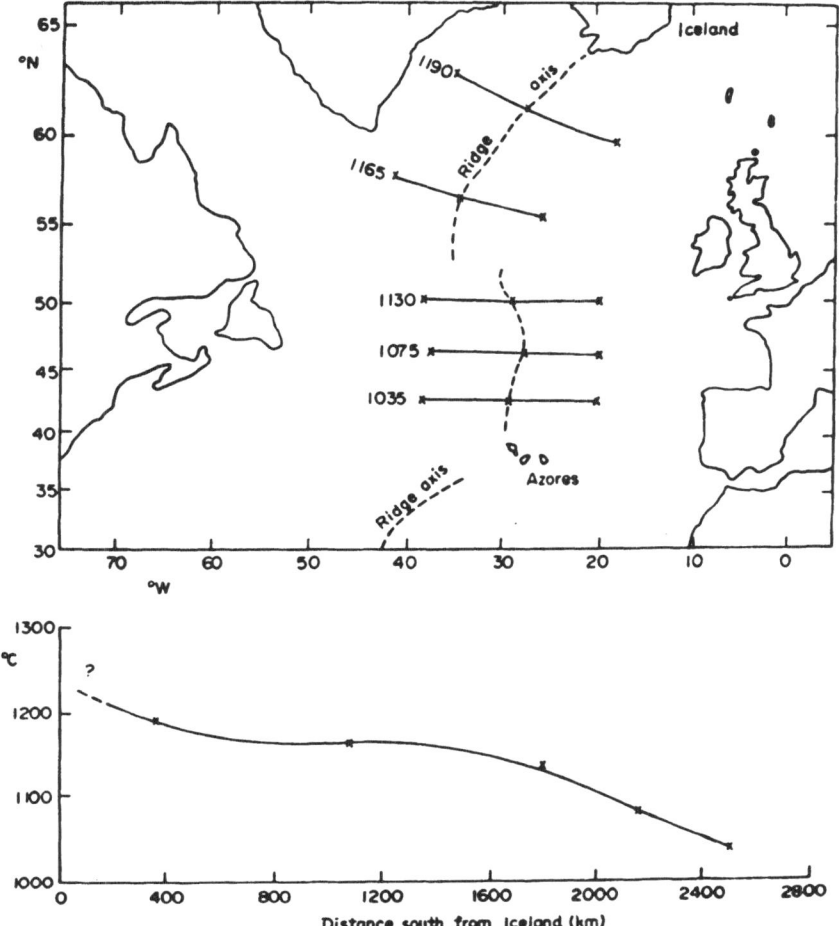

Fig. 4. Temperature (°C) at 65 km depth calculated by modelling
Atlantic bathymetry, shown as a map and a profile. Reproduced with
permission from Haigh [39].

to affect the density or seismic velocities greatly.

What produced the high upper mantle temperatures beneath the
Icelandic transverse ridge over the last 60 My? One suggestion is
that there is an underlying plume rising from the lower mantle
[40,41]. A problem of the plume hypothesis is to understand how a
narrow plume initially rising about 60-65 My ago could cause such
widespread early Tertiary volcanism stretching from Canada to
Britain. An alternative explanation suggested by the author [10]

is that a convective overturn occurred at this time, giving rise
to widespread partial fusion in the asthenosphere. This explains
the near contemporaneity and wide lateral extent of the volcanism,
and provides a mechanism for initiation of the new split between
Greenland and North Europe by uplift and dyke intrusion. The
Icelandic transverse ridge then forms above the hottest part of
the upper mantle resulting from the convective overturn. The
apparent decay in the intensity of the hot spot indicated by
crustal thickness variation along the transverse ridge is
agreeable to this hypothesis.

REFERENCES

1. G. L. Johnson, J. R. Southall, P. W. Young and P. R. Vogt,
    J. Geophys. Res., 77, 5688, 1972.
2. P. H. Serson, W. Hannaford and G. V. Haines, Science, N.Y.,
    162,3555, 1968.
3. G. L. Johnson and B. Tanner, Jökull, 21, 45, 1972.
4. M. H. P. Bott and A. Ingles, Geophys. J. R. Astr. Soc., 30,
    55, 1972.
5. U. Fleischer, F. Holzkamm, K. Vollbrecht and D. Voppel,
    Deut. Hydrogr. Z., 27, in press 1974.
6. M. H. P. Bott, C. W. A. Browitt and A. P. Stacey, Marine
    Geophys. Res., 1, 328, 1971.
7. J. R. Heirtzler, X. Le Pichon and J. G. Baron, Deep-Sea Res.,
    13, 427, 1966.
8. P. R. Vogt, N. A. Ostenso and G. L. Johnson, J. Geophys. Res.,
    75, 903, 1970.
9. P. R. Vogt, G. L. Johnson, T. L. Holcombe, J. G. Gilg and
    O. E. Avery, Tectonophysics, 12, 211, 1971.
10. M. H. P. Bott, The evolution of the Atlantic north of the
    Faeroe Islands, in: Implications of continental drift to the
    Earth Sciences (1), ed. by D. H. Tarling and S. K. Runcorn,
    Academic Press, London and New York, 175, 1973.
11. M. Båth, J. Geophys. Res. 65, 1793, 1960.
12. G. Pálmason, Soc. Sci. Islandica, 38, 67, 1967.
13. G. Pálmason, Crustal structure of Iceland from explosion
    seismology, Science Institute, University of Iceland, 239 pp,
    1970.
14. E. Tryggvason, Bull. Seism. Soc. Am., 52, 359, 1962.
15. E. Tryggvason, Bull. Seism. Soc. Am., 54, 727, 1964.
16. R. E. Long and M. G. Mitchell, Geophys. J. R. Astr. Soc., 20,
    41, 1970.
17. Tr. Einarsson, Soc. Sci. Islandica, 30, 22 pp, 1954.
18. G. Pálmason, The insular margin of Iceland, in: The geology
    of continental margins, ed. by C. A. Burk and C. L. Drake, in
    press 1974.
19. G. L. Johnson and E. D. Schneider, Earth Planet. Sci. Lett.,
    6, 416, 1969.

20. E. J. W. Jones, M. Ewing, J. I. Ewing and S. L. Eittreim,
     J. Geophys. Res., 75, 1655, 1970.
21. M. H. P. Bott and J. Sunderland, in preparation, 1974.
22. U. Fleischer, Marine Geophys. Res., 1, 314, 1971.
23. A. D. Ingles, The interpretation of magnetic anomalies between
     Iceland and Scotland, Ph.D. thesis, University of Durham, 1971.
24. G. Bodvarsson and R. P. Lowell, J. Geophys. Res., 77, 4472,
     1972.
25. J. D. A. Piper, Earth Planet. Sci. Lett., 12, 199, 1971.
26. R. A. Scrutton, Geophys. J. R. Astr. Soc., 27, 259, 1972.
27. D. G. Roberts, D. A. Ardus and R. Dearnley, Nature Phys. Sci.
     244, 21, 1973.
28. M. H. P. Bott and A. B. Watts, Inst. Geol. Sci. Rep. 70/14,
     89, 1971.
29. M. H. P. Bott, J. Sunderland, P. J. Smith, U. Casten and
     S. Saxov, Nature, 248, 202, 1974.
30. T. J. G. Francis, Geophys. J. R. Astr. Soc., 17, 507, 1969.
31. G. Pálmason, Soc. Sci. Islandica, 38, 111, 1967.
32. J. F. Hermance and L. R. Grillot, Phys. Earth Planet. Inter.,
     8, 1, 1974.
33. M. H. P. Bott, Geophys. J. R. Astr. Soc., 9, 275, 1965.
34. J. G. Sclater, R. N. Anderson and M. L. Bell, J. Geophys. Res.
     76, 7888, 1971.
35. P. R. Vogt and O. E. Avery, J. Geophys. Res., 79, 363, 1974.
36. X. Le Pichon, R. D. Hyndman and G. Pautot, J. Geophys. Res.,
     76, 4724, 1971.
37. P. L. Ward, Bull. Geol. Soc. Am., 82, 2991, 1971.
38. P. R. Vogt, Nature, 239, 79, 1972.
39. B. I. R. Haigh, Geophys. J. R. Astr. Soc., 33, 405, 1973.
40. W. J. Morgan, Nature, 230, 42, 1971.
41. P. R. Vogt, Earth Planet. Sci. Lett., 13, 153, 1971.

MORPHOLOGY OF THE MID-OCEAN RIDGE BETWEEN ICELAND AND THE ARCTIC BASIN

G.L. Johnson

U.S. Naval Oceanographic Office
Washington, D.C. 20373, U.S.A.

ABSTRACT. The Mid-Ocean Ridge in the environs of Iceland is anomalously elevated. Morphologically the ridge crest is a rugged arch lacking a central rift. With distance from Iceland the rift valley reappears and the ridge subsides to more normal depths.

1. INTRODUCTION

1.1. Oceanic Crustal Processes

The Mid-Ocean Ridge is one of the principal active structures of the earth. This globe-encircling submarine ridge extends through the Arctic, Atlantic, Indian, Antarctic, South Pacific, and the eastern extremity of the North Pacific Ocean for a total length of more than 75,000 km [1]. The concept that sea floor is generated along the crestal region of the Mid-Ocean Ridge has passed from mere speculation to essentially proven fact and is today widely accepted. Dietz [2] and Hess [3] proposed a hypothesis for evolution of the ocean basins by spreading of the sea floor away from the crest of the Mid-Ocean Ridge system. They suggested that new ocean crust is formed at the axis of an oceanic ridge by upwelling mantle derivatives. This new crust accretes to the edge of the plate and thence is carried laterally away from the ridge crest in a conveyor belt fashion. A more detailed review of the oceanic crust and its generation is found in reference [4].

Kristjansson (ed.), Geodynamics of Iceland and the North Atlantic Area. 49-62. *All Rights Reserved.*
Copyright © 1974 by D. Reidel Publishing Company, Dordrecht-Holland.

1.2 Morphology

With time the crust of the ocean floor becomes mantled by a sedimentary blanket which buries the original igneous surface. The
morphology of the sea floor is therefore the result of the generation of igneous topography as a result of faulting and volcanism
at the crest of the Mid-Ocean Ridge and its slow burial by sedimentary processes.

Morphologically the Mid-Ocean Ridge is a broad fractured arch
rising 1-3 km above the adjacent sea floor with an average width
of 1500 km in most areas. The shape of the ridge is apparently a
function of a variable rate of spreading and a relatively constant
rate of crustal subsidence with time. In detail the morphology
seems to depend on spreading rate. "Fast" ridges such as the East
Pacific Rise appear as broad, topographic arches with small (180 m
amplitude, 9 km wavelength) relief superimposed on the arch [5].
The crestal region is marked by a slight arch or horst [5] and no
axial valley is present. Slower spreading rates, such as prevail
in the Atlantic, usually create an uplifted blocky crestal zone,
rift valley and rugged flank topography (300-800 m amplitude, 15
km wavelength). There is approximate symmetry in topographic profiles with respect to bottom roughness, regional slope, sediment
thickness and with respect to the ocean basins across the Atlantic
and other branches of the Mid-Ocean Ridge.

Iceland is geographically part of the ridge system; however,
it is anomalous in that it is subaerial although it may be no
more atypical than other segments of the ridge. The Mid-Ocean Ridge
in the environs of Iceland is also atypical in that it does not
fit the slow-fast category. After examination of the morphology
with increasing distance from Iceland, I will present a possible
influencing factor.

2. MORPHOLOGY OF THE MID-OCEAN RIDGE NORTH OF ICELAND

I will limit my discussion to the Mid-Ocean Ridge north of Iceland.
For information on the Reykjanes Ridge see references 6 to 9.

2.1 Kolbeinsey Ridge

The branch of the Mid-Ocean Ridge between Iceland and Jan Mayen
is called Kolbeinsey Ridge. The connection between the spreading
axis of Iceland and Kolbeinsey Ridge is complex (Fig. 1).

Saemundsson [13] has proposed a 75 km wide fracture zone
between Husavik and the southern extent of Kolbeinsey Ridge. He
suggests an eastward migration of the spreading axis from Skagi

(20°W) 4 mybp to 18°W thence ∿1 mybp to 17°W. These shifts of the spreading axis are postulated by Saemundsson to be the result of the ridge axis' preference to remain near the melting anomaly (hot spot) of central Iceland [14]. Saemundsson predicts that Axar-fjordur (the submarine trough to the north-northeast of Husavik) is now undergoing ductile thinning and will become the next spreading center.

Geophysical data presented in Fig. 1 generally support Saemundsson's conclusions. Just to the south of Kolbeinsey the

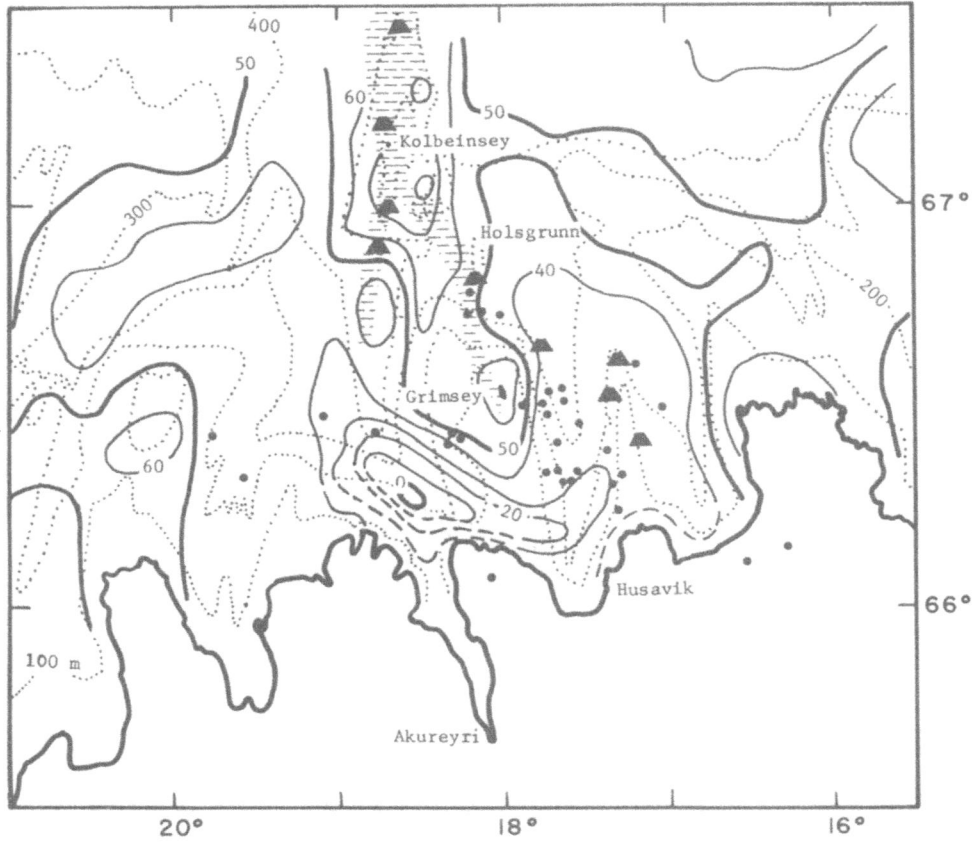

Fig. 1. Geophysical parameters north of Iceland. Dots, earth-quakes [10]; heavy lines, free-air gravity anomaly [11]; dotted lines, bathymetry (m); horizontal dashed lines, positive magnetic anomaly (Project MAGNET data); and truncated triangles, fresh basalt samples [12].

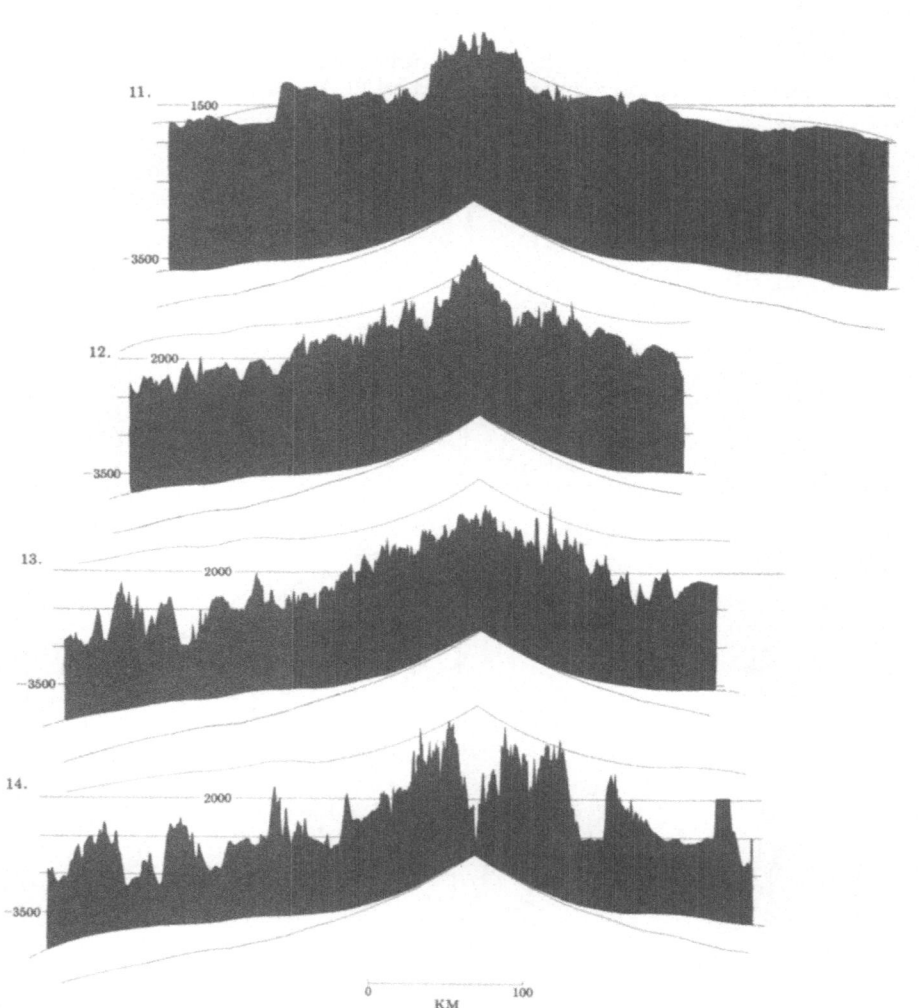

Fig. 2. Reykjanes Ridge profiles [8]. Thin line is the empirical
curve for northeast Pacific subsidence [24]. The profile base
line assumes an average sedimentation rate and thence subtracts
for loading and subsequent depression of the crust by sediments.
The uppermost curve is the Pacific curve with 2000 m added.
Spreading rates of 0.98 cm/yr to 10 mybp; 0.80 cm/yr 10 to 21 mybp;
0.70 cm/yr 21 to 38 mybp; 0.90 cm/yr 38 to 49 mybp and 1.15 cm/yr
49 to 60 mybp were assumed [31]. Depth in meters. Indexed on Fig. 4.

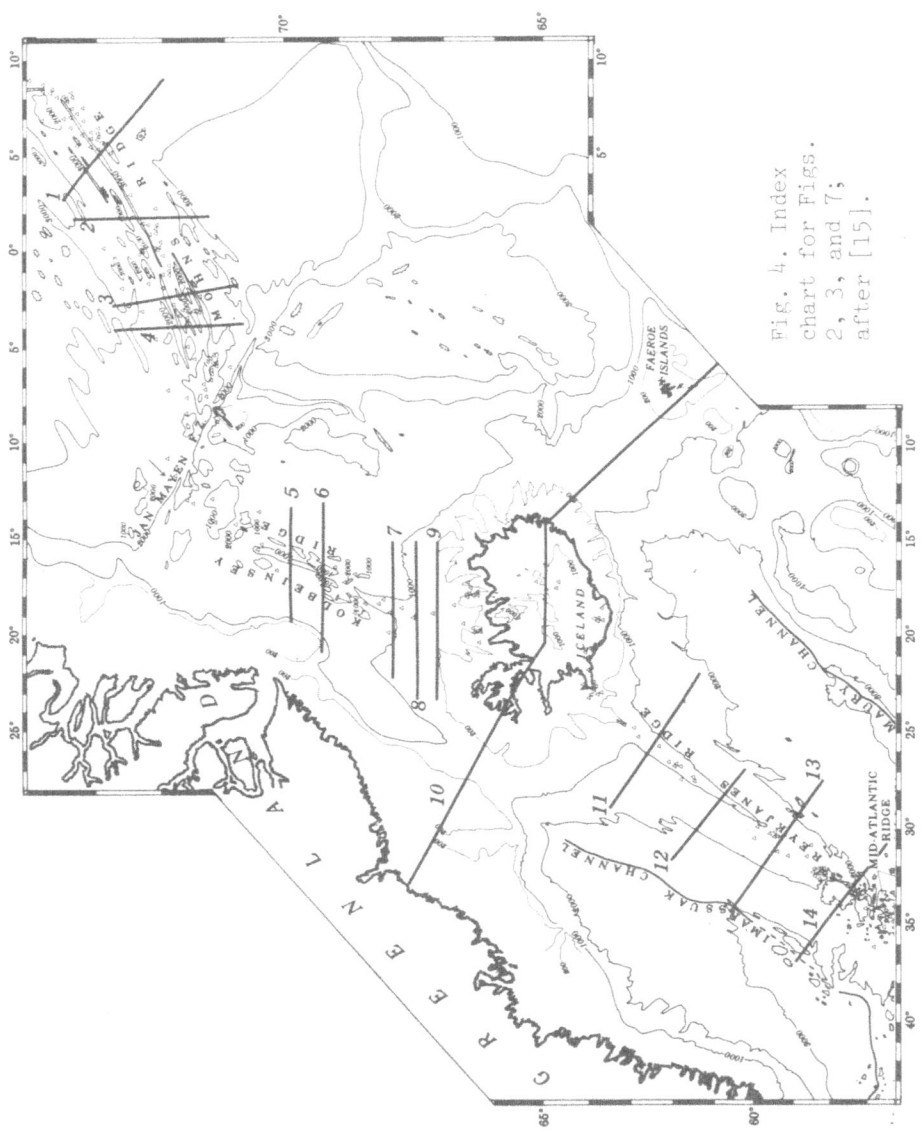

Fig. 4. Index chart for Figs. 2, 3, and 7; after [15].

axial magnetic anomaly bifurcates. The western branch continues
southward along a submarine valley to 66°40'N. The southern part
of this valley contains 100 meters of transparent sediment
suggesting volcanism has not extended this far south [15], or conversely is retreating northward. Fresh basalt pillows have been
recovered at 67°54'N [16]. The southeastern branch is defined by
a positive magnetic anomaly which strikes south-southeast to just
west of Grimsey and dies out at 66°28'N. Earthquake epicenters
to the east of Grimsey (Fig. 1) and along the Manareyjar chain
(north of Husavik) suggests this region is tectonically active.
The sea floor here is pierced by numerous volcanic peaks [15]
from which fresh basalt has been dredged (Fig. 1).

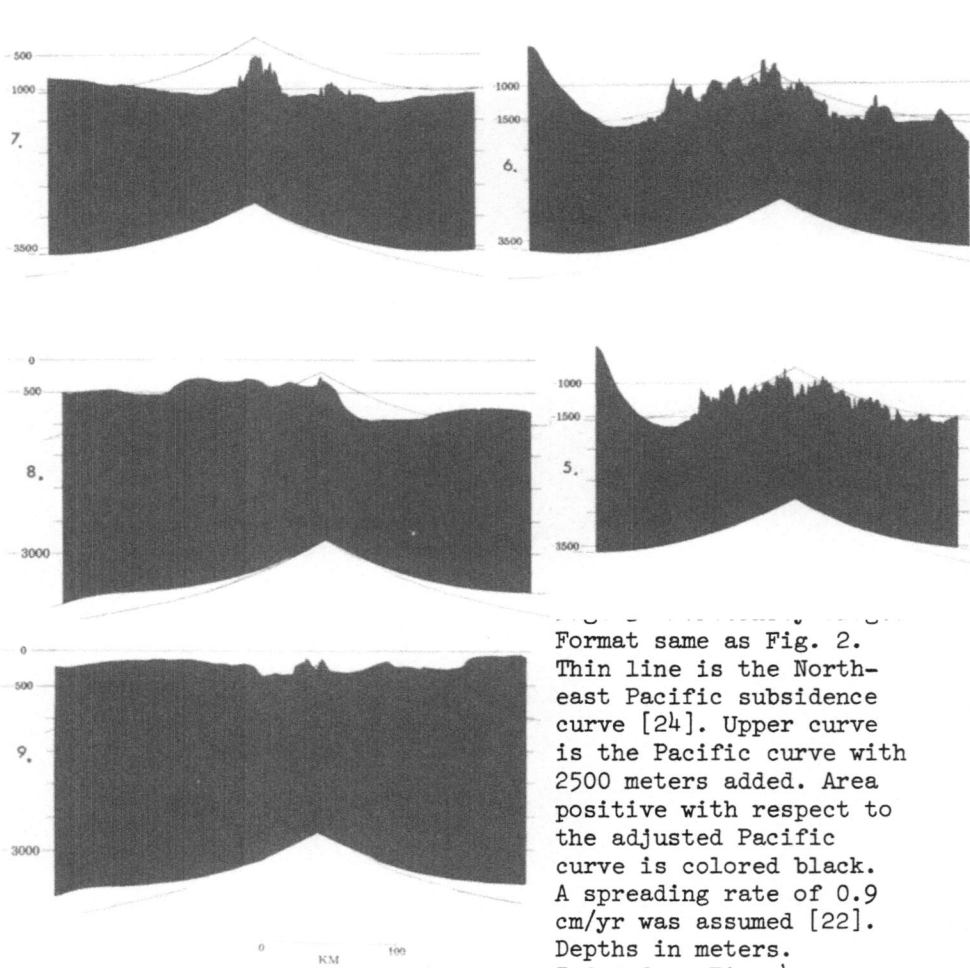

Format same as Fig. 2.
Thin line is the North-
east Pacific subsidence
curve [24]. Upper curve
is the Pacific curve with
2500 meters added. Area
positive with respect to
the adjusted Pacific
curve is colored black.
A spreading rate of 0.9
cm/yr was assumed [22].
Depths in meters.
Indexed on Fig. 4.

The Husavik fracture zone is clearly defined by a gravity low
[11] and by the bathymetric contours. The northern extent of the
Grimsey shoal area as marked by epicenters and topography (Fig. 1)
may represent the northern margin of the Tjörnes Fracture Zone.
There is, however, no offset to the magnetic anomalies in Fig. 1.

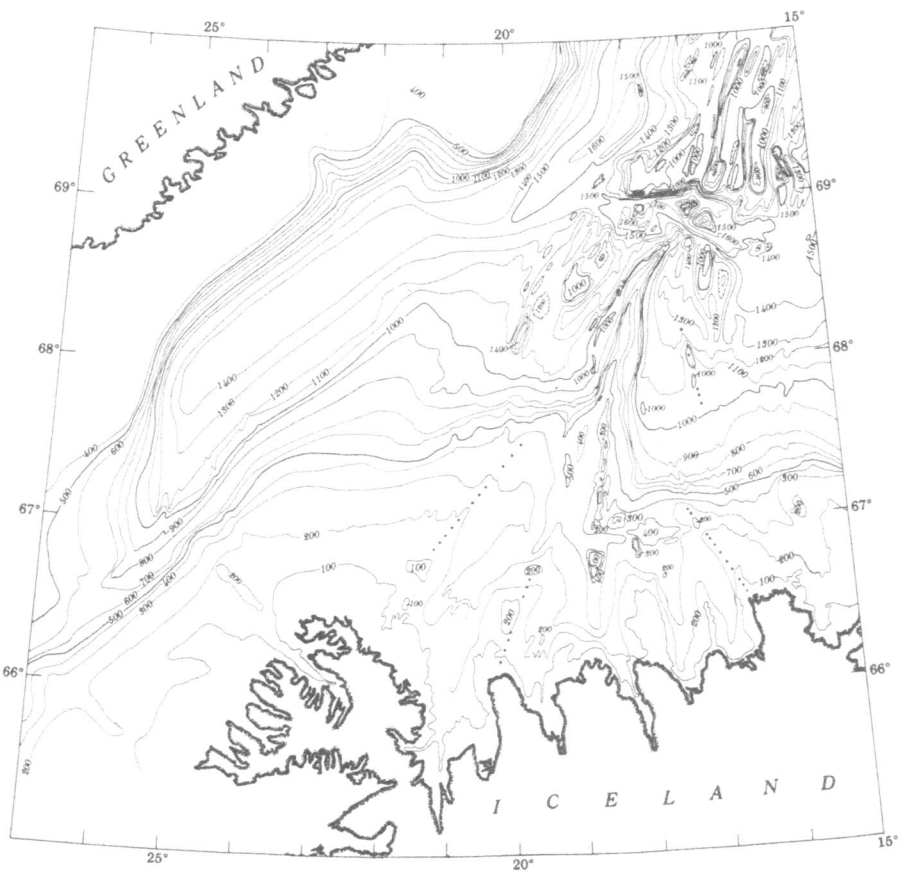

Fig. 5. Bathymetric chart of Kolbeinsey Ridge in meters. Based
on data from [17,22]. Dotted lines are possible time-transgressive
basement ridges.

Kolbeinsey Ridge does not have a well developed axial valley
such as is found farther south in the Atlantic (Fig. 2, Profile 14)
[17]. However, north of Spar Fracture Zone at 68°30'N an irregular
depression of greater than 100 m along the axial province of the

Fig. 6. Chart of N. Kolbeinsey Ridge. Based on data from LYNCH 1973 and LDGO. Depths in meters.

ridge seems to be a rift valley (Fig. 3, Profile 5, Figs. 4-6).
There is an increase in width and relief of the ridge to the north.
(Fig. 3); in an opposite sense it is morphologically similar to
the Reykjanes Ridge (Fig. 2) which widens to the south. This may
be largely due to burial of the flanks at a high rate of sedimentation near Iceland.

Southward pointing V-shaped (time-transgressive) basement
ridges have been documented on the Reykjanes Ridge south of Iceland
[18]. Vogt [19] has postulated that these ridges are trails left by
southward flowing magma from the postulated Iceland mantle plume
(hot spot) [14,20,21]. Careful examination of the bathymetric con-

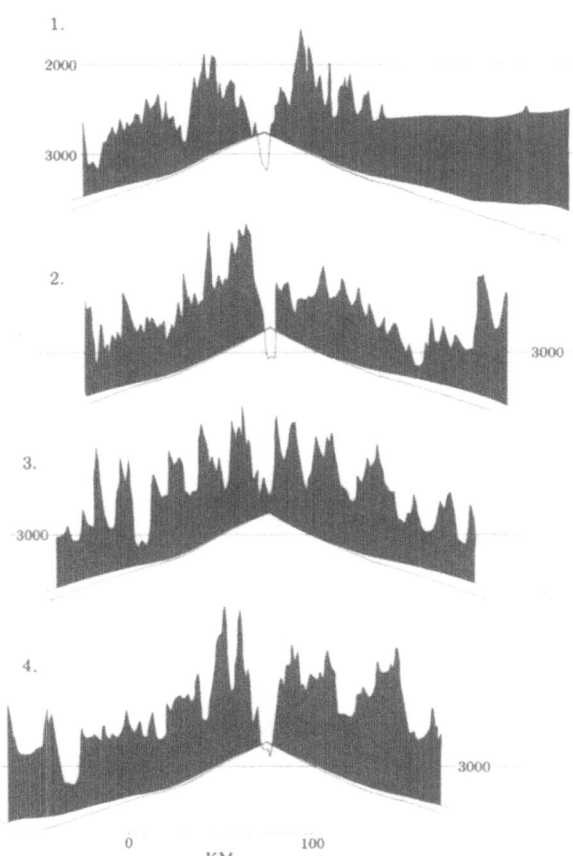

Fig. 7. Mohns Ridge. Format same as Fig. 2. Spreading rates of 0.88
cm/yr for 0-10 mybp; 0.70 cm/yr for 10-21 mybp; 0.62 cm/yr for
21-38 mybp; 0.80 cm/yr for 38-49 mybp and 1.01 cm/yr for 49-60 mybp
were assumed [31]. Indexed on Fig. 4.

tours north of Iceland suggests a similar phenomena; in that the
topographic lineaments on Fig. 5 are time-transgressive and cut
obliquely across the reported magnetic anomalies [17]. This
pattern implies that a variable northward migration of ultrabasic
partial melts may also occur along Kolbeinsey Ridge in response
to the mantle plume. Petrochemical evidence of rocks from the
present ridge crest, however, suggests that Tjörnes Fracture Zone
acts as a dam [12].

3.   MOHNS RIDGE

That portion of the Mid-Ocean Ridge which strikes from northeast
of Jan Mayen toward Bear Island is commonly called Mohns Ridge.
Mohns Ridge (Fig. 7) resembles in form a typical mid-ocean ridge.
The crestal zone, which averages from 900 to 1800 m shoaler than
the rift valley, is about 15 km in width, with the unburied flanks
extending more than 130 km away from the crest. A deep axial rift
is present on all profiles. A small offset is apparently present
in the rift at $1\frac{1}{2}°$E (Fig. 8); however, no unusual earthquake
activity is noted in this region.

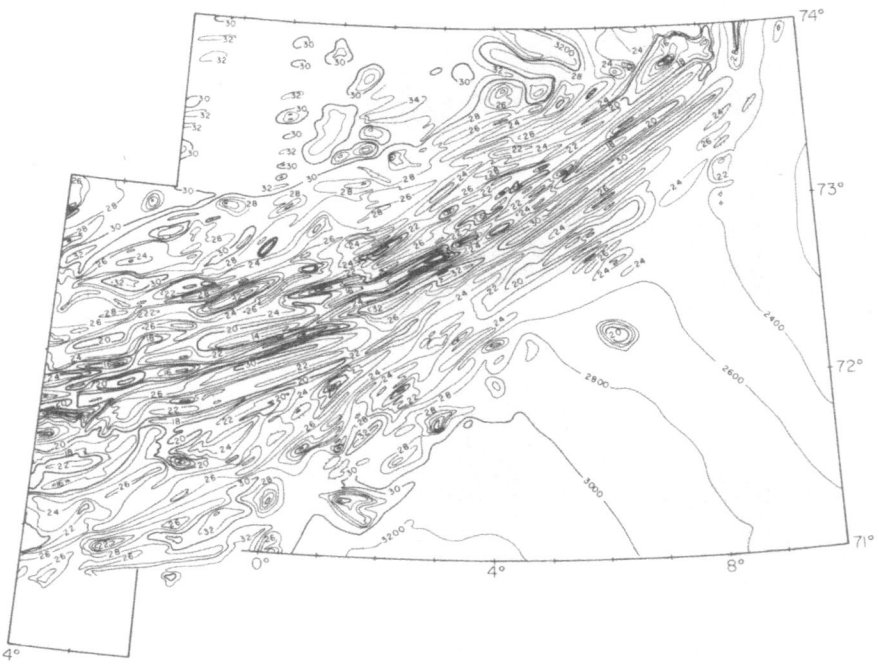

Fig. 8. Bathymetric chart of Mohns Ridge. Contours in meters with
alterations from [23].

4. KNIPOVITCH RIDGE

Knipovitch Ridge abuts Mohns Ridge and strikes north-south toward
Svalbard, where it is offset by fracture zones prior to the Mid-
Ocean Ridge entering the Arctic Basin [25]. Morphologically the
ridge is atypical in that in addition to the expected rough arch
its eastern flank is bordered by a continuous linear trench
(Fig. 9). This trench has generally been accepted as an axial
rift valley or/and a fracture zone [23] or trench [25]. The earthquake epicenters tend to lie in or very near to the axis. The
rift is narrow, generally 4 to 5 km wide at the 3200 m isobath and
has well developed structural benches on its walls. The rift
mountains generally lie at depths less than 2200 m and, to the
east, are barely discernible, due to burial. The trench abruptly
terminates at 78°30'N as it impinges upon the continental rise of
Svalbard.

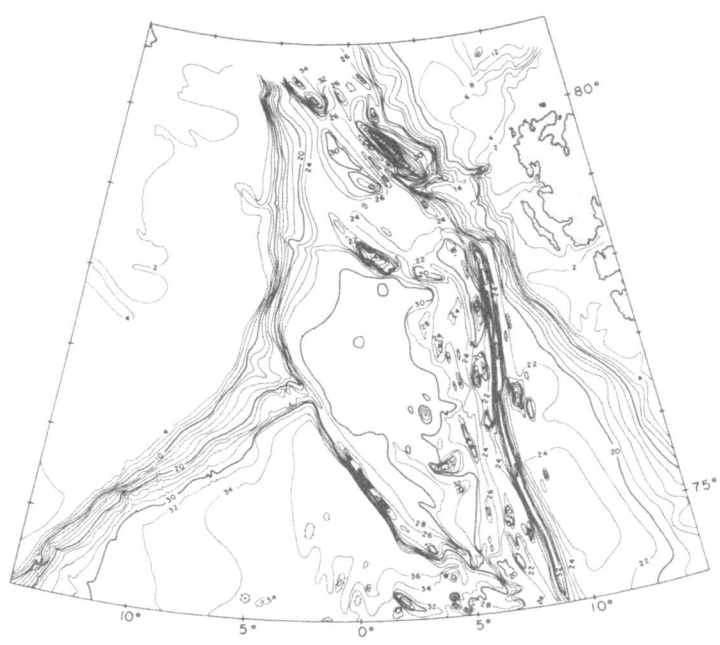

Fig. 9. Bathymetric chart of Knipovitch Ridge based on [23 and
25] and recent LDGO data. Contours in meters.

5.   ANOMALIES IN BASEMENT ELEVATION

It has been known for some time that basement subsides rather
uniformly after formation at axial depths of about 2.5 to 3 km
[24]. The subsidence seems to be due largely to thermal contraction of the cooling lithosphere with isostatic loading by the
increasing water and sediment overburden amplifying the thermal
effect. An alternative view is that plates thicken continuously
with crustal age, and are slightly denser than the immediately
underlying asthenosphere. The plates will therefore sink further
into the asthenosphere as they thicken and cool [26].

Sclater et al. [24] have documented the age/depth relationship
for many oceanic areas; use of their data makes it now possible
to determine any variation from the norm. To investigate any
anomalies in the depth versus crustal age relationship in the
region near Iceland, we subtracted the average North Pacific subsidence law, as derived empirically by Sclater et al. [24], from
the basement level. Because of the relatively thick sediment
cover in the northern Atlantic, it is necessary not only to strip
off the sediment, but to correct for the isostatic depression of
the basement caused by the sediment load. To make the isostatic
correction we assumed an average sediment density of 2.0 gm/cc
and an average compensating density of 3.3 gm/cc.

Inspection of Figs. 2, 3, and 7 shows that almost the entire
region is positive with respect to the North Pacific. The greatest
topographic anomaly is in the region of Iceland (Figs. 2 and 3)
and it progressively decreases to both the north and south [27].

5.1   Conclusions

Both the Reykjanes, Kolbeinsey and Galapagos Ridges are elevated
crestal blocks with no rift valley despite a slow spreading rate
of about 1 cm/yr for the first two and 2.2 to 3.6 cm/yr for the
latter [6,28].

The Mid-Ocean Ridge in the vicinity of Iceland with the excep-
tion of the fracture zones is relatively aseismic for some 900 kilometers north and south of the postulated Iceland hot spot [29,32],
Francis [30] suggests that the elevated temperatures along the
northern end of the Reykjanes Ridge have increased the ductility
of the mantle and therefore depressed the level of seismic activity.
Vogt et al. [4] suggested that under fast spreading ridges the
melting surface for basalt reaches the base of the oceanic layer
under the axis ensuring a continuous injection. At slower spreading rates there is discontinuous injection. A thermal anomaly
beneath Iceland would therefore alter the model of slow spreading
ridges to that of the faster ridges [4]. The petrochemistry of

the Mid-Ocean Ridge lavas near and on Iceland [11,13,15] suggests a mixing of depleted asthenosphere and a deeper mantle source. Additionally the high amplitude of magnetic anomalies reflecting a high Fe/Ti content, near postulated hot spot centers such as Iceland has only just been recognized [29].

All these lines of reasoning suggest that the Mid-Ocean Ridge in the environs of Iceland has been influenced in regard to morphology, elevation, petrochemistry, magnetization and seismicity by a thermal anomaly centered in Iceland.

REFERENCES

1. B. C. Heezen and M. Ewing, The Mid-Oceanic Ridge, in: The Sea, ed. by M. N. Hill, 3, Interscience Publ., New York, 1963.
2. R. S. Dietz, Nature, 190, 854, 1961.
3. H. H. Hess, History of the Ocean Basins, in: Petrologic Studies, Geol. Soc. Amer., Denver, 1962.
4. P. R. Vogt, E. D. Schneider and G. L. Johnson, The Crust and Upper Mantle Beneath the Sea, in: The Earth's Crust and Upper Mantle, ed. by P. J. Hart, Am. Geophys. Un., Washington, D.C., 1969.
5. R. N. Anderson and H. C. Noltimier, Geophys. J. R. Astr. Soc., 34, 137, 1973.
6. M. Talwani, C. C. Windisch and M. G. Langseth, J. Geophys. Res., 76, 473, 1971.
7. U. Fleischer, A. Korschunov, G. Schulz and P. R. Vogt, "Meteor" Forsch. - Ergebnisse, 13, 64, 1973.
8. J. Ulrich, Kieler Meeresforschungen XVI, 155, 1960.
9. U. Fleischer, The Reykjanes Ridge, (this volume).
10. L. R. Sykes, Bull. Seism. Soc. Am., 55, 501, 1965.
11. G. Palmason, The Insular Margin of Iceland, in: The Geology of Continental Margins, ed. by C. A. Burk and C. L. Drake, in press, 1974.
12. J. G. Schilling, D. G. Johnson and T. H. Johnston, Trans. Am. Geophys. Un., 55, 294, 1974.
13. K. Saemundsson, Geol. Soc. Am. Bull., 85, 495, 1974.
14. S. P. Jakobsson, Lithos, 5, 365, 1972.
15. G. L. Johnson, The Morphology and Structure of the Norwegian-Greenland Sea, Ph. D. thesis, University of Copenhagen, 1974.
16. P. R. Pinet, J. G. Schilling and R. L. McMaster, Trans. Am. Geophys. Un., 55, 284, 1974.
17. O. Meyer, D. Voppel, U. Fleischer, H. Closs and K. Gerke, Deut. Hydrogr. Z., 25, 13, 1972.
18. P. R. Vogt and G. L. Johnson, Earth Planet. Sci. Lett., 15, 248, 1972.
19. P. R. Vogt, Earth Planet. Sci. Lett., 13, 153, 1971.
20. J. T. Wilson, Roy. Soc. London Phil. Trans., 258, 145, 1965.

21. W. J. Morgan, Nature, 230, 42, 1971.
22. G. L. Johnson, J. R. Southall, P. W. Young and P. R. Vogt,
     J. Geophys. Res., 77, 5688, 1972.
23. G. L. Johnson and B. C. Heezen, Deep-Sea Res., 14, 755, 1967.
24. J. G. Sclater, R. N. Anderson and M. L. Bell, J. Geophys.
     Res., 76, 7888, 1971.
25. G. L. Johnson, Jokull, 22, 65, 1972.
26. R. L. Parker and D. W. Oldenburg, Nature Phys. Sci., 242,
     137, 1973.
27. B. I. R. Haigh, Geophys. J. R. Astr. Soc., 33, 405, 1973.
28. G. L. Johnson, P. R. Vogt, R. Hey, J. Campsie and A. Lowrie,
     Mar. Geophys. Res., in press, 1974.
29. P. R. Vogt and G. L. Johnson, Nature, 245, 373, 1973.
30. T. J. G. Francis, Earth Planet. Sci. Lett., 18, 119. 1973.
31. P. R. Vogt and O. E. Avery, J. Geophys. Res., 79, 363, 1974.
32. P. R. Vogt and G. L. Johnson, Earth Planet. Sci. Lett., 18,
     49, 1973.

THE AZORES-GIBRALTAR PLATE BOUNDARY

A.S. Laughton and R. B. Whitmarsh

Institute of Oceanographic Sciences,
Wormley, Surrey, England.

1. INTRODUCTION

Between the Azores and Gibraltar there is a belt of earthquakes
which, with its eastward extension into the Mediterranean, marks
the boundary between the African and Eurasian plates. There is
no offset of the Mid-Atlantic Ridge at the western end of this
belt, which goes through the East Azores Islands, and no indication of a continuation of the belt westwards into the western
Atlantic.

Various authors [1,2,3] have shown that the Azores-Gibraltar
plate boundary results from the different spreading rates of the
Eurasian and African plates relative to the North American plate
thus partially transforming the spreading of the Mid-Atlantic Ridge
north of the Azores into the complex compressional region of the
Mediterranean. The Azores are believed to be the result of spreading
from a zone extending from the junction of the Eurasian, African and
North American plates [4]. From an analysis of the gross movement
of the Eurasian and African plates, and from earthquake fault plane
solutions, McKenzie [1] concluded that today the western end of the
Azores-Gibraltar plate boundary is dilatational, the central section
is dextral shear and the section near to Spain and Africa is compressional.

For some time the region between the Azores and Gibraltar
was known as the Azores-Gibraltar Ridge, although the evidence
for an east-west ridge was rather poor. An analysis of all available bathymetric data now shows that there is no continuous ridge
and that the region is considerably more complex. The principal

Kristjansson (ed.), Geodynamics of Iceland and the North Atlantic Area. 63-81. *All Rights Reserved.*
Copyright © 1974 by D. Reidel Publishing Company, Dordrecht-Holland.

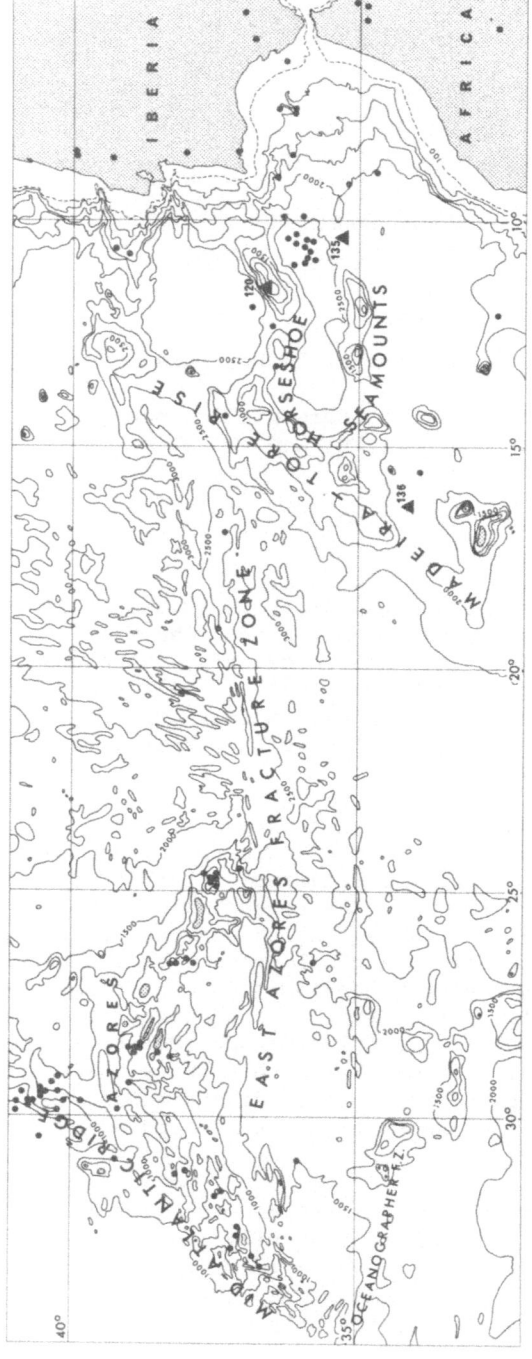

Fig. 1. Bathymetry of the region between the Azores and Gibraltar showing earthquake epicentres 1961-1970 and the position of DSDP holes 120, 135 and 136. Depths in corrected fathoms, contour interval 500 fathoms.

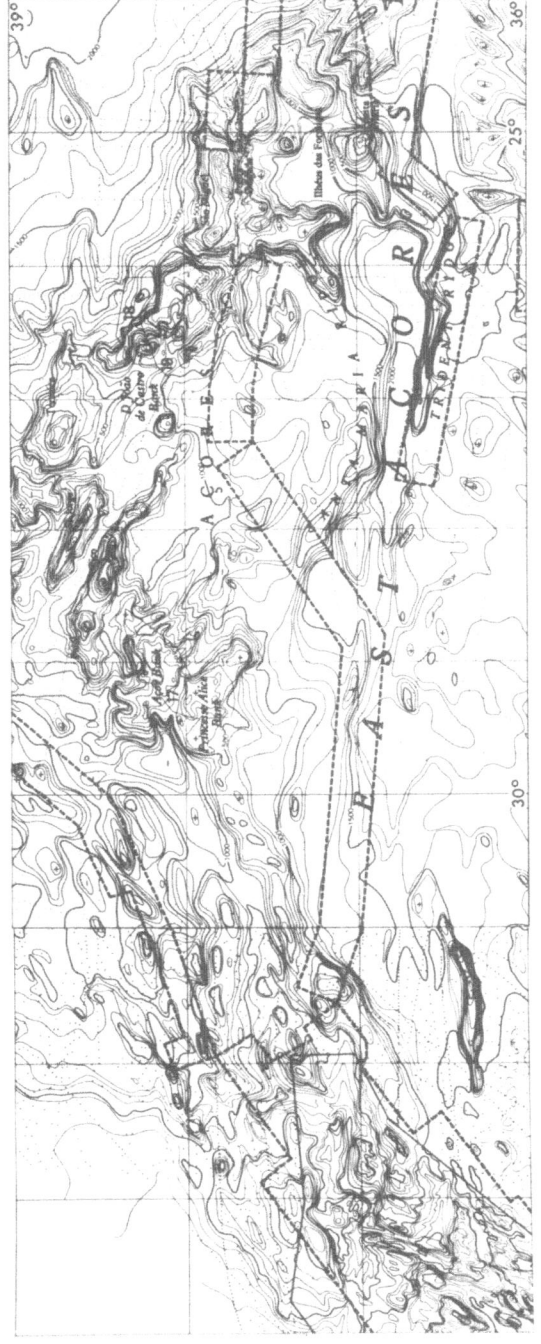

Fig. 2. Bathymetry of the Mid-Atlantic Ridge and the southern part of the Azores Plateau showing area covered by long range side-scan sonar. Depths in corrected fathoms, contour interval 100 fathoms. (Figs. 2, 3 and 5 are part of unpublished chart by Laughton, Roberts and Graves).

elements (Fig. 1) are the East Azores Fracture Zone between 15°W and 31°W, and further east the complex of ridges and seamounts which includes the Horseshoe Seamounts. The western end of the East Azores Fracture Zone diverges from the seismic belt and runs south of the Azores Plateau to approach the Mid-Atlantic Ridge axis at 37°N. West of the Ridge, the Pico Fracture Zone is collinear with the East Azores Fracture Zone and extends to at least 40°W [5].

In 1969, 1971 and 1973 sections of the plate boundary between 15 and 30°W were examined in R.R.S. "Discovery" using GLORIA, a long range side-scan sonar [6], together with other geophysical techniques. Also in 1973, sonar studies were made in the FAMOUS area of the Mid-Atlantic Ridge (at 37°N) where a westward extension of the East Azores Fracture Zone would intersect the median valley. The following study, which incorporates the data obtained on these three cruises, is also based on compilations of all available bathymetric and magnetic data, some of which are as yet unpublished and have been kindly made available for this paper.

2. MORPHOLOGY

The morphology can be described in three sections starting at the Mid-Atlantic Ridge.

(1) The Azores section (Fig. 2)

The active plate boundary (trend 115°) passes through the Azores Islands [4] but south of the Azores Plateau there is the inactive part of the East Azores Fracture Zone. In the FAMOUS area a number of E-W fracture zones offset the median valley. The largest offset is 75 km at 37°N. However even this fracture zone does not appear to penetrate a very prominent NE-SW ridge which continues NE to the Princesse Alice Bank. East of the ridge there is an E-W scarp forming the southern side of this bank south of which there are relatively thick sediments giving a smooth topography. Between 28 and 25°W, a prominent sediment filled trough and a ridge (Trident Ridge) lie just south of the Azores Plateau [4,7]. The east end of the ridge is linked with Santa Maria Is., a volcanic feature of Middle Miocene age. Thus in this sector the East Azores Fracture Zone is not continuously well developed, but there are strong parallel almost collinear lineations which suggest a common origin.

Between 27 and 28½°W the Santa Maria Ridge, which links up with Princesse Alice Bank, parallels the Terceira spreading axis and is not therefore part of the East Azores Fracture Zone.

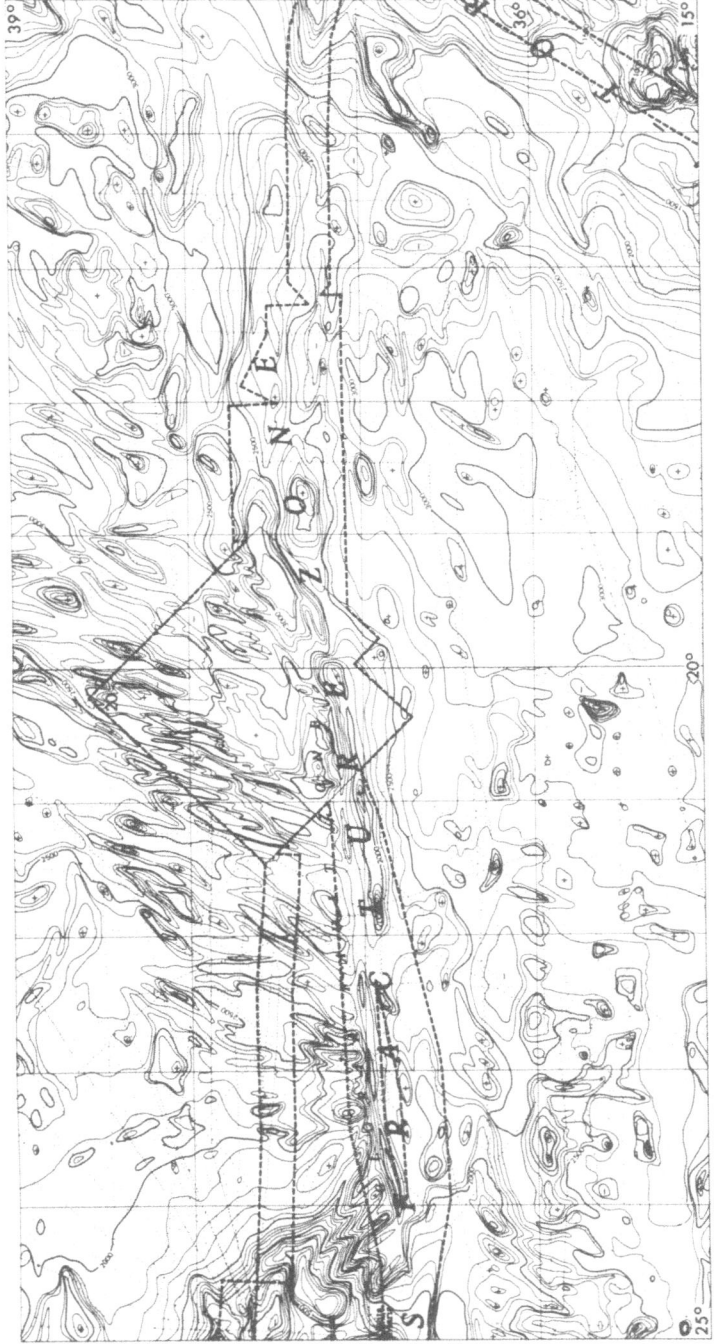

Fig. 3. Bathymetry of the central part of the Azores-Gibraltar plate boundary showing area covered by long range side-scan sonar. Depths in corrected fathoms, contour interval 100 fathoms.

(2) The central section (Fig. 3).

    The southern slope of Trident Ridge can be traced south of
Santa Maria Is. and links with the very narrow and well defined
valley of the Gloria Fault discovered in 1971 [8]. This valley
truncates NNE-SSW trending topography to the north and separates
it from a more thickly sedimented region to the south. There is
an E-W ridge immediately south of the valley. These features were
particularly clearly shown by a sonar mosaic representing 25,000
sq. km between 19 and 21°W [9]. In the northern half of the mosaic,
echoes were obtained from NNE-SSW ridges (Fig. 4) which parallel
the sea-floor spreading magnetic anomalies, indicating that the
ridges are the relics of the original sea-floor spreading topography. In the southern corner the Gloria Fault truncates these
ridges at the same position where the linear magnetic anomalies
end.

    In the eastern part of the mosaic more E-W or ENE-WSW scarps
appear and the pattern becomes more complex. An eastward extension of the mosaic area by 17,000 sq. km was made in 1973. These
data together with data from the mapped topography and from the
magnetic anomaly map show that between 15 and 19°W the plate
boundary probably lies along 38°N since south of the boundary
NE-SW trends can be associated with the African plate whereas the
NNE-SSW ridge trends found in the northern part of the first
mosaic can be associated with the Eurasian plate. A large E-W
ridge dominates the topography between 16 and 19°W but the sonar
showed this to consist of several large blocks bounded by NW-SE
or NE-SW scarps. However the northern face of the ridge was not
surveyed by sonar and it is here that the continuity of the

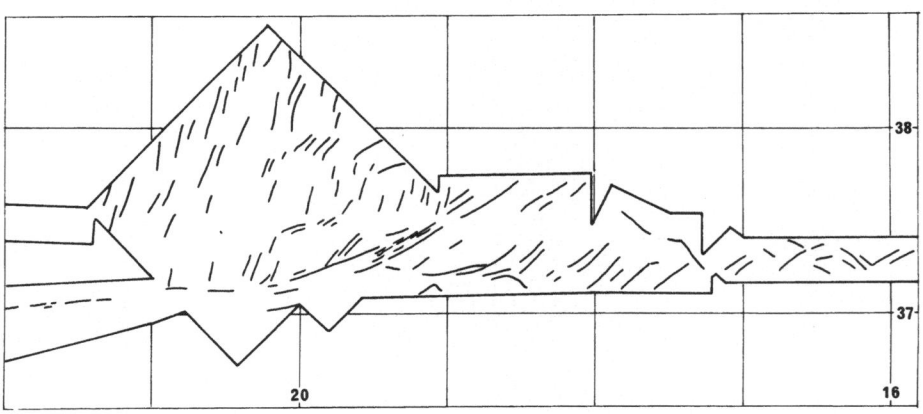

Fig. 4. Topographic trends derived from sonographs of the central
part of the Azores-Gibraltar plate boundary.

oceanic magnetic anomalies from the north is interrupted. From
the bathymetric chart it appears that there may also be a
significant valley north of this ridge with a trend of ENE-WSW
leading towards Tore Seamount.

(3) The eastern section (Fig. 5)

The East Azores Fracture Zone does not continue as a clear
bathymetric feature east of 16°W. Between this longitude and 10°W
the topography changes completely and is dominated by a number of
large seamounts strung out along NE-SW and E-W ridges in the form
of an italic E. A third prominent trend of ENE-WSW can be recognised on scarps or individual ridges. A broad rise stretching
from Madeira to Tore Seamount (the Madeira-Tore Rise) forms the
western side of the E and carries several large seamounts separated
in places by relatively deep and sediment covered cols. The ENE-
WSW trend is particularly well developed in Gorringe Ridge, the
south side of which is collinear with the north side of Ampère
Seamount. The blocky nature of this whole region has been attributed to compressional tectonics which has both uplifted oceanic
crust to form seamounts and depressed it in regions now below the
thick sediments of the intervening plains [10,11,12; Purdey, private
communication]. The earthquake belt runs approximately along the
central arm of the E through Gorringe Ridge and into the east end of
the Horseshoe abyssal plain, but it is not well defined and earthquakes have occurred 400 km either side of this line.

3. MAGNETIC ANOMALIES

The magnetic anomalies adjacent to the Azores-Gibraltar plate
boundary were studied because they were expected to give independent evidence of the past and present position of the shear zone
in the crust. The position of this zone at the time of formation
of the crust producing the magnetic anomalies should be indicated
by a discontinuity in the pattern of the anomalies, due to differing spreading rates, and by a change in trend from the NNE trending Eurasian plate anomalies to NE trending African plate anomalies.
Offsets within one of these sets of anomalies should indicate either
old fracture zones or currently active shear zones.

Ships' tracks between 33° and 40°N were compiled and anomaly
profiles drawn with their vertical axes normal to the tracks.
From these data it was possible to draw a contoured chart from
which a simplified version depicting areas of positive and negative anomaly was prepared (Fig. 6). The density of tracks on this
figure is variable, being greatest in the middle longitudes and
less west of 28°W and south of 35°N. Correlations between tracks
are based on bathymetric trends, on large course changes and on

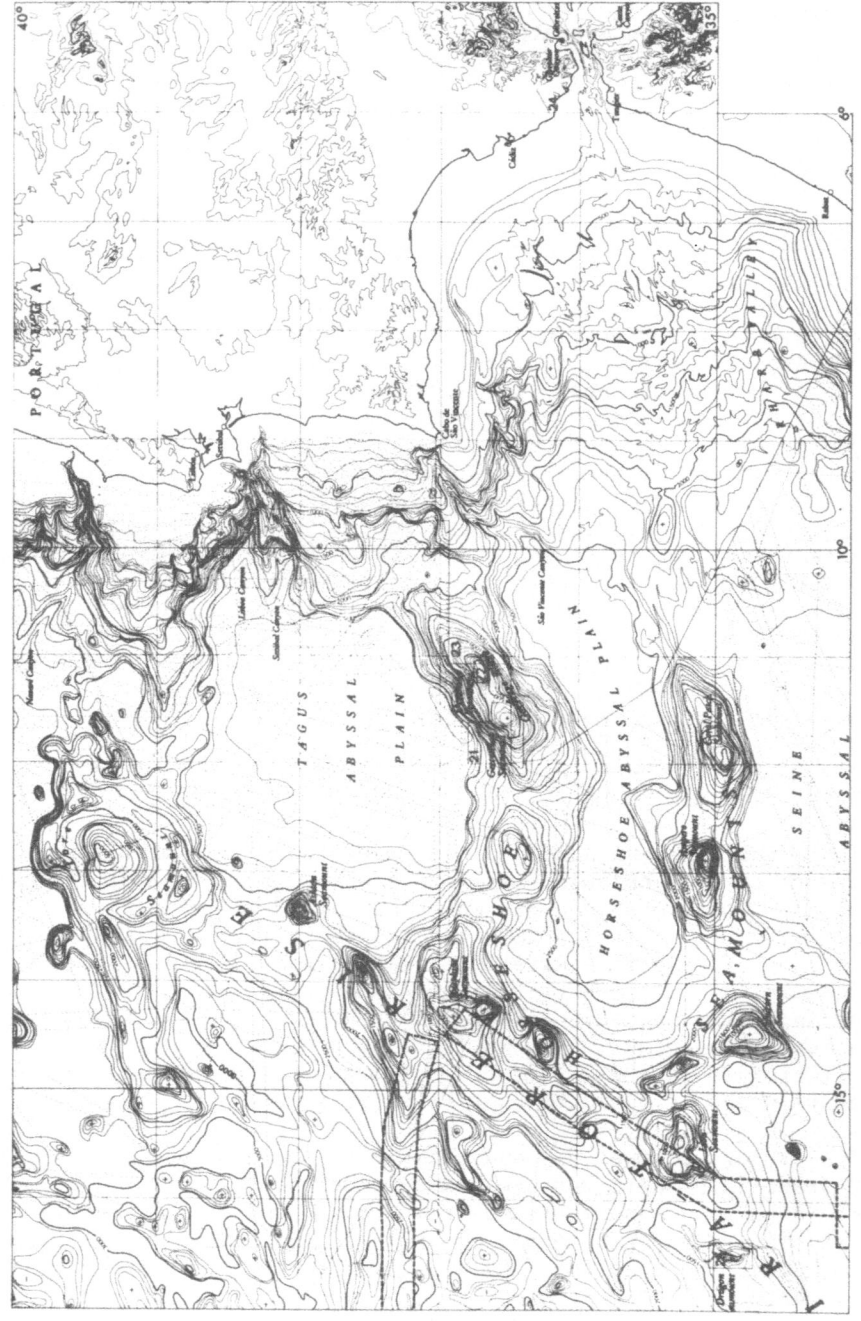

Fig. 5. Bathymetry of the eastern part of the Azores–Gibraltar plate boundary showing area covered by long range side-scan sonar. Depths in corrected fathoms, contour interval 100 fathoms.

Fig. 6. Magnetic anomaly chart, using IGRF, based on data from along the ships' tracks shown. Numbered heavy tracks show location of profiles illustrated in Fig. 7. Positive anomalies are stippled. Anomaly identifications are derived from comparison with models shown in Fig. 7. Boundary of the quiet magnetic zone (dotted) determined by inspection of profiles. (The authors are grateful to Lamont-Doherty Geological Observatory, Woods Hole Oceanographic Institution, University of Rhode Island, the US Naval Oceanographic Office, US Naval Research Laboratory, Bedford Institute of Oceanography, Centre National pour l'Exploitation des Océans, Vening Meinesz Laboratorium, and Cambridge University for supplying unpublished magnetic profiles, and to Krause and McGregor for permission to use their contoured magnetic chart around the Azores).

a study of selected profiles projected along a fixed direction
known to be approximately normal to the regional anomaly trends.
Correlations were established between these latter profiles on
the basis of anomaly shape and wavelength and were subsequently
used to identify sea-floor spreading anomalies and to determine
spreading rates. Near the Azores contours from other surveys or
compilations have been used to supplement the data [4,14].

North of the Azores and of latitude 37°N, the 025° trending
magnetic lineations are well developed as described by Williams
and McKenzie [13]. Along the line of the East Azores Islands Krause
and McGregor [14] have shown that there are lineations
parallel to the Terceira Rift which extend about 70 km either side
of the axis. South of 37°N the gross trend of the anomalies is
quite different from that to the north being mainly 045° although
this may consist in detail of a pattern of more north-south segments en echelon. The steps in this pattern can be ascribed to a
series of E-W fracture zones for which there is evidence also in
the bathymetry, although they are not well surveyed.

The present Eurasia-Africa plate boundary, as defined by
recent seismic activity, extends from a triple junction, where it
meets the American plate, near 39°N, 30°W [1,14] eastsoutheastwards
through the Azores in the form of a spreading axis and then eastwards
as an active transform fault. This fault appears to coincide with
the sharp change in trend of the magnetic anomalies between 25° and
15°W at latitude 37°N. Between 25° and 20°W the bathymetric and
sonar data have already demonstrated that a fault exists along this
latitude.

$\longrightarrow$

Fig. 7. Projected magnetic anomaly profiles and calculated profiles.
Profiles 1-9 were projected at 115°, the remainder at 130°. Model
anomaly profiles were generated along the direction 115° for a
point at 37°N, 22°W. Mesozoic anomalies assumed to have been originally generated in direction 130°. Source layer was between 5 and
7 km below sea-level with an effective susceptibility of 0.01 c.g.s.
units. Before anomaly 20 time the original latitude was taken as
22°N. The thick dashed line marks the proposed plate boundary.
Anomalies J [2] and R [15] are also shown. Spreading rates are the
values which result from being projected onto the direction 086°
(south, before anomaly 28 time; north, all profiles) or 097° (south,
after anomaly 28 time). Directions obtained from fracture zone
trends.

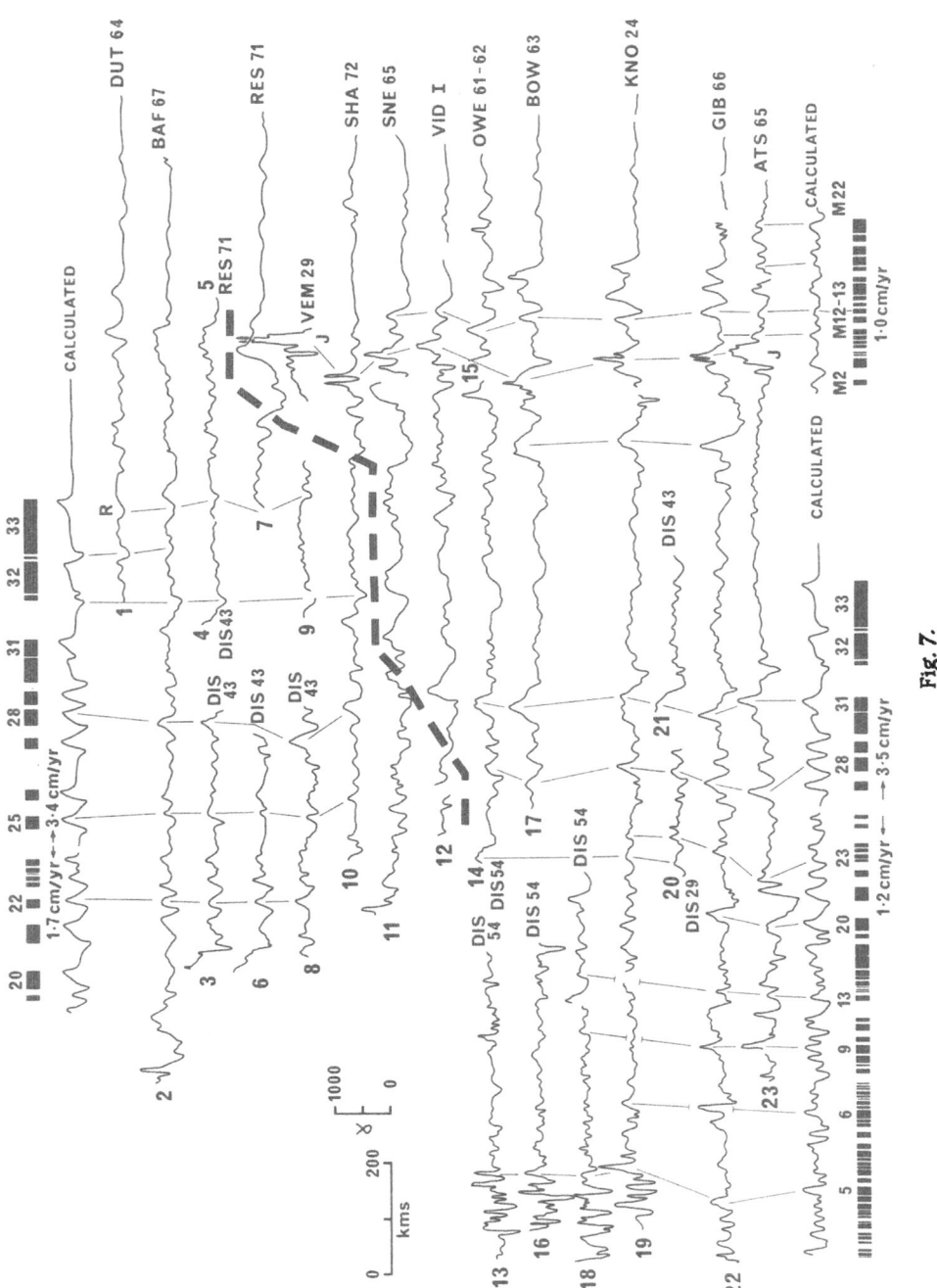

Fig. 7.

The position of the plate boundary east of 20°W can be determined by a combination of the sonar and magnetic data. Large negative anomalies (up to 500 gamma) appear to be associated with the
boundary. Similar large fracture zone anomalies were found slightly
further south in the Atlantic by Vogt, Anderson and Bracey [15]
and Schouten [16]. A large anomaly follows the sonar trend of 065°
between 18° and 20°W suggesting a plate boundary that lies close
to an earthquake epicentre and truncates anomaly 31. An E-W negative
anomaly at $37\frac{1}{2}$°N, $15\frac{1}{2}$°W, which is associated with the north side of
the large topographic ridge, and a similar anomaly at 38°N, 14°W,
which separates Ashton Seamount from Josephine Bank, may both
indicate the position of the plate boundary, although the latter
lies 70 km north of the nearest epicentre.

A number of anomaly profiles obtained along suitably oriented
tracks were projected onto lines approximately perpendicular to
the main lineations and oriented at 115° (north of the plate
boundary) or at 130° (south). The advantage of this presentation
is that when the projected profiles are stacked and correctly
aligned anomalies of the same age should lie one above the other
(providing also that the chosen directions are correct and that
fracture zones have not displaced the anomalies). Some of the
key correlations which could then be made are shown in Fig. 7.

North of the fault and east of the Azores firm anomaly
identifications can be made up to the 31-32 trough. The anomalies
have a well-developed 025° trend unbroken by fracture zones. A
sharp change in spreading rate is required around anomaly 24-25
time (63 my) as suggested by Pitman and Talwani [2] and Williams
and McKenzie [13]. A plot of anomaly age versus distance along the
spreading direction suggests that in fact there may have been a
gradual slowing down of spreading between anomalies 27 and 21
(Fig. 8). East of the 31-32 trough the anomaly amplitudes are
reduced and only tentative correlations of the 32-33 trough (77 my)
and an earlier trough can be made. The latter trough, which is not
predicted by the magnetic model, can be correlated with anomaly
R noted by Vogt et al. [15] in the north-west Atlantic. The region
between 16° and 13°W therefore corresponds to all or part of the
mid-Cretaceous constant polarity interval [17]. However within
this zone there are broad less well defined, but nevertheless quite
clear, anomalies trending 025°. East of 13°W the magnetic quiet zone
correlates with the western edge of the Tagus abyssal plain.

South of the fault the anomalies appear to have an 045° trend
but the sinuosity of the contours here and the well established 020°
trends south of the Azores and in the FAMOUS area suggest that in
fact the 045° trend consists of 020° trending anomalies that have
been offset by closely spaced east-west fracture zones. Some high
amplitude roughly east-west anomalies may correspond to such zones.
The distinctive peak of anomaly 31 (72 my) and its eastern flanking

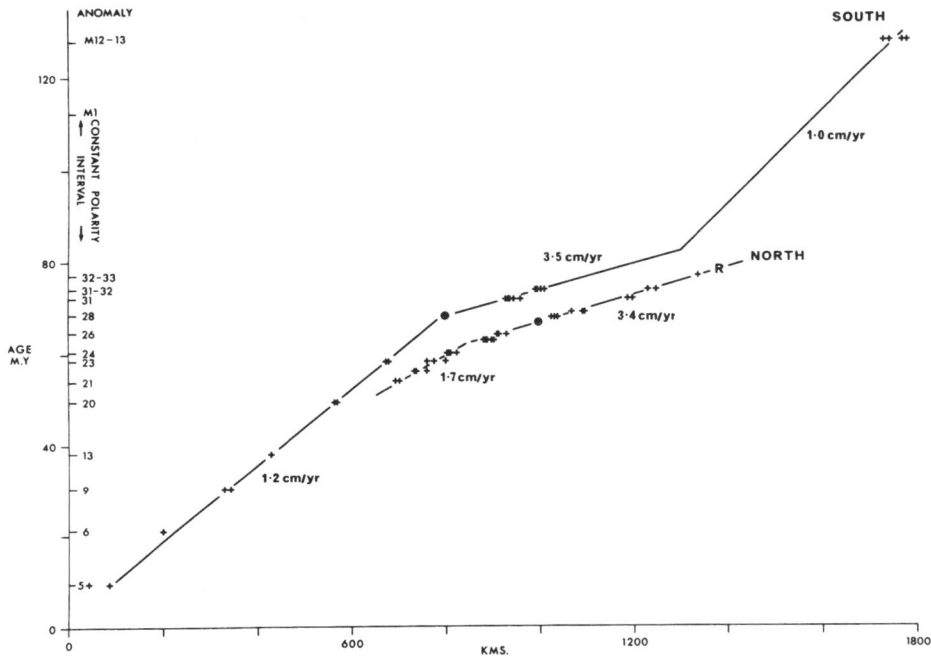

Fig. 8. Age of observed anomalies plotted against distance in
directions given in caption to Fig. 7. North of the plate boundary
distances are relative to anomaly 27 arbitrarily placed at 1000 km.
South of the boundary, distances are relative to anomaly 28 placed
at 800 km. Spreading rates per limb are in the directions parallel
to fracture zones.

trough are easily identified in this region (Fig. 7). It is more
difficult to pick anomalies to the west because prior to anomaly
25 the spreading was rather slow. However the present interpretation is broadly in agreement with that of Pitman and Talwani [2].

The plot of anomaly age versus distance along the spreading
direction (Fig. 8) indicates that a fairly abrupt change in
spreading rate occurred at about 68 my ago, 5 my later than on
the Eurasian plate. The differential spreading rate of 2.2 cm/yr
between the plates during these 5 my accounts for the relatively
larger separation of anomalies 25 and 28 (63 and 68 my) on the
Eurasian plate. East of anomaly 31, the anomalies correspond to
the mid-Cretaceous constant polarity interval and are subdued
until the foot of the Madeira-Tore Rise is reached. It is
interesting that, as to the north with anomaly R, a distinctive

trough is seen just east of the expected position of anomaly 33 although the data are inadequate to contour it with confidence.

Over the Madeira-Tore Rise, where igneous basement rocks have been uplifted closer to the surface, there is a band of complex high amplitude anomalies which were called anomaly J by Pitman and Talwani [2]. Since anomaly J closely follows the east side of the Madeira-Tore Rise it might be supposed to be a purely topographic effect. However it is present on Snellius profile I at 34°N which crosses a col on the Rise north of Madeira and also on tracks SW of the Rise shown in [2]. North of 38°N it is absent in spite of the elevated topography. The anomaly is therefore considered to be due to sea-floor spreading and to be fortuitously coincident with the Rise over which it is amplified by the small depths to basement. At 35°N an offset in the topography and a large E-W anomaly suggest that there is a fracture zone here which may link up eastwards with the Ampère Bank – Gorringe Ridge 060° trending lineation and possibly westwards with the fracture zone at 35°N, 22°W.

It is expected that east of anomaly J there exists older crust which should contain Late Jurassic-Early Cretaceous sea-floor spreading anomalies [17]. It is difficult to identify such anomalies in the region of the Horseshoe Seamounts due to the great sea-floor relief. However profiles south of these seamounts in the deep water north of Madeira should reveal such anomalies. Our only two profiles in this region may be interpreted in terms of the model in Fig. 7 based on the time scale of Larson and Pitman [17]. Although the correlations between anomalies M10 and M22 are quite convincing the data are very scant. Correlations of the positive anomaly M12-13 (128 my) can be seen in Figs. 6 and 7. It is striking however that Lattimore, Rona and DeWald [18] identified anomalies M4 and M22 at 22°N in the eastern Atlantic the same distance apart as predicted by our model. On the basis of our model anomaly J is related to the younger part of the Mesozoic sequence of reversals and is estimated to be 120 my old in contrast to 135 my estimated by Pitman and Talwani [2]. This interpretation can be combined with the subsequent spreading history on the anomaly age/distance plot (Fig. 8) using the identifications of the peak M12-13 on Fig. 7 and the spreading rate of 1.0 cm/yr. Figure 8 suggests therefore that the Late Cretaceous fast spreading period was short-lived covering just the period 82 to 68 my.

DSDP hole 136 [19] was drilled near to the M12-13 anomaly. The oldest sediments cored were Late Aptian (104 my) although these were 19 m above basaltic basement. A crustal age of 108 my was inferred by extrapolating the sedimentation rate. However this assumes that sedimentation started immediately the crust was formed, whereas it is possible now to find exposed basement 20 my old without sediment cover. Other larger hiatuses in sedimentation of several tens of millions of years were observed in hole 136 between the Late Cretaceous and the Miocene. These data are not therefore in conflict with the age of 128 my from the magnetic anomalies. The oldest sediments 19 m above basement recovered at DSDP hole 120 [11] were Barremian (118 my) although the magnetic anomalies would imply an age of 143 my. Again the difference may be due to an early hiatus in sedimentation.

The western boundary of the magnetic quiet zone is shown in Fig. 6 and in the profiles in Fig. 7. On the basis of these figures it is difficult to argue, as Pitman and Talwani [2] have done, that this boundary represents an isochron. The boundary is not parallel to anomaly J, even when the disruptions caused by the Horseshoe Seamounts are neglected. If one assumes that there was a change of spreading rate north of the fault at the same time (82 my) as one occurred south of the fault, and that prior to this time the spreading rates were equal (1.0 cm/yr), then the age of the quiet zone boundary to the north is about 100 my and to the south is 145 my. The above evidence leads us to believe that the correct explanation for the quiet zone is unrelated to the magnetic reversal sequence.

4. DISCUSSION AND CONCLUSIONS

On the basis of bathymetry, sonographs and magnetic anomalies a linear fracture in the crust has been traced along latitudes 37°N and 38°N from longitude 31°W to 13°W (Fig. 9). Only east of 16°W where bathymetric evidence is inconclusive and sonar evidence is not available is there the possibility that the fault, whose position was chosen on the basis of magnetic data only, is a fossil feature which is inactive today. Since epicentres at this end of the fault are scattered, current seismicity does not help us to choose the present plate boundary. It is probably unreasonable to expect to be able to pick a single boundary east of 14°W since the motion between Eurasia and Africa becomes compressional in this region [12]. It has been found in the eastern Mediterranean, for example, that the lithosphere has broken up into many small blocks or platelets [1,3].

An important result of this study is the discovery that the magnetic anomalies south of the Azores-Gibraltar plate boundary have an 020° to 030° trend not very different from that seen north of the fault. Just as in the FAMOUS area of the Mid-Atlantic Ridge at 37°N the crust south of the plate boundary is cut by a number of fairly closely spaced fracture zones separating anomalies with right-handed offsets for which bathymetric, as well as magnetic, evidence exists as far east as 23°W. Here the bathymetric evidence is obscured by the sediments of the Madeira Abyssal Plain and magnetic correlations disappear in the mid-Cretaceous constant

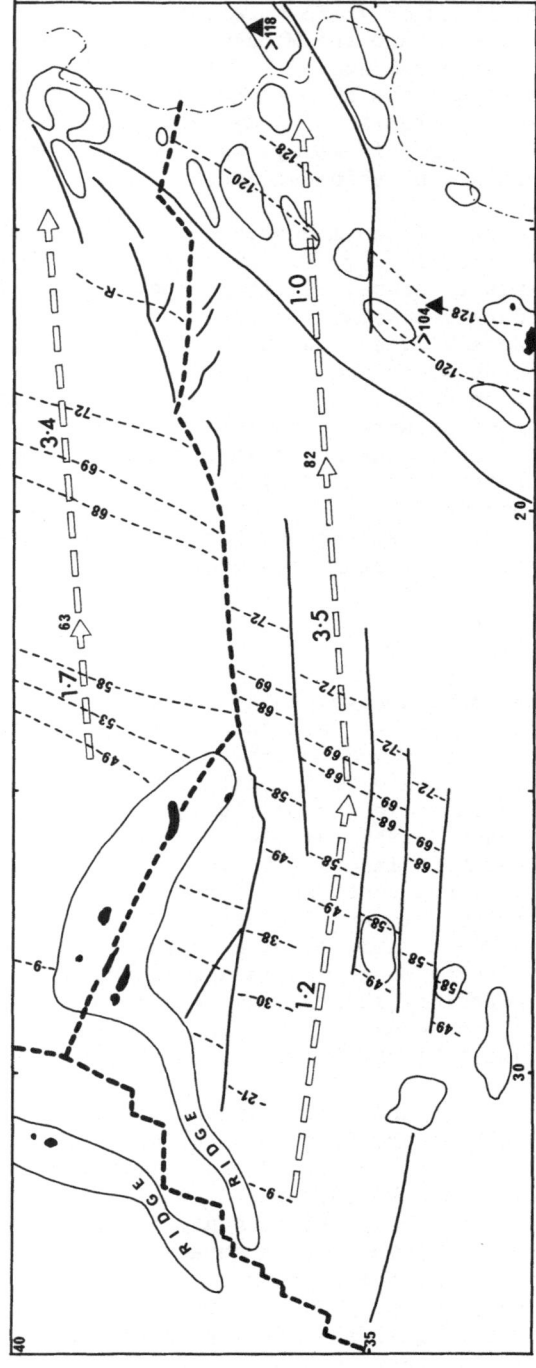

Fig. 9. Outline of the main features and isochrons (in m.y.) relating to the evolution of the Azores–Gibraltar region. Dotted line shows presently active plate boundary. Solid lines are fracture zones or prominent linear features. Enclosed areas are the principal topographic highs. Dot–dash line is boundary of the magnetic quiet zone. Arrows show the direction, duration and rate (cm/yr) of spreading per limb. Triangles show DSDP holes and minimum ages of basement.

polarity interval. Since the spreading rate may have increased markedly at the end of this interval (Fig. 8) it is possible that the fracture zones came about as a result of this acceleration. Conversely there is some slight evidence for a fracture zone passing close to Lion Seamount which may represent the eastward extension of the 35°N fracture zone. The eastward continuity of these fracture zones can probably only be established by means of seismic reflection profiling.

The fracture zone that forms the westward extension of the presently active transform fault section of the plate boundary passes south of the Azores. Between this zone and the 140 km wide region of Terceira Rift magnetic anomalies, there are well developed and unambiguously mapped 020° trending magnetic anomalies. It is not possible to date these anomalies on the data available, but their position and trend place severe restraints on an evolutionary model of the Azores triple junction and are not consistent with those proposed by Krause and Watkins [4] and by McKenzie [1].

There is an interesting association of the series of seamounts and ridges between 33°N, 29°W and 35°N, 27°W with the 58 my isochron. Vogt and Avery [20] note a similar association of an extended ridge at 51°N, 22°W with the 60 my isochron and postulated that it was the result of excessive asthenosphere discharge resulting from the major rearrangement of plate geometry at this time [21]. No such ridge however is found northeast of the Azores.

Some conclusions can be drawn concerning the relative motion across the transform fault section of the plate boundary. From the position of anomaly 20 the mean spreading rate since 49 my north of the boundary is 1.0 cm/yr whereas south it is 1.2 cm/yr. This must have resulted in a net sinistral relative motion, i.e. in the opposite sense to that derived from post-10 my rates and from the dextral strike slip earthquakes along the fault [1]. Assuming symmetrical spreading during this period and noting that anomaly 20 immediately north and south of the plate boundary is collinear, the offset of anomaly 20 west of the Mid-Atlantic Ridge must be twice that of the ridge axis today. This is consistent with the offset of the 4000 m contour across the Pico Fracture Zone [5]. Between 49 and 63 my ago the north side of the Azores-Gibraltar plate boundary was spreading faster than the south by 0.5 cm/yr and between 63 and 68 my by 2.2 cm/yr resulting in dextral shear across the boundary.

Thus the boundary between the Eurasian and African plates has suffered several changes of differential movement which in the above discussion has been assumed to be mainly strike slip. Differences in spreading directions will substantially modify the topographic expression of the plate boundary. Any offsets in a strike slip fault, such as at 15° and $18\frac{1}{2}$°W, will result in local

compression or extension and subsequent fracturing of the crust
into separate blocks. The complex nature of the boundary must
reflect such changes in relative plate movement.

The spreading history between 49 and 128 my ago south of the
boundary is independent of the existence of the E-W fracture zones
between 34° and 37°N, so that these must reflect original offsets
in the spreading axis. A different spreading history to the north
of 37°N has been identified between 49 and 78 my ago. We therefore conclude that the boundary proposed between 13 and 24°W has
been an essentially permanent feature since the earliest
Cretaceous.

ACKNOWLEDGEMENTS

In addition to thanking those who supplied magnetic data (see
Fig. 7), the authors are grateful to the Experimental Cartography
Unit of NERC for assistance in preparing the bathymetric charts
and to M. T. Jones of the British Oceanographic Data Service and
D. G. Roberts of this Institute for assistance in compiling magnetic
profiles.

REFERENCES

1. D. McKenzie, Geophys. J. R. Astr. Soc., 30, 109, 1972.
2. W. C. Pitman and M. Talwani, Geol. Soc. Am. Bull., 83,
    619, 1972.
3. J. F. Dewey, W. C. Pitman, W. B. F. Ryan and J. Bonnin,
    Geol. Soc. Am. Bull., 84, 3137, 1973.
4. D. C. Krause and N. D. Watkins, Geophys. J. R. Astr. Soc.,
    19, 261, 1970.
5. E. Uchupi, Woods Hole Oceanographic Institution Tech. Rep.
    71-72 Bathymetric atlas of the Atlantic, Caribbean and
    Gulf of Mexico, 1971.
6. J. S. M. Rusby, Int. Hyd. Rev., 47, 25, 1970.
7. R. B. Whitmarsh, Deep-Sea Res., 18, 433, 1971.
8. A. S. Laughton, R. B. Whitmarsh, J. S. M. Rusby, M. L.
    Somers, J. Revie, B. S. McCartney and J. E. Nafe, Nature,
    237, 217, 1972.
9. R. B. Whitmarsh and A. S. Laughton, Trans. Am. Geophys. Un.,
    53, 1123, 1972.
10. X. Le Pichon, J. Bonnin and G. Pautot, Upper Mantle Comm.
    Symp., Flagstaff, Arizona, (abstract), 1970.
11. W. B. F. Ryan, K. J. Hsu et al., Initial reports of the
    Deep-Sea Drilling Project, 13, Pt. I., Washington (US Govern-
    ment Printing Office), 514 pp, 1973.
12. Y. Fukao, Earth Planet. Sci. Lett., 18, 205, 1973.

13. C. A. Williams and D. McKenzie, Nature, 232, 168, 1971.
14. D. C. Krause and B. A. McGregor, Geol. Soc. Am. Bull.,
     in press, 1974.
15. P. R. Vogt, C. N. Anderson and D. R. Bracey, J. Geophys.
     Res., 76, 4796, 1971.
16. H. Schouten, Trans. Am. Geophys. Un., 55, 232, 1974.
17. R. L. Larson and W. C. Pitman, Geol. Soc. Am. Bull., 83,
     3645, 1972.
18. R. K. Lattimore, P. A. Rona and O. E. Dewald, J. Geophys.
     Res., 79, 1207, 1974.
19. D. E. Hayes, A. C. Pimm et al., Initial Reports of the
     Deep Sea Drilling Project, 14, Washington (US Government
     Printing Office), 975 pp., 1972.
20. P. R. Vogt and O. E. Avery, J. Geophys. Res., 79, 363,
     1974.
21. A. S. Laughton, Nature , 232, 612, 1971.

DEEP DRILL INVESTIGATIONS OF THE OCEANIC CRUST IN THE NORTH ATLANTIC

F. Aumento and K. D. Sullivan

Department of Geology, Dalhousie University, Halifax, Nova Scotia, Canada.

ABSTRACT. Deep drilling projects undertaken in 1971-1973 in the median valley of the Mid-Atlantic Ridge and on the islands of Sao Miguel, Azores, and Bermuda have enabled us to construct quite detailed models of the structure and evolution of these three tectonically and chronologically distinct regions of the North Atlantic ocean floor.

Deep Drill 1971 yielded six oriented rock cores from the median valley of the MAR at 45°N. In the resultant model for 45°N we envision lopoliths of basaltic liquids emplaced at shallow depths beneath the median valley floor. Differentiation of these magmas produces the commonly observed peridotite-gabbro-basalt stratigraphy. The peridotite-gabbro transition constitutes the Moho which displays quite a variability of depth and definition in the crestal region. Stress build-up after the lopoliths solidify eventually results in a new phase of fissuring and intrusion. Upon the basic pattern of a median valley developed by block-faulting of a simple three-layered oceanic crust are superimposed complexities due to hydrothermal and higher grade metamorphism, serpentinization and diapiric serpentinite intrusion.

Deep Drill 72 on Bermuda penetrated 772 m of igneous rocks beneath a 36 m limestone capping. Of the ca. 1,000 igneous units cored 64% by volume were altered ocean floor tholeiites. These were dated to be at least 90 my old, but may be as old as the surrounding seafloor (100-110 my). A second igneous event 33 my ago intruded 36% by volume of very basic, titanium-rich sheets.

The Azores hole, drilled in 1973, penetrated 981 m of subaerial

Kristjansson (ed.), Geodynamics of Iceland and the North Atlantic Area. 83-103. *All Rights Reserved.*
Copyright © 1974 by D. Reidel Publishing Company, Dordrecht-Holland.

and submarine flows, pyroclastics and volcanogenic sediments.
The subaerial-submarine transition is marked by a 107 m thick
igneous-sedimentary sequence found at 631-737 m below sealevel.
The entire sequence is normally magnetized, suggesting a maximum
age of 0.69 m.y. Initial results of K-Ar work currently in progress suggest, however, that the oldest rocks in the core are
ca. 0.25 m.y. old.

The vertical motion of neither island is consistent with
models based on thermal contraction of a lithospheric plate. The
scarcity of intrusives in the Azores core suggests that intrusives
play only a minor role in the principle phase of island building.
Both lava piles have been extensively hydrothermally altered over
a short period of time.

1. INTRODUCTION

Deep Drill is a continuing programme investigating the nature
and history of the oceanic crust in a variety of chronological and
tectonic environments. To date three drilling operations have
been completed: In 1971 in the median valley of the MAR, at 45°N,
in 1972 on the older Bermuda Seamount and in 1973 on the quite
young island of Sao Miguel in the Azores. At the time of this
writing, the JOIDES Deep Drill 74 Project is proceeding successfully at a drill site near the crest of the MAR at 36°30'N. (Fig. 1).

Deep Drill 71 (DD 71) was part of the Hudson Geotraverse
project [1]. The detailed survey and sampling results of the Geotraverse and the oriented drill cores of the DD 71 operation allowed construction of a very detailed model of the creastal region
of the MAR at 45°N.

The Atlantic oceanic islands rank second in importance only
to the ridge crest as sites of volcanism. In the past few years,
numerous workers have attempted to explain the evolution and composition of volcanic islands and seamounts within the framework of
the seafloor spreading hypothesis [1-5]. All of these studies
have been hampered by the lack of complete stratigraphic sections
and of chemical, mineralogical and age data for the volumetrically
predominant submarine portions of the lava pile. In 1972 and 1973,
Deep Drill recovered cores from Bermuda and Sao Miguel, Azores,
which provide us with unique samples from an ancient and a youthful
Atlantic island.

Deep Drill 74 will attempt to penetrate the oceanic crust
down to Layer 3. At the time of this writing, the "Glomar Chal-
lenger" is drilling through basaltic basement material at a site
ca. 20 km west of the MAR axis, near 36°30'N.

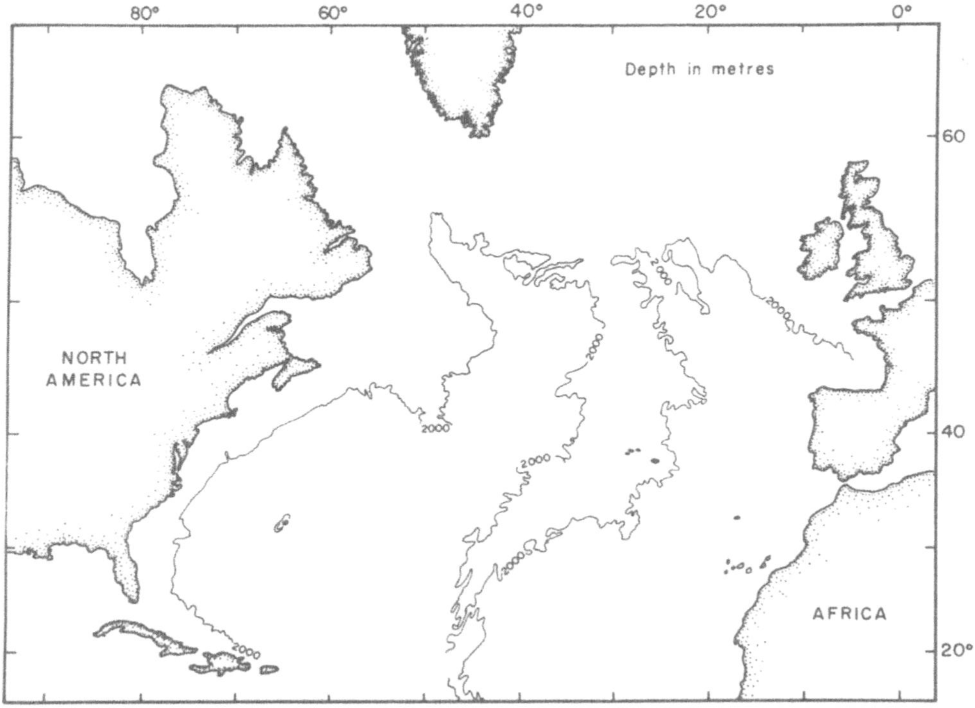

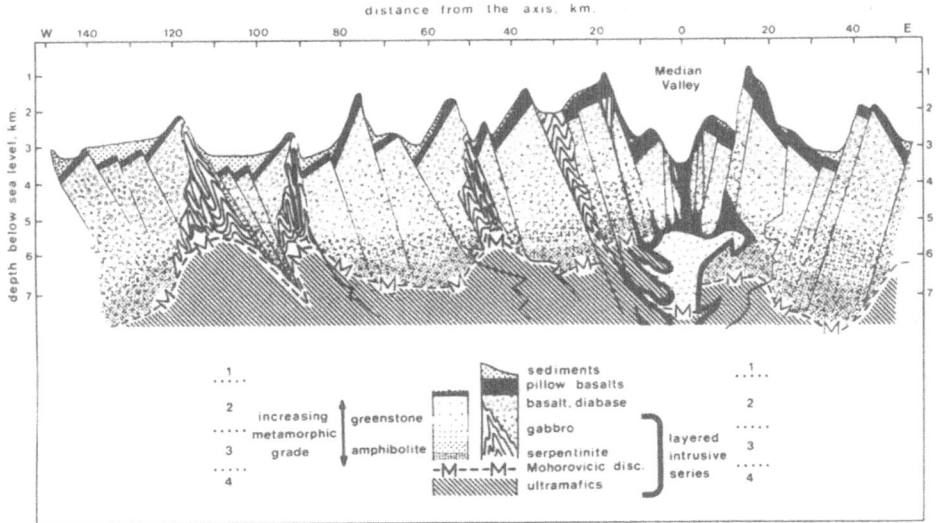

Fig. 1. Location map of the North Atlantic.

Fig. 2. Schematic cross-section of the Mid-Atlantic Ridge, 45°N.

2. DEEP DRILL 1971:   Mid-Atlantic Ridge

Deep Drill 71 constituted a final phase of the Hudson Geotraverse
Project, which surveyed to some extent nearly all of a 1° wide
strip across the North Atlantic between 45° and 46°N. The central
area straddling the axis of the MAR received the most attention,
and was the site for the 1971 drilling operation. Geotraverse
collected continuous bathymetric, magnetic and gravimetric data
over some 30,000 km of survey lines controlled by strict satellite
and radar navigation. A total of some 400 successful sampling stations were undertaken, as well as numerous observational stations
and detailed seismic reflection and refraction experiments [6,8].
The DD 71 project employed an autonomous sea-floor drill deployed
on the floor of the median valley [9]. Six oriented rock drill
cores were recovered and studied for paleomagnetics, petrology and
geochemistry.

2.1 Structure

The data collected during the Hudson Geotraverse Project made it
possible to draw a detailed picture of the sea-floor and oceanic
crust in the crestal zone of the MAR, a slowly spreading ridge.
Fig. 2 is a schematic cross-section of the crestal zone of the
MAR at 45°N compiled from actual measurements. While details in the
figure are obviously due to artistic liberties, parameters such as
layering thicknesses, attitude of layers and topography are factual.

     The crust in this region is believed to consist of ca. 5 km
of discontinuously layered igneous and metamorphic rocks (Layers
2 and 3) overlain by a veneer of sediment (Layer 1). Layer 2 is
thought to be composed of an upper section of fresh pillow basalts
and weathered massive basalt and a lower section of massive basalts,
sheeted diabase and fresh serpentinite. Layer 3 consists of the
metamorphosed equivalents of Layer 2 rocks. The uppermost upper
mantle, Layer 4, is peridotitic material. Layers 1 and 3 thicken
away from the ridge axis, while Layer 2 thickens towards the axis.
Mean seismic velocities for each layer are as follows: Layer 1=
2.2 km/sec; Layer 2a=4.1 km/sec; Layer 2b=4.7 km/sec; Layer 3=6.7
km/sec; Layer 4=8.1 km/sec. The crust in this region displays a
0.25 km/sec compressional wave velocity anisotropy; the azimuth
of low velocity is parallel to the ridge crest. Hess [7] and other
workers have attributed such anisotropies to olivine orientation
or other crustal fabric elements. We feel that the anisotropy might
also reflect the presence of elongate zone of low rigidity (perhaps magma chambers) beneath the median valley.

     The layering of the igeneous and metamorphic rocks of Layers
2 and 3 is disrupted by diapiric serpentinite intrusions at dis-

tances of 30 km or more from the axis, and by steep normal faults striking parallel to the ridge axis. These faults, which are responsible for much of the topographic relief at the crestal zone, have inward-facing scarps as steep as 45° and vertical throws of up to 1.5 km. Ages of rocks from the median valley floor and the top of the first fault block above the valley range from 0-25,000 and 150,000-170,000 years respectively, indicating that extrusion and faulting may be cyclic events separated by 100,000 to 300,000 years.

The thickness of the crust has been observed to vary locally by as much as 3 km in the crestal region as a result of block faulting and variability of the Moho near the axis (see below) [8,10].

Fig. 3 gives a more detailed cross-sectional view of the median valley at 45°N. The factors responsible for the asymmetrical form and the small trough along the eastern side are not understood. Eruption of new basaltic material occurs periodically through fissures in the valley floor; batches of magma which do not succeed in breaking through to the sea-floor solidify as dykes and small lopoliths beneath the valley floor.

We propose the following model for the evolution of the oceanic crust at the ridge crest: Large, discontinuous lopoliths of mafic magma are emplaced beneath the axis of the ridge following a phase of major lateral fracturing due to the forces responsible for ocean-floor spreading. As indicated in Fig. 2, the roofs of these lopoliths may be within 1-2 km of the median valley floor; the lopolith floors are less than 24 km below sealevel.

Magmatic differentiation by gravity crystal settling takes place under quiescent conditions after emplacement, producing an upward dunite-peridotite-gabbro-extrusives stratigraphy such as is observed on the fault scarps facing the median valley. The gabbros and intrusives accrete to the base of the crust, whereas the ultramafics beneath them accrete to the top of the upper mantle. The gabbro-peridotite interface thus formed represents the local Mohorovicic discontinuity between Layers 3 and 4 which, by nature of its origin, exhibits considerable variations in depth and definition. Failure to observe the Moho at some places on the crest may be due to insufficiently advanced differentiation.

Metamorphism due to burial, subsequent intrusions and extrusions, percolating sea water and generally high heat flow affects these rocks in a manner approaching that of continental regional metamorphism, transforming gabbros and diabases into greenschists and amphibolites, and at the same time possibly concentrating the dispersed sulphide ores found in these rocks. This upwardly encroaching metamorphic horizon is thought to represent the junction between Layers 2b and 3.

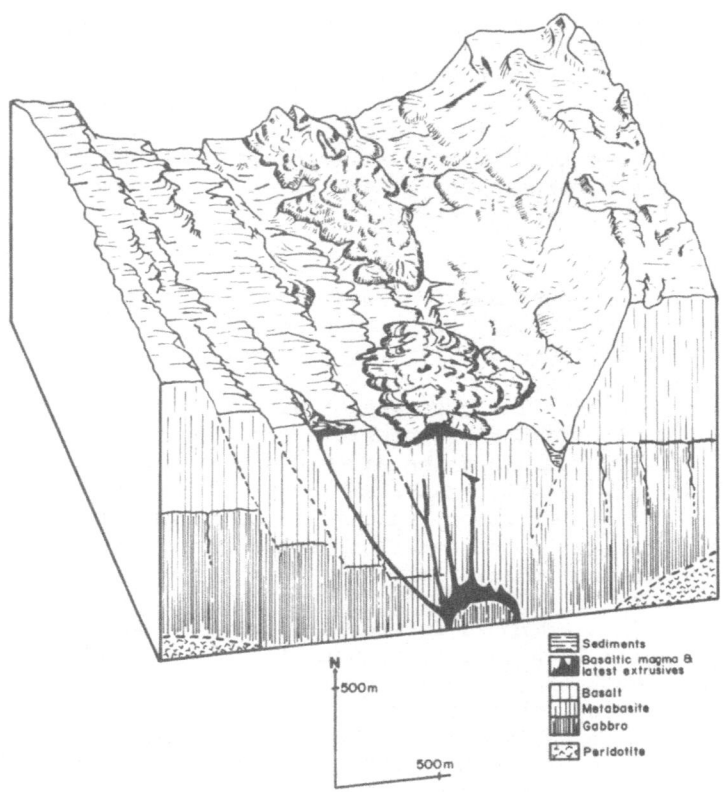

Fig. 3.   Physiography and structure of the median valley at 45°N.

        The final complete crystallization and solidification of
these lopoliths restricts further plastic stress release, and con-
tributes to a new cycle of tensional stress build-up, resulting
in renewed fracturing beneath the axis, and the intrusion into the
voids thus formed of a new generation of liquids. The solid rocks
formed during the previous cycle are deformed cataclastically as
they are moved apart; however, this tectonic damage may often be
repaired by renewed solid-state recrystallization under a hot,
high stress condition. Cataclasis is followed by amphibolitization;
rodingitizing metasomatism appears contemporaneous with the last
stages of amphibolitization and heralds incipient serpentinization.
The bulk of the serpentinization results from later, more extreme
periods of hydration. Diverse serpentinization processes are evident here: the initial phases are the results of very hot, perhaps
juvenile waters; the final phases are produced under tensional
stress, by the hydrating effects of cooler meteoric waters, and
are possibly contemporaneous with diapiric emplacement through the

oceanic crust.

2.2 Magnetics

The magnetic anomaly pattern of the ocean floor at 45°N, as elsewhere on the axes of ridge systems, is dominated by a strong
positive lineation (over 800 gammas) coincident with the median
valley. Bhattacharya [11] has shown that the confused pattern on
either side of the central anomaly is due to the fact that the overall 017° anomaly trend actually is composed of an en echelon arrangement of short north-south anomaly segments offset right-lat-
erally by east-west striking faults.

Some 120 km west of the axis there appears a continuous anomaly of 350 gamma maximum amplitude, corresponding to an anomaly
which is readily identifiable over other ridge axial regions, and
which has been designated as oceanic anomaly 5. The relative position of anomaly 5 with respect to the ridge axis (anomaly 1) indicates that the ocean floor has probably been spreading at an overall
average rate of 1.28 cm/yr over the last 10 m.y.

Paleomagnetic tests on the rocks from this area have shown
that only the eruptives belonging to the uppermost 200 m or so of
oceanic Layer 2 are likely contributors to the regular magnetic anomaly pattern over the ridge [12]. High intensities of natural remanent magnetism found in the serpentinites (avg. 40×10^{-4} cm^{-3} cgs)
produce substantial disruptions of the linearity of the surface anomaly patterns coincident with diapiric intrusions of serpentinized
ultramafics. These high n.r.m. intensities also suggest that the
oceanic Layer 3 is not composed primarily of serpentinites.

The six oriented rock cores drilled during DD 71 revealed both
normal, reversed, and shallow remanent magnetic inclinations. Fission track dates on the glassy rinds of the cores gave ages of less
than 25,000 years for all six cores [13]. These orientations and
young ages can be explained either by the operation of a mechanism
of self-reversal of polarity [14] or by the recent existence of a
rapid, short-lived reversal of the earth's magnetic field, or by
local magnetic effects within the floor of the median valley.

2.3 Chronology

Potassium-argon and fission track ages [13] show that the spreading
rate of the ocean floor may have changed from 1 cm/yr during the
period 3-16 m.y.b.p. to 4 cm/yr during the period from 3 m.y.b.p. to
the present. This change in spreading rate cannot be detected in the
magnetic anomaly patterns because of their confused nature. The
physiographic boundary between the Crest Mountains and the High

Fractured Plateaus, and possibly also increases in sediment and
ferromanganese coating thicknesses on rocks at that boundary may
reflect the change in spreading rate. Correlative changes in
spreading rate have been suggested for other ocean ridge systems.

2.4 Petrology

Mafic rocks: These vary in texture from glassy, sometimes spheru-
litic pillow basalts, to more massive diabasic and gabbroic rocks.
Vesicles are common in the pillow lavas; no correlation between
vesicle size and depth of extrusion has been found. Resorbed
calcic plagioclase phenocrysts, indicative of differentiation at
shallow levels beneath the ridge axis, also characterize many of
the pillow basalts.

Chemically, the mafic rocks vary from quartz-normative tholeiites to nepheline-normative alkali basalts, the more alkaline
rocks being associated with volcanism at greater distances from
the ridge axis.

Meta-basites: Zeolite facies metamorphism of ridge basalts
and diabases takes place under very shallow burial conditions, possibly of the order of a few tens of metres. Plagioclase phenocrysts
are altered to analcite, and the groundmass to other zeolites. In
the greenschist facies the mafic rocks still retain their igneous
textures; here calcic plagioclase alters to albite, augite and
actinolite; olivine and glass alter to chlorite. In the higher
metamorphic grades epidote, tremolite, quartz, calcite, talc, titano-
maghemite and green hornblende also occur. Amphibolite facies meta-
basites occur at greater crustal depths (well into Layer 3). These
rocks have lost all igneous textures and most relict minerals, and
exhibit strong fabric lineations. Two mineral assemblages are common: A lower grade assemblage of quartz, plagioclase, biotite,
hornblende, epidote, sericitized orthoclase, magnetite and sphene,
and a higher grade assemblage of hornblende, diallagic diopside,
plagioclase, sericitized orthoclase and minor biotite.

Serpentinized ultramafics: These rocks include dunites, harz-
burgites, lherzolites, amphibole peridotites, wehrlites and troc-
tolitic gabbros. Many show evidence of cumulate crystal layering,
of subsequent mylonitization and other mechanical deformations, as
well as metasomatic changes such as amphibolitization and rodingi-
tization, and ubiquitous, almost total serpentinization.

Sialic intrusives: Hornblende rich quartz-diorites, grading
into trondhjemites or albite granites, occur as small pockets in
intimate association with the ultramafic rocks. These contain
xenoliths of basalt, serpentinites and metabasites, and are reminis-
cent of the albitized diorites characteristic of the late stages

of alpine intrusive complexes.

3. DEEP DRILL 1972: Bermuda Seamount

The Bermuda Islands occupy the northernmost of three platforms
located 1100 km east of Cape Hatteras, or approximately one-third
of the way to the MAR axis. The only lithology exposed on the emer-
gent seamount is limestone. Seismic reflection results [15] indicate that the top of the volcanic seamount material lies an average of 76 m below sealevel and is shallowest in the area which includes the Bermuda Biological Station, where the DD 72 site was
located (Fig. 4). The drill encountered the first solid volcanics
at a depth of 26 m below sealevel, and penetrated in all some 920
igneous units beneath the 36 m limestone capping. Subaqueously
erupted ocean floor tholeiites comprised 64% of the igneous units
and lamprophyric/limburgitic sheets a further 32%. The entire core
was extensively altered by hydrothermal activity.

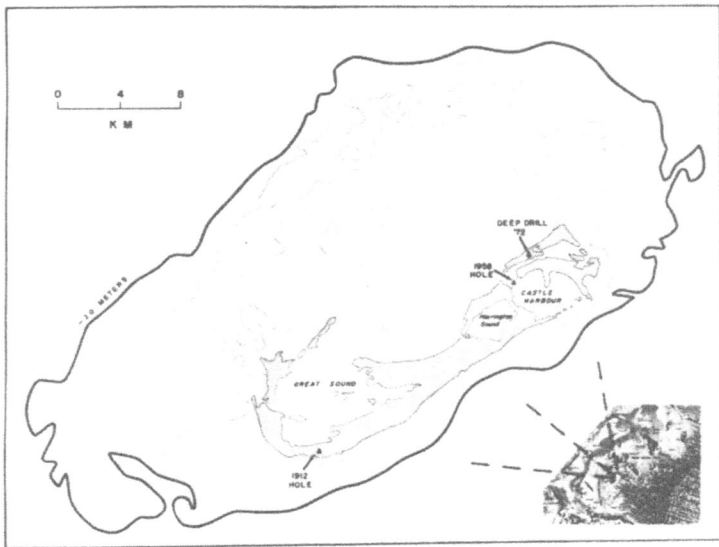

Fig. 4. Location of Deep Drill 1972, Bermuda.

    The seafloor in the Bermuda area has been dated by Larson and
Pitman [16] as 100-110 m.y. old, on the basis of Mesozoic magnetic
anomalies. Potassium-argon studies on the Bermuda core rocks revealed two igneous events, one dated with some uncertainty as at
least 91 m.y.b.p. and the other firmly dated as 33 m.y.b.p. [17].
The 91 m.y. event may actually be coincident with the origin of the
sea-floor, and has been interpreted as marking the construction of

the Bermuda Platform and its pedestal. The seamount was reactivated at 33 m.y.b.p. by the intrusion of the very basic, alkalic sheets.

3.1   Petrology

The igneous rocks recovered were as follows (Fig. 5): 431 suba-
queous "lava flows", 493 intrusive sheets and three horizons of
fragmented lava with calcareous cement. Some of the so-called flows
may actually be individual pillows from a single flow lying one on
top of another; such a distinction could not be made in the core.

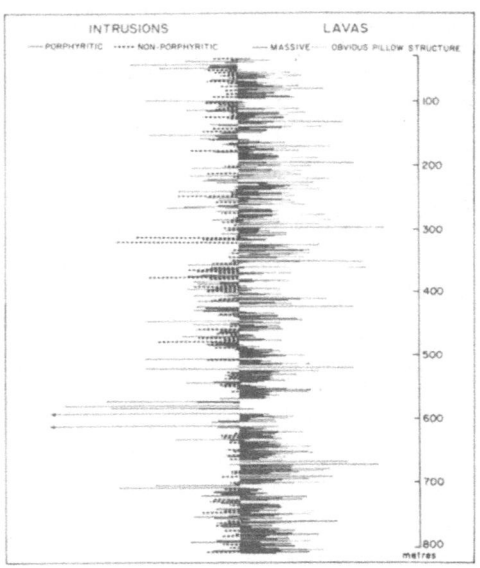

Fig. 5. Distribution of lavas and sheets in the Bermuda drill core.

The extensive alteration and unusual chemistry of the Bermuda
rocks make a petrologic classification very difficult.

Flows: The flows are altered ocean floor tholeiites with an
average thickness of one meter. They characteristically consist of
a topmost interflow material of green and black volcanic debris
with a calcareous matrix, which is followed downward by pillow sections separated by chloritic zones. Massive, slightly amygdaloidal
(zeolitized) lavas and, sometimes, a second pillow section, underlie
the upper pillow lava section. Pillowed flows are extremely abundant below 600 m, common between 150 and 600 m and absent between

30 and 150 meters. This sequence may represent the gradual emergence of the seamount during its growth. The distribution of the chlorite-carbonate cemented interflow breccia parallels the pillow distribution pattern.

Mineralogically, the flows contain andesine-labradorite laths set in variolitic, hypocrystalline or pilotaxitic matrix matter.

Sheets: By volume, 32% of the lava pile consisted of steeply-dipping intrusive sheets of 0.5 m average thickness. The following three types have been defined:

a. Porphyritic sheets with phenocrysts of calcite pseudo-morphs of olivine (ca.40%), clinopyroxene (0-60%) and biotite (0-30%) in a matrix of highly altered plagio-clase and devitrified, microlitic or totally chlori-tized glass.

b. Fine- to medium-grained amygdaloidal sheets, with vesi-cle concentrations in zones parallel to the contacts.

c. Uniformly fine-grained sheets.

All of the intrusives show a marked preference for the inter-flow zones of the igneous sequences, where they tend to concentrate by multiple intrusion. The unique chemistry and effects on the lavas of exposure to seawater are discussed in some detail by

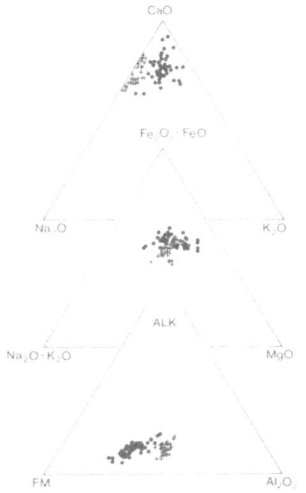

Fig. 6. Ternary variation diagrams for Bermuda lavas (+) and sheets (·).

Aumento and Gunn [18], but the unique basic, high-titanium composition of the sheets should be noted here (Figs. 6 and 7).

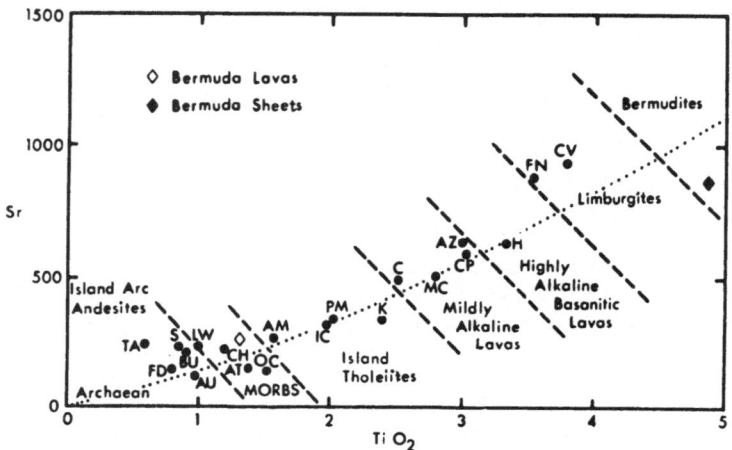

Fig. 7. TiO_2 vs. Sr for suites of oceanic basalts. Black diamond: avg. of 49 Bermudites; open diamond: avg. of 31 spilitised Bermuda basalts. CV=Cape Verde: FN=Fernando de Noronha; H=Hawaiian basanites; AZ=Azores; C=East Is. (Crozet); K=Kilauea Iki (Hawaii); IC=Iceland; AM=Amsterdam Is. tholeiites; OC=Mid-Atlantic tholeiites; AT=JOIDES drill cores; TA=Tongan Is. arc andesites; AV=avg. Australian Archean metabasalts.

3.2 Magnetics

Eight major remanence groups were defined within the lava succession on the basis of upward and downward remanent magnetic inclinations. The inclinations were found not to be antiparallel (-42° and 21°) and to differ in intensity (after 200 oe demagnetization) by a factor of 3.2, the positive inclination group having the higher intensity.

The cleaned remanence is thought to consist of two components with comparable intensities and coercivities. The first component consists of two groups of antiparallel directions, which represent the initial remanence (TRM) acquired during Paleogene polarity reversals. Superimposed upon this component is a second TRM with a shallow downward inclination which was acquired during the extensive hydrothermal alteration.

3.3   Heat flow

Heat flow in the Bermuda core is close to the mean value for ocean basins of 1.36 μcal/cm^{2}sec; this is approximately 0.16 μcal/cm^{2}sec higher than the value predicted for sea-floor of Cretaceous age. The excess heat flow can be accounted for by radioactivity in the 7 km of "excess" basaltic material which makes up the seamount.

The observed Bermuda heat flow is much lower than the values predicted for 30 m.y. old sea-floor by Sclater and Francheteau's relationship [19]. However, good agreement is found with the values predicted for 100-110 m.y. old sea-floor, indicating that the present seamount consists of volcanics erupted through older oceanic crust during events of limited extent. The same conclusion is reached when data on the subsidence of Bermuda are compared with various theoretical age-vs-depth curves for oceanic crust migrating away from active ridge crests [20,21].

3.4   Conclusions

The Bermuda basalts were extruded through the oceanic crust close to the spreading centre of the MAR some 110 m.y. ago, much like the basalts of the median valley at 45°N described in the preceeding section. The initial constructive phase of the seamount was followed by a 60 m.y. long period of dormancy during which hydrothermal alteration by seawater, ocean floor erosion and subsidence (possibly even submergence) of the island occurred. Hydrothermal alteration of the basalts by seawater began shortly after extrusion and continued for an unknown length of time.

Approximately 33 m.y.b.p. the island was "reactivated" by an event of unknown cause. Intrusion of the sheets and further alteration of the basalts took place at this stage. Perhaps as much as 40% of the present mass of the seamount was added with this event.

Emergence of the seamount prior to the intrusive episode, at the time of its formation near the MAR axis, is suggested by the distribution of pillow lavas and chlorite-carbonate cemented interflow breccias. Subaerial erosion during such a period of emergence might explain the lack of subaerial extrusive units in the core.

4.   DEEP DRILL 1973: Sao Miguel, Azores

The Azores archipelago is an elongate group of nine islands which lies astride the MAR at about 38° north latitude. Fig. 8 is a location map, on which can be seen several noteworthy features of the islands' regional setting:

a. The 35° change in trend of the MAR.
b. The broadening of the MAR to form the Azores Platform.
c. The linear trend of the island chain and of individual islands.
d. The East Azores Fracture Zone, which the diagram erroneously
    shows passing to the north of the islands; it in fact runs
    to the south of the archipelago, and is seismically active
    from just east of Santa Maria to its intersection with the
    MAR.
e. The bathymetric depressions near Terceira and Sao Miguel, which
    mark approximately the Terceira Rift.

        Krause and Watkins [32] have proposed a plate tectonic model
for the evolution of this system. They envision evolution in five
stages, beginning with a right-lateral offset of the ridge crest.
Faster spreading rates to the south of the offset, plus a change
of spreading direction on the southern ridge segment, produce a
"leaky transform fault" [23] and lead to the development of the
Terceira Rift as a secondary spreading center. Watkins and Ridley
[24] present a much more detailed synthesis of available geologic
and geophysical data from the Azores region, and specifically
relate the origin of the islands to volcanism associated with the
secondary spreading centre. The testing and refinement of this
and other models for the evolution of the Azores system will require
more geophysical data and much more detailed investigation of the
vertical and horizontal chemical variations, both within individual
islands and across the entire archipelago.

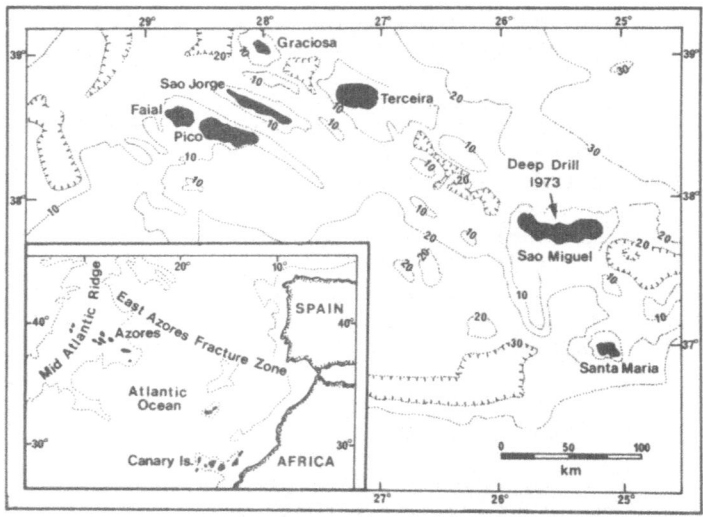

Fig. 8. Location of Deep Drill 1973, Azores.

Sao Miguel, the largest of the Azorean islands (Fig. 9),
lies near the eastern end of the archipelago, 400 km from the MAR
crest. The surface geology is dominated by four large calderas,
three of which have erupted during historic times, producing an
extensive blanket of trachytic pyroclastic material [25]. Surface
rocks range in composition from alkali basalt to trachyte [26,27],
and are unusually potassic [28].

The DD 73 site was located on the flanks of the volcano Agua
da Pau near the town of Ribeira Grande (25°31.4'W, 37°48.9'N), at
a distance of 5 km from the caldera wall. Drilling commenced at an
elevation of 72 m above sealevel and continued to a depth of 981 m,
at which point steam erupted through the drill pipe, forcing termination of the hole.

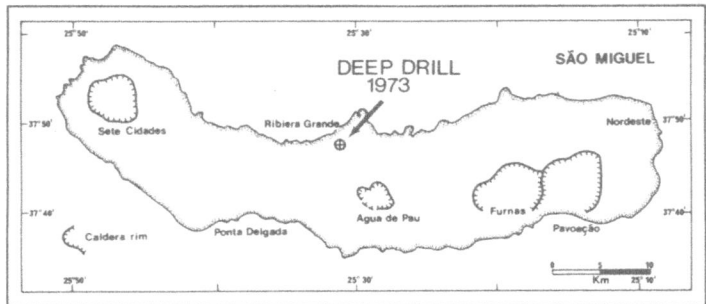

Fig. 9. Sao Miguel: DD 73 site and major volcanic centres.

Core recovery was continuous after the first 148 m of unconsolidated pyroclastic and mudflow deposits had been penetrated.
A synoptic core log is given in Fig. 10.

Extrusive lavas constitute 72% of the drill core and occur in
140 separate flow units of 4.8 m average thickness. Alkali basalts,
hawaiites and mugearites predominate and only three trachyte flows
were encountered. A subaerial origin for the flows is suggested by
the massive nature of the flow centres; the complete absence of
pillow structures; the intercalation of numerous pyroclastic units
that lack any indication of aqueous re-working or stratification;
the development of some lateritic horizons; the occurrence of thick,
vesicular and auto-brecciated flow tops. Pillowed basaltic rocks
with altered glassy margins were first met at 880 m in the core,
where they occur interbedded with massive basaltic flows in a
sequence devoid of pyroclastics.

In sharp contrast to the Bermuda drill core, intrusive units
constitute an insignificant fraction of the Sao Miguel core, and

basaltic flows were encountered only at 662 m and 738 m in the
core.

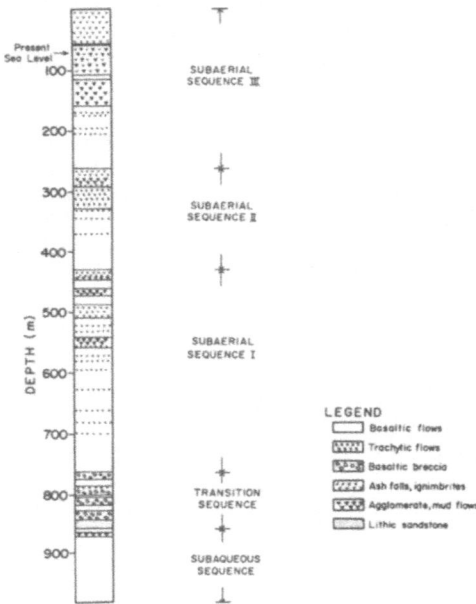

Fig. 10. Schematic graphic log, DD 73 core.

    Pyroclastic deposits, chiefly trachytic in character, account
for 22% of the core and range from fine-grained tuffs to pumice
deposits, agglomerates and mud flows. Ignimbrite cooling units
from 6 cm to 3.8 m in thickness are common in the 262-508 m depth
interval. Only two such units have previously been described on
the island [28]. Individual pyroclastic units are frequently trachytic in character at the top but grade downward into more basic
compositions.

    A 107 m thick igneous-sedimentary sequence marks the transi-
tion from subaerial to submarine volcanism. Basaltic breccias with
a chloritic matrix (altered glass or ash) are interbedded with massive flows in the upper part of the sequence and overlie 15 m of
bedded coarse lithic sandstones. One 7 m thick bed consisting of
angular basaltic fragments embedded in a matrix of black lithic
quartz-carbonate cemented sand is strikingly similar to the black
beach sands found along today's coastline.

4.1   Paleomagnetic measurements

Paleomagnetic measurements on lava flows in the core indicate
that the subaerial, transitional and submarine sequences are all
normally magnetized. This probably indicates magnetization during
the Brunhes polarity epoch, although the possibility of remagnetization during hydrothermal alteration certainly cannot be discounted completely. If magnetization does reflect geomagnetic field
polarity changes, the time scale of polarity reversals provides an
upper age limit of 0.69 m.y. for the entire volcanic-sedimentary
sequence. Preliminary K-Ar determinations on material from the
uppermost and near bottom sections of the core suggest, however,
that the actual age range may in fact be much narrower, perhaps
110,000-ca. 250,000 years.

4.2   Temperature measurements

Bottom hole temperatures were measured to 914 m depth with maximum
thermometers at intervals during drilling when water circulation
was stopped temporarily. They probably represent in situ temperatures to ±2°C [30]. The results show a nearly constant temperature
of 20-25°C to 100 m depth, followed by a sudden jump to over 100°C
at 100-125 m, then a uniform 250°C/km gradient and, finally, a
gradient of less than 10°C/km to the bottom of the hole (Fig. 11).

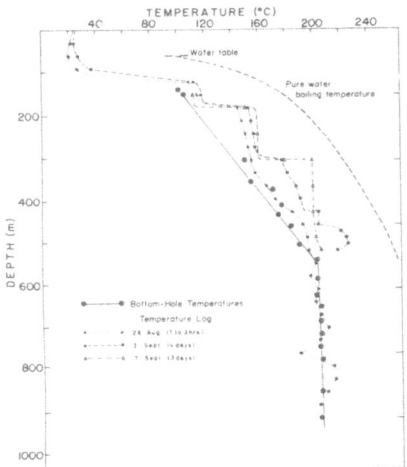

Fig. 11. Temperatures recorded in the DD 73 borehole as a function
of depth.

     In contrast to the general type of temperature profile in a
hydrothermal area (high gradients and conductive heat transfer observed in impermeable rocks; low gradients and convective heat transfer in permeable rocks) the highly porous subaerial section above
550 m in the Sao Miguel core has a high thermal gradient, while the
less permeable rocks below this level have a very low gradient. It
has been suggested [30] that vertical convection is not possible in
the complete section drilled, nor even in the more porous sections
above 550 m because of the frequency of horizontal, impermeable
horizons that limit circulation. The lack of cellular convection
in the section is seen in the rapid rise of temperature in the hole
recorded over a period of four days after drilling had ceased,
when convection was possible through the borehole itself. The temperatures show an upward migration of 200°C water from 550 m depth
to 290 m depth at the time of the last measurement. However, water
flow parallel to the bedding (nearly horizontal) is probably not
restricted in the pyroclastic horizons. High temperature water
(205°C) is believed to flow parallel to the bedding at about 550 m,
and 100°C water to flow just beneath the impermeable layer at 110-
120 m depth. The low gradient below 550 m suggests that there is
no volcanic heat source located directly beneath the drill site.

4.3 Discussion

Deep drilling on Sao Miguel has disclosed several important aspects
of the island's geological evolution which are not evident from an
examination of its surface geology. Three distinct subaerial
eruptive sequences totalling 762 m in thickness were found to
overlie a complex 107 m transition zone marking the change-over
from a submarine to a subaerial environment. Each subaerial eruptive sequence began with a quiescent phase of basaltic lava extrusion which was followed by increasingly explosive activity. Inter-
calated tuff beds increase in frequency and thickness, and grade
into a terminal phase of trachytic pumice eruption with rare trachyte
extrusion. Compositional zonation within individual pyroclastic
units further supports strong compositional zonation in the magma
chamber at this stage. Each eruptive sequence may signal the ar-
rival of a fresh magma batch from depth into a sub-crustal magma
chamber. Shallow-level differentiation processes then produce the
increasingly volatile, silicic liquids which follow the basalts.

     The depth at which submarine lavas were first encountered
(786 m) indicates substantial subsidence of the island. Paleomagnetic dating of the core at younger than 0.69 m.y. yields a minimum average subsidence rate of 0.1 cm/yr. If the preliminary K-Ar
date of 0.25 m.y. for a basalt flow near the bottom of the core is
correct, then this subsidence may be on the order of 1.0 cm/yr.
However, the exposure of submarine lavas and sediments of Miocene
to Quaternary age on the nearby island of Santa Maria [31] indicates

that that island is being uplifted, and implies that the subsidence of Sao Miguel is not a regional phenomenon.

Post-glacial sealevel rise following the last ice advance can account for only 130 m of the subsidence [32]; theoretical models for crustal subsidence due to thermal contraction [19,20, 21] predict subsidence an order of magnitude less than the minimum value of 0.1 cm/yr. Neither of these factors would seem to be dominant in the case of Sao Miguel.

Moore [33] has reported a correlation of decreasing subsidence rates with increasing age and decreasing volcanic intensity in Hawaii. Quite a similar pattern is exhibited by Santa Maria and Sao Miguel in the Azores. In spite of the more limited data available in the case of the Azores, a similar cause of the subsidence is suggested by the magnitude of the subsidence rate and the apparent correlation between island age and activity.

Swanson [34] suggests that isostatic readjustment in response to crustal loading is sufficient to account for the subsidence of the island of Hawaii, provided a large volume of sub-surface intrusives is taken into account. Evaluation of such a model for the Azores must await more data on the volume of ejecta and the role of intrusives.

Unless the upward transfer of material from the upper mantle towards the surface is sufficiently compensated for by lateral mass transfer at depth, thermal contraction may be expected to occur and to be accentuated by rapid crustal loading (extrusion). Upon termination of volcanic activity, heat loss is confined largely to slower conductive processes. The island's vertical motion is then determined by the balance between thermal contraction, accumulation of lava and erosion plus isostatic uplift in response to unloading.

5. CONCLUSION

Our understanding of the structures and processes of mid-ocean ridges has increased tremendously in the three years since Deep Drill 1971. Many features of mid-ocean ridges which had then only recently been discovered and were in the process of being defined are now common knowledge among geologists concerned with oceanic provinces. Further extension of the frontier of knowledge in this field will come through projects such as the International Program for Ocean Drilling (IPOD), for which Leg 37 of DSDP (Deep Drill 74) is an important trial project [35].

On oceanic islands, deep drilling is proving to be an excellent means of overcoming poor exposure to obtain samples of some of the most critical units of the lava succession. Upon completion

of initial chemical and mineralogical studies of the Sao Miguel core
we will have much important new data concerning the structure and
evolution of the volcanic islands of the ocean basins.

ACKNOWLEDGEMENTS

Financial support for this work was provided by the Faculty of
Graduate Studies at Dalhousie University, the National Research
Council of Canada, the National Science Foundation and Research
Corporation of America and Lamont-Doherty Geological Observatory.
We are grateful for the constant support of our colleagues at
Dalhousie, Lamont, and Bedford Institute and to Tau Rho Alpha,
J. G. Moore (U.S.G.S.) and the U.S. Naval Oceanographic Office
for permission to reproduce one of their illustrations. Deep
Drill is especially indebted to the late Dr. Maurice Ewing, with-
out whose help the 1972 and 1973 programs would not have been possi-
ble.

REFERENCES

1.   F. Aumento et al., Phil. Trans. Roy. Soc. London, 268A,
     623, 1971.
2.   A. McBirney and I. Gass, Earth Planet. Sci. Lett., 2,
     265, 1967.
3.   P. Gast, Geochim. et Cosmochim. Acta, 32, 1057, 1965.
4.   J. Gilluly, Bull. Geol. Soc. Am., 82, 2382, 1971.
5.   E. A. K. Middlemost, Lithos, 6, 123, 1973.
6.   M. J. Keen and K. Manchester, Can. J. Earth Sci., 7,
     735, 1970.
7.   H. H. Hess, Nature, 223, 629, 1964.
8.   C. E. Keen and C. Tramontini, Geophys. J. R. Astr. Soc.,
     20, 509, 1970.
9.   J. Brooke and R. L. G. Gilbert, Deep-Sea Res., 15, 483,
     1968.
10. C. E. Keen and D. Barrett, EOS, Trans. A. G. U., 50, 241,
     1969.
11. P. J. Bhattacharya, Unpubl. Ph.D. thesis, Dalhousie Univ.,
     1970.
12. E. Irving, Can. J. Earth Sci., 7, 1528, 1970.
13. F. Aumento, Can. J. Earth Sci., 10, 679, 1973.
14. P. Ryall, Ph.D. thesis in prep., Dalhousie Univ., 1974.
15. R. A. Gees and F. Medioli, Maritime Seds., 6, 21, 1970.
16. R. Larson and W. C. Pitman, Bull. Geol. Soc. Am., 83,
     3645, 1972.
17. F. Aumento and P. Reynolds, EOS, Trans. A. G. U., 54,
     485, 1973.
18. F. Aumento and B. G. Gunn, in prep., 1974.
19. J. G. Sclater and J. Francheteau Geophys. J. R. Astr.

Soc., 20, 509, 1970.
20. H. W. Menard, Earth Planet. Sci. Lett., 6, 275, 1969.
21. J. G. Sclater et al., J. Geophys. Res., 76, 7888, 1971.
22. D. Krause and N. D. Watkins, Geophys. J. R. Astr. Soc., 19, 261, 1972.
23. H. W. Menard and T. Atwater, Nature, 219, 463, 1968.
24. N. D. Watkins and I. Ridley, in: The Ocean Basins and Margins, Vol. II, Plenum Press, N.Y., 1974.
25. F. S. Weston, Boletim. do Museum e Lab. Mineral. e Geol. Lisboa, 10, 3, 1964.
26. P. Esenwein, Zeitschr. Vulkanologie, XII, 108, 1929.
27. T. de Assuncao, Comm. Serv. Geol. Portugal, XLV, 81, 1961.
28. H. U. Schmincke and M. Weibel, N. Jb. Mineral. Abh., 117, 253, 1972.
29. A. Cox, Science, 163, 237, 1969.
30. G. K. Muecke et al., Submitted to Nature, 1974.
31. G. Zbyszewski, Comm. Serv. Geol. Portugal, XLV, 1961.
32. A. L. Bloom, Quaternaria, 12, 145, 1971.
33. D. G. Moore, Sympos. on volcanoes and their roots, Oxford, 1969, Vol. abstracts, 67, 1970.
34. D. A. Swanson and D. W. Petersen, U.S.G.S. Prof. Paper 800-C, C1, 1972.
35. Int'l. Program Ocean Drilling, Proposal UCSD 6297, Scripps Inst. Oceanography, La Jolla, Calif.

THE ICELAND PHENOMENON: IMPRINTS OF A HOT SPOT ON THE OCEAN
CRUST, AND IMPLICATIONS FOR FLOW BELOW THE PLATES

Peter R. Vogt

Ocean Floor Analysis Division
U. S. Naval Oceanographic Office
Washington, D. C. 20373, U.S.A.

ABSTRACT. Iceland is the summit of a 4,000 km long regional bulge
of the Mid-Oceanic Ridge. Although the associated gravity high
is of similar extent, anomalies in seismicity and crestal morphology extend only along 2,000 km of the Mid-Oceanic Ridge. The
geochemical anomaly covers less than 1,000 km, and anomalous
crustal thickness appears confined to the Iceland platform itself.
In detail various parameters suffer abrupt discontinuities across
major transform faults. Within 500 km of Iceland time-transgressive basement trends are apparent from bathymetric and seismic
surveys. These trends form nested arrows pointing north and
south away from Iceland. Since the latest Cretaceous the evolution
of the northeast Atlantic by plate tectonics has evidently been
closely linked to the time-space history of the Iceland regional
phenomenon.

An initial attempt to account for the above observations involves a rising convection plume whose core is presently located
under southeast Iceland. The source depth of this mantle upwelling is
probably at least a few hundred kilometers, and possibly as deep as
the core/mantle boundary. The central part of the plume is geochemically anomalous; on the flanks normal (depleted) mantle may also be
entrained in the rising flow. Asthenospheric partial melts flow
"downhill" away from Iceland primarily in a conduit below the axis
of the Mid-Oceanic Ridge; flow rates of a few to a few tens of
cm/yr are believed measurable from time-transgressive trends. Major
transform faults interrupt this mid-oceanic plumbing system, thereby
explaining discontinuities across the faults.

Kristjansson (ed.), Geodynamics of Iceland and the North Atlantic Area. 105-126. All Rights Reserved.
Copyright © 1974 by D. Reidel Publishing Company, Dordrecht-Holland.

1. INTRODUCTION

It has been known for many years that Iceland forms the summit of a vast regional bulge of the Mid-Oceanic Ridge (Fig. 1). The nature and origin of this bulge, and similar features elsewhere, is a major problem in the study of the solid earth. Palmason and Saemundsson [1] have recently reviewed most of the relevant data. With the advent of sea-floor spreading and plate tectonics it became apparent that volcanic and tectonic activity on Iceland, distributed as it is over a broad, complex zone, departs from the classic simplicity of the typical Mid-Oceanic Ridge. The Icelandic crustal structure was found to differ--mainly in thickness--from typical ocean crust [2], and its composition chemically and mineralogically distinct [3]. Until recently little attention was given to the question of how far the anomalous characteristics of Iceland extended north and south along the Mid-Oceanic Ridge, and whether the transition is sharp or gradational. In the last few years the southward [4] and northward [5] extent of the Iceland petrochemical anomaly [6] has been charted along the zero age isochron. Transitions in axial topography (presence or absence of a median rift valley), accompanied by changes in crestal seismicity, have been identified on both sides of Iceland [7,8] (Fig. 2). The spreading axis has jumped to new locations within 900 km north of Iceland [9], and on Iceland itself [10]. Time-transgressive basement structures occur on Reykjanes [11] and Kolbeinsey [12] ridges, their existence suggested by detailed geophysical surveys [9,13-15].

Any hypothesis that attempts to explain why there should be an Iceland on the Mid-Oceanic Ridge must somehow accommodate a variety of geophysical and geochemical observations extending over a vast oceanic region, of which subaerial Iceland is merely the exposed summit. In this paper I wish to consider the problem by examining the various parameters that measure the longitudinal extent of the Iceland phenomenon. The most comprehensive portrait can be developed for the present spreading axis, or zero age isochron. Accordingly, emphasis is herein placed on late Tertiary and younger events, but only because they are better documented.

I shall assume the basic validity of plate tectonics; clearly there is in the Iceland area some major perturbation of normal plate accretion processes and/or processes in the mantle below the plates. As a working hypothesis I shall assume this aberration to arise from a relatively stationary, persistent upwelling of mantle material below the plates [16]. Certainly the coincidence of regional free-air and topographic highs along the Mid-Oceanic Ridge suggests rising convection [17]. The term "plume" is appropriate for such localized upwelling irrespective of the depth from which the material rises. Other kinds of hypotheses have been advanced for "hot spots" or "melting spots" such as Iceland [18-20]; there is no room here to discuss their merits.

   The identification of hot spots as <u>deep</u> mantle plumes [21,22]
is an attractive but unproven hypothesis. Certainly there is an
extensive region of partial fusion below Iceland [1,23-26], but
it is still unknown whether this region extends only a few
hundred kilometers down or if it reaches all the way to the core-
mantle boundary [21,22]. Geochemical and other crustal information

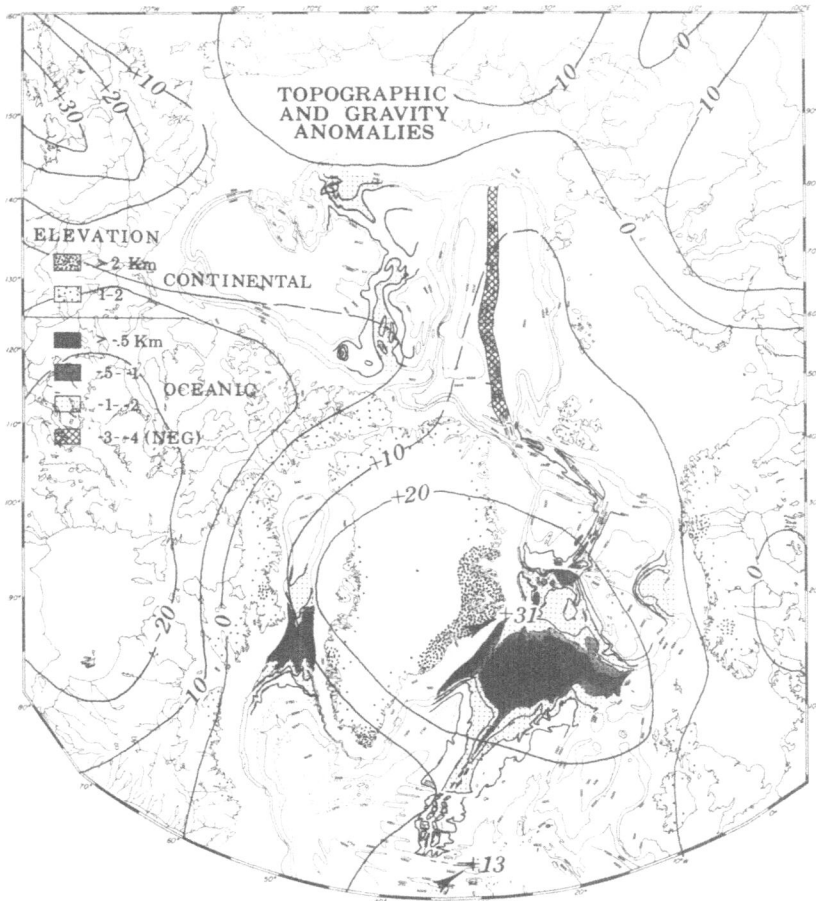

Fig. 1. Regional topographic [27] and satellite derived free-air
gravity anomalies in milligals [28], of the Arctic region. Isobaths
at 500 m interval, with elevated areas emphasized. Stippling on
adjacent continents shows maximum elevations. Note broad extent
of Iceland gravity and topographic anomalies, and rough correlation
between Icelandic and Davis Straits elevations and elevations on
adjacent continental margins.

have so far been of little help in demonstrating a very deep
origin; teleseismic methods seem most promising. For the purposes
of this paper the precise depth of origin may remain moot. The
same applies to the "thermal," "nonthermal," or mixed nature of
the density deficits that drive the plume. If there is any kind
of mantle upwelling below Iceland, however, there should be radial
outflow of partial melts moving away from Iceland in the
asthenosphere, i.e., below the plates [11]. Such outflow should
be particularly channeled below the spreading axis, where the
asthenosphere is thick and at the same time of low viscosity [4,11].

In this paper I consider first the regional picture and its
implications for mantle convection, followed by a more specific
model for flow below the spreading axis at high levels of the
asthenosphere. This subject will be treated more exhaustively
elsewhere [12].

2.   THE ICELAND "HOT SPOT" AS A REGIONAL PHENOMENON

The broad extent of the Iceland hot spot and its lower Tertiary
cousin in Davis Straits is shown in Fig. 1. Late Cretaceous/
early Tertiary igneous activity was distributed over a similiarly
large area [29]. Detailed studies of crustal age versus basement
depth suggests that crust formed at anomalously shallow depths
within 2,000 km of Iceland subsides like normal ocean crust, more
or less maintaining the elevation anomaly it had at the spreading
axis [29]. There is even a suggestion that the highest elevations
of coasts bordering the Davis Straits and Iceland hot spots occur
in the general vicinity of ocean crust anomalies (Fig. 1).
Although late Tertiary uplift is often invoked, some of the
continental elevations may well be relics of anomalously high
uplifts dating from initial igeneous activity and sea-floor
spreading about 60 mybp.

The Iceland topographic bulge as seen in cross-section along
the present spreading axis is shown with seismicity, spreading
angle (obliqueness), and spreading rate in Fig. 2. Evidently neither
the spreading angle nor spreading rate correlate systematically
with elevation and seismicity. The elevation anomaly, although
regionally symmetric about Iceland, is broken into sections of
distinct slopes; some sections are separated by step-like discontinuities. Most sections also exhibit distinct levels of
seismicity. A region of generally low seismicity, correlating
with the absence of a median rift valley, extends from the 57°N
bight through Iceland to the Jan Mayen fracture zone [7,8]. High
seismicity characterizes Mohn and southern Reykjanes ridges
(to 52°N); more moderate activity exists still further from Iceland.
The overall symmetry of earthquake activity and regional topography

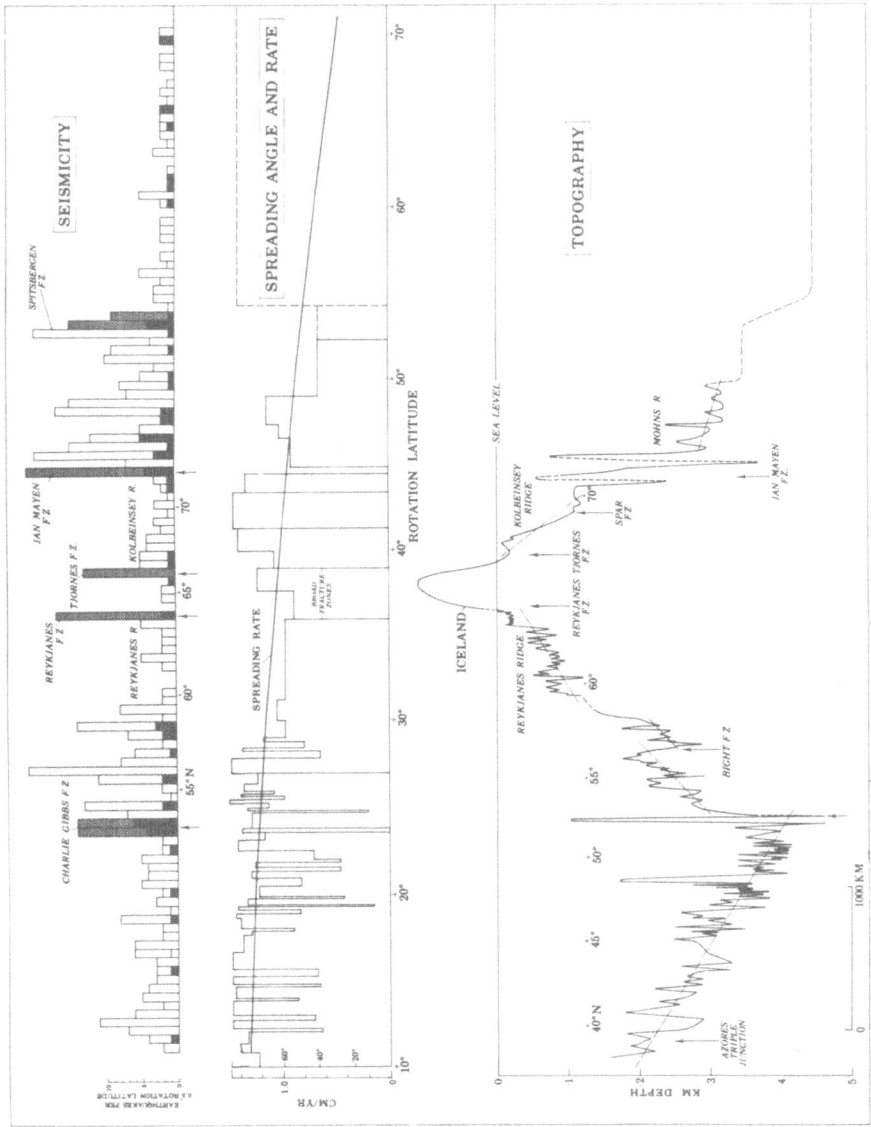

Fig. 2. Seismicity [33] (top), spreading rate and angle (middle) and topography (bottom) profiled along present spreading axis, plotted against plate rotation latitude. Magnitudes greater than 5.0 are black; crosshatching indicates seismicity of major fracture zones. Spreading half-rate, in direction of plate motion, is cosine curve fitted to rate measurements [9]. Region without prominent central rift valley extends from 58°N to Jan Mayen F. Z.

about Iceland is remarkable. Interpretated in terms of a rising
mantle convection cell, this symmetry must mirror the symmetry of
flow at depth. However, detailed geochemical [4,5] and morphological
[9,12,13,15] differences between the Kolbeinsey and Reykjanes
ridges indicate flow asymmetries located at shallow depths.

Major transform fractures often separate spreading axes of
distinct depth, slope, seismicity, crestal morphology, or other
characteristics. This is seen not only near Iceland (Fig. 2) but
elsewhere [30]. One hypothesis for such abrupt differences across
transform faults is that flow of partial melts in a pipe-like region
below the spreading axis is more or less blocked at the faults. Such
blockage must occur if lithosphere thickens with crustal age [31].
Melts of one state or composition would be unable to reach a neigh-
boring axis; if state or composition influences seismicity, basement
morphology, crustal thickness, and composition, such parameters
would be offset across faults [30]. Longitudinal flow and damming
at transform faults would thus modify an otherwise smooth regional
variation in upper mantle properties [32]. The partial melts
producing Iceland's exceptionally thick crust may be partially
dammed by the Tjörnes and Reykjanes fracture zones; similarly the
crestal profile (Fig. 2) suggests that the Charlie Gibbs Fracture
Zone is a "double dam " that prevents northward flow from the
Azores area and southward flow from Iceland. In any case it is
clear that the shape of the Iceland topographic bulge, and therefore any mantle processes responsible for it, have been influenced
by the shape of the accreting plate boundary.

A significant feature of the Iceland hot spot is that different
"anomalies" terminate at different distances from Iceland. Low
seismicity and the lack of a median rift valley are features that
occur only within 900 km of Iceland [7,8]. The spreading axis in
this central region resembles fast-spreading ridges like the East
Pacific Rise. Presumably the lack of a median valley around Iceland
indicates the existence of a thermal anomaly; the more elevated
isotherms--and possibly thicker crust-- would permit continuous
injection of low-viscosity ultrabasic slushes, making the crust
"Pacific" in character [34]. The lack of a graben-like rift valley
and the associated faulting seems to explain the reduced seismicity [8].

Perhaps more enigmatic is the limited extent of the geochemical
anomaly [4,5] compared to the less seismic province and the entire
topographic bulge. If the upwelling brings up a primary mantle
material [4], then either this material merely forms the core of
a convection cell, or else it spreads out as a "megalaccolith" at
some depth below the level of melt segregation (Fig. 3). The
former case would require that "depleted" mantle [4] is entrained

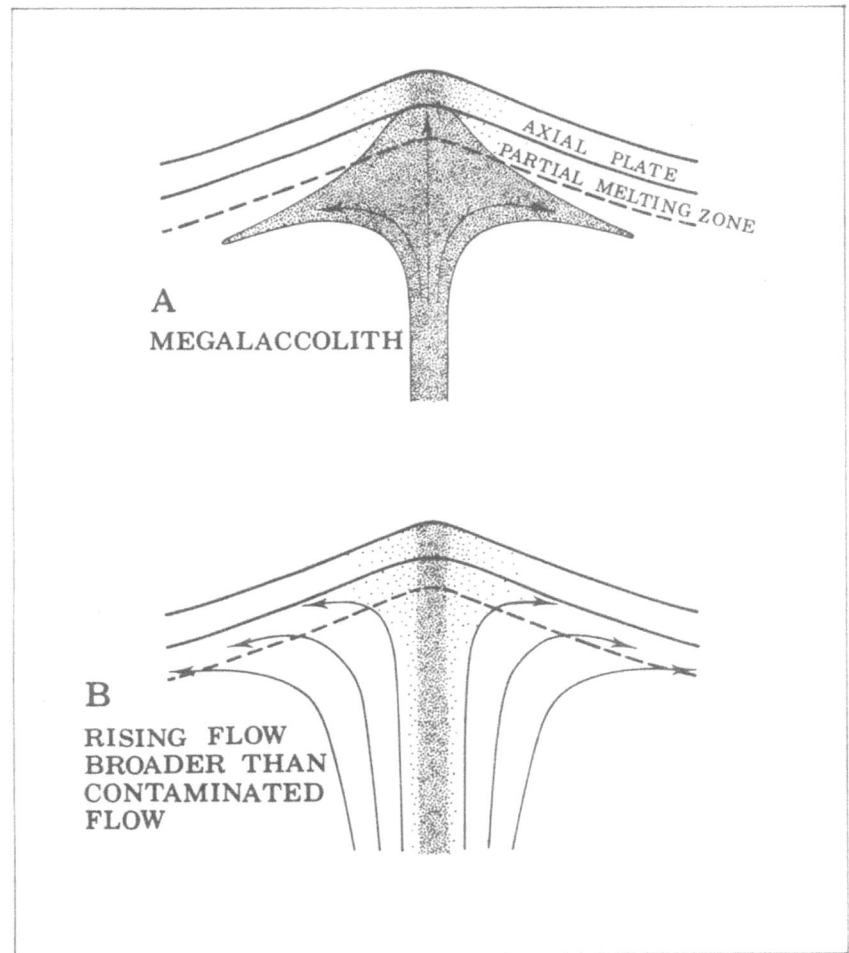

Fig. 3. Two hypothetical <u>longitudinal</u> profiles through the
Iceland topographic bulge showing possible explanations why
topographic, gravity, and seismicity anomalies are broader than
geochemical anomaly (stippled). Basalt melts segregate only in
partial melting zone between axial plate and dashed line. Both
"A" and "B" assume distinct mantle sources [4]. More detailed
models, describing flow in partial melting zone, are shown in
Figures 7, 8 and 9. "A" resembles model b of Schilling [4].

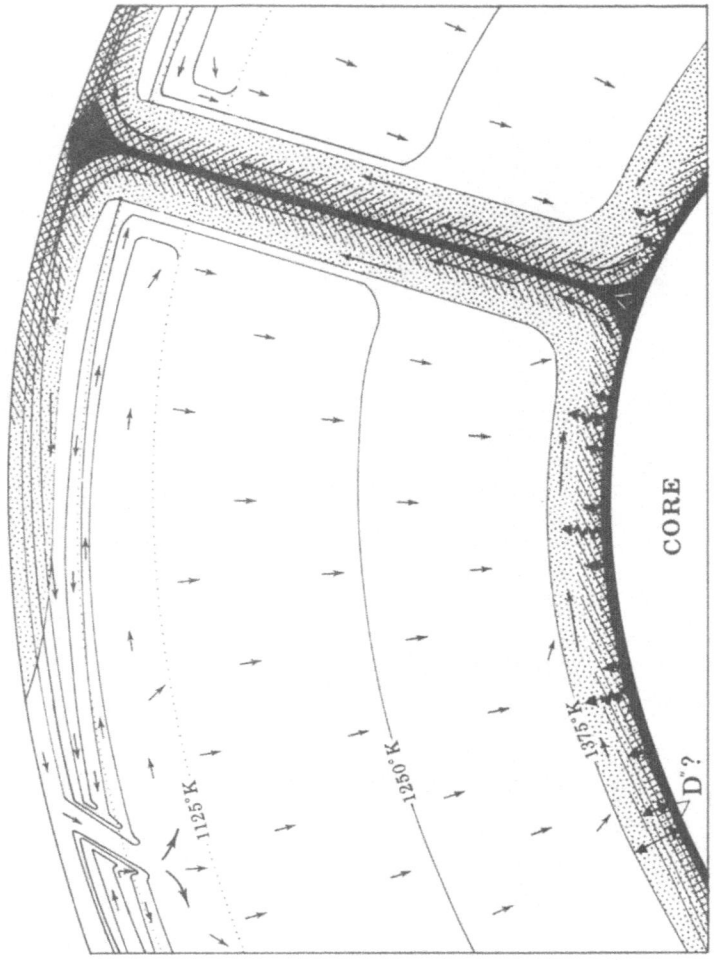

Fig. 4. More speculative extension of model B (Fig. 3) is based on deep mantle plume hypothesis of Morgan [21] and Deffeyes [22]. Chemically anomalous mantle originates by diffusion from core (or an anomalous region elsewhere in the mantle!) and ends up in the center of the convection cell. Thermally anomalous but already depleted mantle is entrained along the margins of cell and yields normal basalts on outer parts of broad hot spot topographic anomaly. Greater thermal than chemical contamination could derive from diffusivity differences. Diagram modified from Deffeyes [22].

in a broad region along the edges of the plume. Should the "deep mantle plume" model [21,22] be demonstrated, this would have the interesting consequence (Fig. 4) that depleted mantle formed in the upper mantle in the geological past has already sunk to the lower mantle, to be recently regurgitated. A speculative possibility (Fig. 4) is that diffusion through the core-mantle boundary (or some other interface higher in the mantle) creates a thin geochemically contaminated layer, overlain by a region that is merely "thermally contaminated," the difference being due to the higher thermal diffusivity. Such a model raises some major problems about the core, however.

3. ASTHENOSPHERE MOTION AT SHALLOW DEPTHS ALONG THE MID-OCEANIC RIDGE

If mantle upwelling occurs under Iceland [21-23] there should be radial asthenosphere motion away from Iceland; in particular there should be motion along the Reykjanes and Kolbeinsey ridges. Such flow should occur, given the low viscosity [36], the thickness of the asthenosphere, and the available horizontal pressure gradient given by the regional topographic slope down the spreading axis away from Iceland [8,30]. Indications that such flow does occur come from diachronous ("time-transgressive," "oblique," or "V-shaped") basement structures on Reykjanes [11,35] and possibly also Kolbeinsey ridges [12,14]. Independently, the progressive changes in basalt chemistry and mineralogy along the Reykjanes Ridge also suggest such flow [4]. Finally, if there is flow below the spreading axis it would be blocked by major transform faults [30]; evidence for such "transform dam" effects constitute still another line of evidence for asthenosphere flow, at least under the ridge axis. In the following discussion simple models are constructed that attempt to incorporate the above observations. A more detailed treatment will be published elsewhere [12].

Diachronous basement features are found both on the Reykjanes Ridge [11,35] south of Iceland (Fig. 5) and on southern Kolbeinsey Ridge [12,14]. They are best revealed when the regional subsidence curve of the ridge is first subtracted (Fig. 5). The structures converge northward north of Iceland and southwestward on the Reykjanes Ridge. If these features of the upper crust are formed at the spreading axis--with no subsequent tectonic or volcanic modification--then propagation of some function of magmatism along and under the spreading axis is implied. Unless the flow has such nonsteady-state features, the crustal morphology would vary smoothly away from Iceland, without any direct registration of flow.

The propagation speeds are of the order 5 to 20 cm/yr under the Reykjanes Ridge [11,35] and 1 or 2 cm/yr along southern Kol-

beinsey Ridge [12,14]. The inferred speed might possibly reflect
the propagation of some magmatic instability, but I shall assume
the actual motion of asthenospheric melts is being measured [11].
Such flow away from Iceland is a predictable consequence of mantle
upwelling hypotheses [21-23] and is not readily explained by other
models [18-20] for the origin of hot spots. It is not possible
from the diachronous trends to discriminate between (1) a largely
radial asthenosphere flow pattern away from Iceland and (2) a
more one-dimensional pattern of flow in a pipe-like region of low
viscosity and high melt concentration below the Mid-Oceanic Ridge
[11]. By the first model the diachrons only measure the component,
along the spreading axis, of asthenospheric variations carried
like circular waves outward from Iceland.

Model (2) is more reasonable, although a relatively sluggish
component of flow east and west from Iceland is also probable.
Such components are required if flow is to help raft the plates
apart by viscous traction, or to create bulges from which the
plates can slide. Although isochrons and diachrons are not as
well defined on the Kolbeinsey as on the Reykjanes Ridge, it seems
probable that flow speeds are an order of magnitude weaker in a
northward direction than southwestwards under the Reykjanes Ridge.
The Icelandic geochemical anomaly also drops sharply from northern
Iceland to the Kolbeinsey Ridge [5], thus supporting the idea that
the northward flow below the spreading axis is less vigorous. A
third difference is the absence, on Kolbeinsey Ridge, of two pairs
of basement "steps" which give the Reykjanes Ridge its blocky
profile [11,13,15] (Fig. 5). The north-south morphologic/geo-
chemical asymmetry might be explained by the Tjörnes transform
fracture zone [10,18] which would act as a dam, impeding subaxial
flow [5,30]. In addition Jan Mayen might mark the site of
another, lesser plume, the flow from which would compete with
flow northward from Iceland. Whatever the reason, it seems clear
that on a regional scale the Icelandic geophysical anomaly displays no marked north-south assymetry (Fig. 2), despite the well-
known northward decrease in spreading rate. I conclude that the
more local asymmetry reflects processes at shallow depths, perhaps
in the range 5 to 40 km below the spreading axis. At greater
depths a more regional flow pattern (Figs. 3 and 4) begins to
ignore the shape of the lithospheric plates above.

What process is responsible for the diachronous trends, and
does this basement topography [11] relate in some way to the observed geochemical anomaly [4,5]? A number of variables could
modulate the instantaneous height of the spreading axis and at the
same time influence the chemistry of basaltic melts. For example,
consider a new batch of asthenosphere moving under the Reykjanes
Ridge south of Iceland. If this batch is less viscous, lower head
losses would allow the ultrabasic mush to climb to a higher level

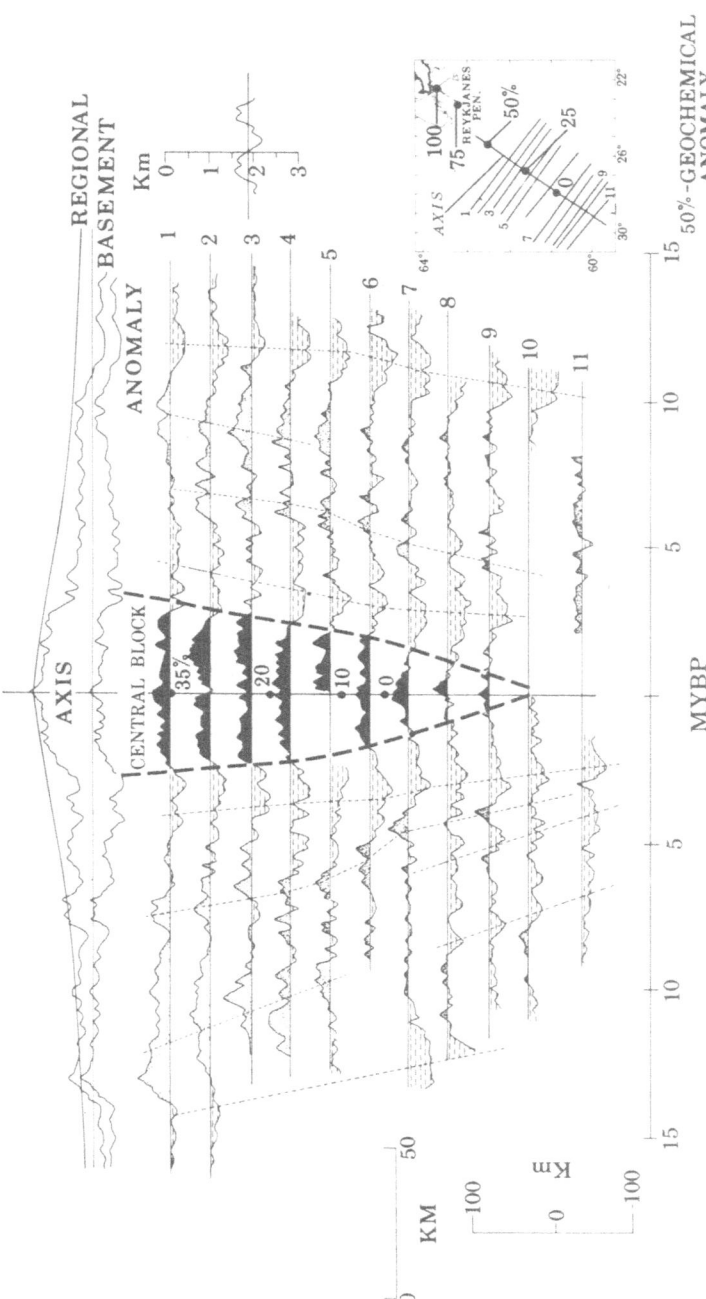

Fig. 5. Basement topography [13] with regional subsidence removed more clearly shows diachronous structures on Reykjanes Ridge. Propagation rates of the order 10-20 cm/yr are implied by thin dashed lines
correlating basement structures, considering well-known crustal isochrons [11]. Geochemical anomaly
(Reykjanes Peninsula =100%) along spreading axis based on linear fits to data [4]. Geochemical anomaly
extends about as far as flat crest of central block, suggesting possible connection.

in the injection zone [37]. A difference in temperature--hence
thermal expansion and viscosity--or in coefficient of thermal
expansion might accomplish this also. The new mush might simply
be less dense. None of these processes necessarily change
crustal thickness, but the basaltic melts would probably differ
in composition. Crustal thickness would also be changed if the
new asthenosphere produces more basalt melt, in which case the
M-discontinuity under the Reykjanes Ridge would not mirror the
observed basement steps. Seismic refraction measurements could
determine whether this is the case. Probably all the listed
factors contribute. In the following simple model I shall assume
that the geochemically anomalous ("primary mantle plume" [4])
ultrabasic mush produces a higher yield of basalt melt. When a
new mass of such anomalous mush, disgorged from the Iceland plume,
arrives at a particular site on the Reykjanes Ridge, crustal
thickness becomes greater and, at the same time, crustal chemistry
becomes more anomalous (Fig. 6). A more explicit model can be
devised that connects crustal thickness and geochemical anomalies. The inspiration for such a model is provided by Fig. 5,
in which the central block of the Reykjanes Ridge (event A-A' [11])
extends about as far from Iceland as Schilling's [4] geochemical
anomaly. (Actually the central block, if measured from the bottom
of the flanking valleys, extends somewhat beyond the geochemical
anomaly (Fig. 5); variables other than composition therefore have
also contributed to making the block. This complication is ignored
here.) I thus explain the time-transgressive central block as the
result of a new pulse of plume discharge which began to deliver
anomalous melts to the northern Reykjanes Ridge about 5-7 mybp;
since then the advancing front has moved southward to about 60°N
(Fig. 5). I do not claim that the crust east or west of the central
block is necessarily normal, depleted [4] ocean crust, but merely
that it is of different composition. Any model for the southward
advance of anomalous mush must account for (a) the gradual southward loss of identity of the central block, and (b) the roughly
linear anomalous geochemical profile [4]. The model discussed
below (Figs. 6 to 9) accomplishes this reasonably well. However,
the actual conditions below mid-oceanic ridges are no doubt
considerably more complex and there is always the possibility that
agreement of data with a simple model could be fortuitous.

In the model I assume there is a pipe-like region below the
spreading axis, extending subhorizontally away from a plume such
as Iceland [4,11]. This mid-oceanic pipe extends from the base
of the axial lithosphere, about 5 or 10 km deep, down to maximum
depths (30 to 50 km?) from which basalt melts segregate and rise.
Tholeiitic fluids would be released from the entire pipe; origin
depths of 23 km for the Mid-Atlantic Ridge and 16 km for the
East Pacific Rise [39] would approximate depth to the center of
the pipe.

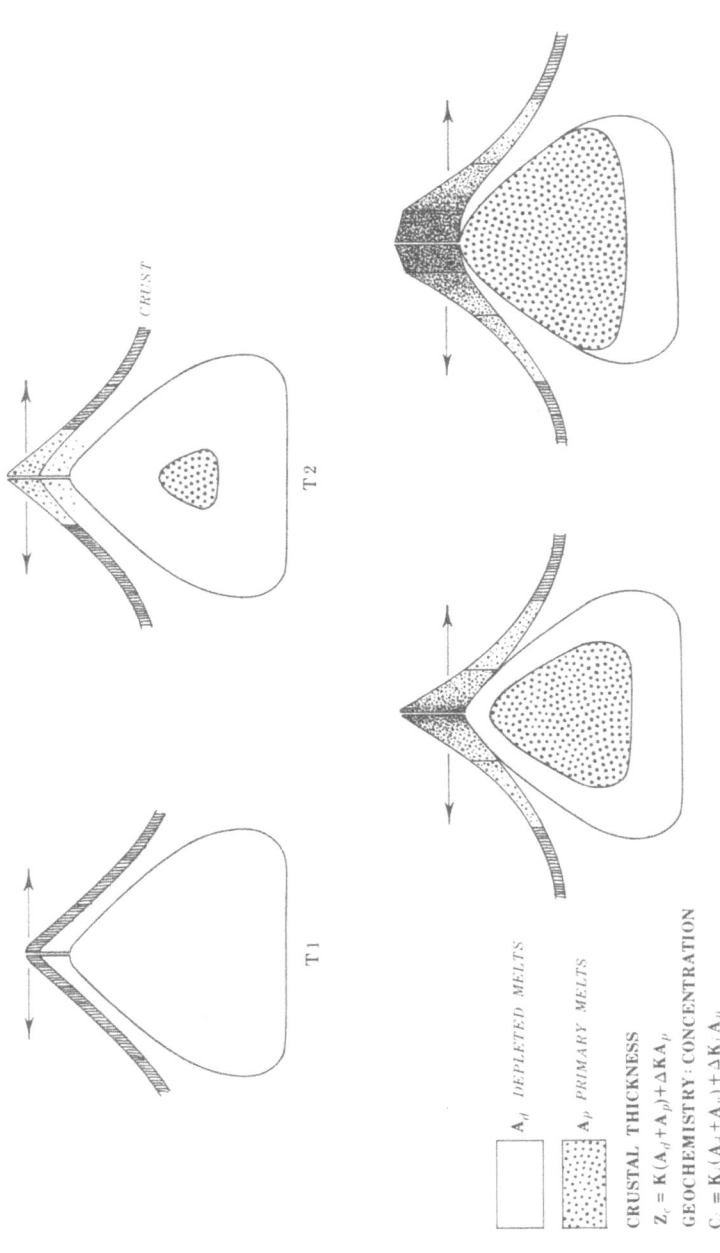

Fig. 6. Hypothesis connecting anomalous crustal thickness with anomalous geochemistry. Cross-sections through subaxial pipe at one point on spreading axis show snout of anomalous composition (stippled) arriving at point and then increasing in area (A_p) and vertical extent. Pipe area occupied by depleted melts (A_d) shrinks accordingly. Average geochemistry (C_i) and crustal thickness (Z_c) might relate linearly to cross-sectional areas (A_p, A_d) and appropriate constants.

The ultrabasic mush in this pipe is assumed to be flowing away from the hot spot at a rate determined by pipe diameter, viscosity, and horizontal pressure gradient [8,30]. The roof of the pipe is probably formed by the plate bottoms which thicken away from the spreading axis at a rate proportional to the square root of crustal age [31]. The actual viscosity is no doubt variable in the pipe. It is not even certain that the flow is Newtonian [40], although the latter is usually assumed to approximate mantle behavior over the time scales considered [17]. Flow with such complex features can be modelled only by numerical techniques, and in any case the details of pipe shape and viscosity distribution can only be guessed at this stage. I therefore make the following simplifying assumptions: (1) the flow is confined either to a circular pipe or, in the other limit, to a sheet-like region between two parallel plates. Seismic data [41] and plate-thickening models [31] suggest that the pipe is more nearly elliptical in cross-section, with the region of extensive fusion greater in width than in thickness. The actual pipe will therefore be intermediate between a circular pipe and sheet flow. (2) Flow behavior is either Newtonian (strain rate proportional to stress) or of the power-law type proposed by Post and Griggs [40] (strain rate proportional to the cube root of shear stress). (3) Viscosity or pseudo-viscosity is constant in time and space within the pipe. (4) Upward motion is assumed to be either zero or a small constant value (in the range a few mm/yr) required to feed the accreting lithosphere [31]. For nonzero upward motion it is assumed that partial melting liberates new material into the bottom of the pipe while accretion removes material from the top. At any instant the profile of horizontal speed with position in the pipe is constant. (5) Horizontal average flow velocity along the pipe is assumed to be given by diachronous structures, viz. about 10 cm/yr [11]. Horizontal inflow towards the ridge crest is neglected. (6) At some time t_o asthenosphere of a different composition (e.g., primary mantle plume material [4]) is allowed to enter the pipe near its connection to the plume. The composition front is initially vertical in the pipe. In the model this new composition is likened to a tracer dye which tags the material without significantly altering its rheological properties. As time progresses the compositional boundary becomes stretched into a progressively longer "snout" because flow is faster at the axis of the pipe than at the margins (Fig. 7). Upward motion required to feed the accreting plates will cause the snout to angle upward (Fig. 8) and eventually reach a limiting range. (7) At any point on the ridge axis, basalt melts are derived from the entire pipe cross-section or from a vertical slice through the pipe below that point (Fig. 6). Anomalous melts contributed by the snout, if present below the point, mix with normal melts from the normal, relatively depleted pipe-fill surrounding the snout. Mixing

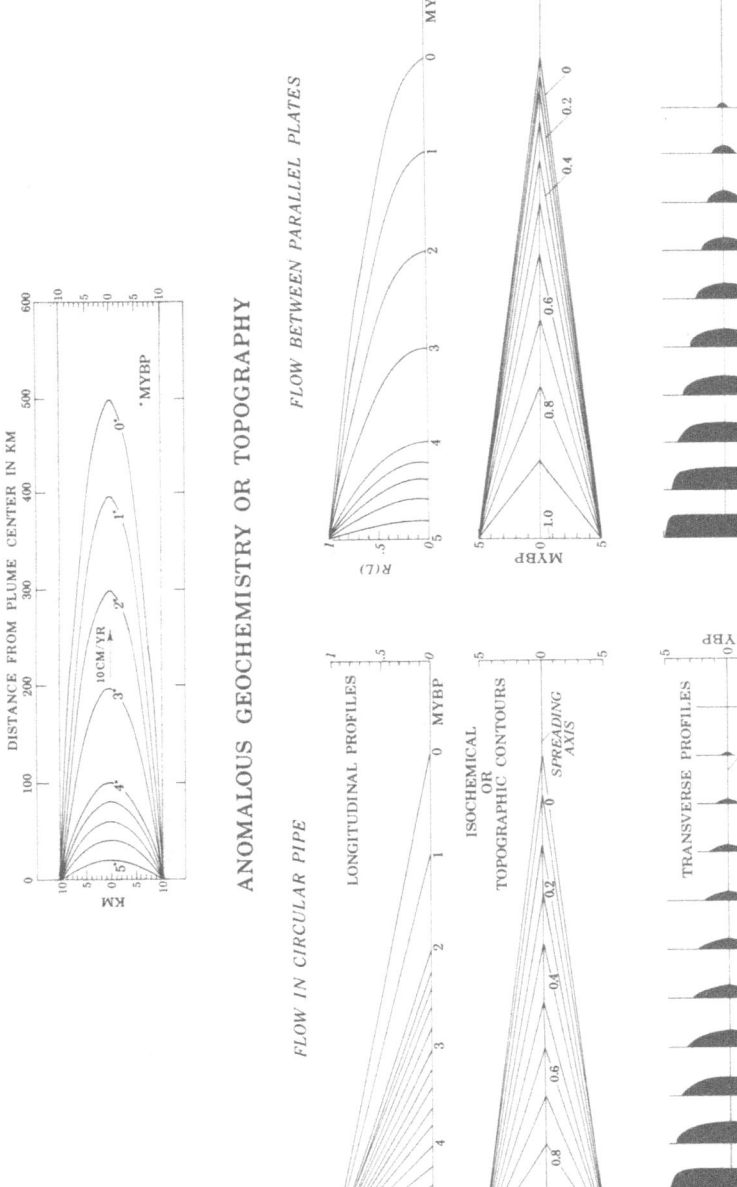

Fig. 7. Simple quantitative model of hypothesis depicted in Fig. 6. Anomalous material is introduced into the pipe 5 mybp, and advances down the pipe, its forward edge progressively stretched due to the velocity gradient. Profiles and contours show R(L), fraction of pipe cross section that is anomalous.

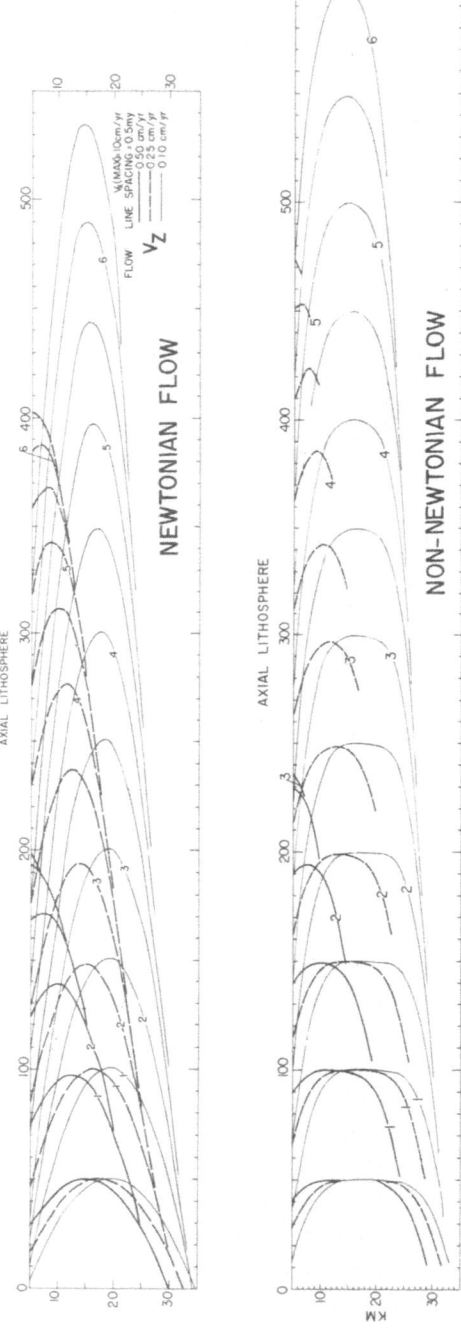

Fig. 8. More complex model allows for slow vertical (V_z) flow (0.1 to 0.5 cm/yr) required to feed accreting lithosphere if published lithosphere models [31] are reasonably correct. Note that a steady-state limit is reached depending on V_x, V_z, and pipe diameter. The pipe is assumed circular, with 30 km diameter. Non-Newtonian flow of type predicted by Post and Griggs [40] produces blunter snout (lower diagram). Depth scale is with respect to ocean floor and flow lines are calculated every 0.5 my.

between two dissimilar mantle materials and their derivative fluids has been discussed in detail by Schilling [4,38]. An anomaly in composition and crustal thickness will form at the point, depending on the fraction of the pipe filled by the anomalous snout. The simplest possible relationship would be for the topographic or geochemical anomaly being formed at a given time on the ridge axis to be linearly proportional to R(L), the ratio of anomalous snout diameter (or area) to the total pipe diameter (or area). For flow between parallel plates R would be the same for diameter as for area. L denotes distance along the spreading axis, measured from the plume.

With the assumptions above, it is possible to follow the compositional boundary as it advances down the pipe (Figs. 7 and 8). Once crustal isochrons have been defined using magnetic lineations, the anomalous parameter R(L) can be constructed along particular isochrons as well as transverse to the spreading axis (Fig. 7).

Construction of the advancing compositional boundary as a function of time is based on an assumed steady-state mantle flow law of the "power-law" form

$$\dot{\varepsilon}_s = B \ (Z) \ \tau^m \qquad [40]$$

where $\dot{\varepsilon}_s$ is strain rate, $B(Z)$ is a depth-dependent parameter, τ is effective shear stress, and m is some positive constant. In the simple model considered here let $B(Z) = $ constant, independent of depth in the pipe. Newtonian flow, often assumed for mantle convection, is given by m = 1. Post-glacial rebound of Fennoscandia suggests a value of m $\simeq$ 3 [40], but there is no assurance that this would apply to the more extensively fused materials below the Mid-Oceanic Ridge. In the following discussion both kinds of flow are considered.

Velocity profiles in a circular pipe of radius r_0 containing a general "power-law" flow are described by setting m = 1 in Eq. 7.55 [42]:

$$V_x \ (r) = \left(\frac{m + 3}{m + 1} \right) \ \overline{V} \ \left[1 - \left(\frac{r}{r_0} \right)^{m + 1} \right] = V_{max} \left[1 - \left(\frac{r}{r_0} \right)^{m + 1} \right]$$

where $V_x \ (r)$ is velocity as a function of radius r, and $\overline{V}$ is mean velocity in the pipe.

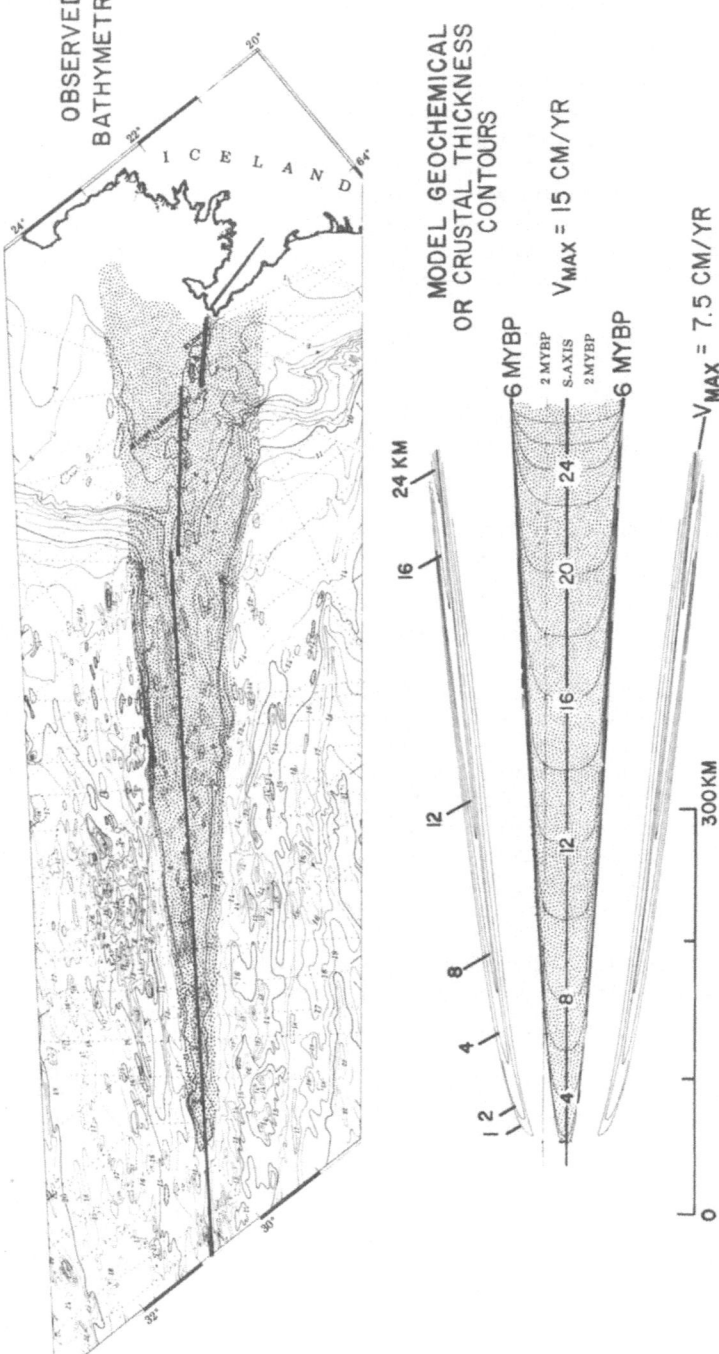

Fig. 9. Observed bathymetry of Reykjanes Ridge, contoured at 100 m interval. Synthetic anomalous bathymetry (or geochemistry) is calculated as in Fig. 7, but using Newtonian flow model of Fig. 8. Contours in km anomalous snout thickness; total pipe thickness is 30 km. Model parameters are $V_z =$ 0.25 cm/yr and V_x = 15 cm/yr to model axial block and V_z = 0.25 cm/yr, V_x = 7.5 cm/yr for a pair of hypothetical diachronous ridges such as B or C [11,35] formed by a 0.5 my long initially cylindrical slug of anomalous mantle.

In Fig. 7, the effects of vertical motion in the pipe are assumed to be negligible out to ranges of several hundred kilometers. Without significant vertical motion the flow would continue indefinitely unless interrupted by a major transform fault [30], and a steady-state condition is not reached. Fluctuations in composition or state would propagate down the pipe, gradually being "smeared out" by the increasingly stretched flow lines. Although neglect of vertical motion is not strictly justified, it might be a good approximation within a few hundred kilometers of the hot spot. Applied to the central block of the Reykjanes Ridge, the model predicts linear or nearly linear geochemical profiles along the present axis (as observed [4]) and older isochrons (Fig. 7), and can therefore be readily tested by sampling of basalts on post-5 mybp crust. In profile and contour form the anomaly R(L) resembles the central block of the Reykjanes Ridge (Fig. 5): The summit of the block is relatively flat; the flanks are steep near the plume and both flank gradients and axial amplitude diminish downstream. Detailed refraction and gravity work, perhaps coupled with deep drilling, could test the hypothesis (Fig. 6) that crustal thickness correlates with bathymetry.

A somewhat more refined flow model incorporates both vertical motion and possible non-Newtonian behavior (Figs. 8 and 9). Vertical motion, V_z, was assumed constant in the pipe; new material is liberated into the pipe bottom at the same rate that melts freeze to the bottom of the accreting plate, thereby being removed from the pipe.

From the rates plates are thought to thicken with age [31], upward motion would be of the order 0.1 to 0.5 cm/yr when averaged over the region 0 to 50 or 100 km from the axis [30]. Composition front positions were calculated by numerical integration at 0.5 million year steps, using several V_z values and both m = 1 and m = 3 (Fig. 8). The non-Newtonian snouts are more plug-like than Newtonian ones (Fig. 8), but the two flow laws produce quite similar consequences. The apparent faster advance of the non-Newtonian snout results from choosing V_{max}, rather than $\overline{V}$, to be the same in both cases. Solutions for other values of $\overline{V}$, V_z and pipe diameter can be derived by appropriate rescaling of horizontal and vertical axes and time increments of Fig. 8. Continuous delivery of anomalous material over several million years results in a steady-state flow pattern; the snout then extends of the order several hundred to 1,000 km downstream before being entirely used up in the accretion process. The steady-state regime resembles model "a" of Schilling [4]. Since geochemical anomalies have in fact been found to extend over such ranges [4,5], steady-state conditions may characterize many hot spots. On the other hand, the central block and other diachronous structures on the Reykjanes Ridge (Fig. 5) must represent variable plume discharge or composition or some other nonsteady-state process. Several lines of other evidence also suggest variable discharge from the Iceland plume, the first-order pattern being high discharge in the early Tertiary or late Cretaceous, and again in the late Tertiary [43]. To construct the model contours matching the Reykjanes central block bathymetry (Fig. 9), a Newtonian compositional front with V_{max} = 15 cm/yr and V_z = 0.25 cm/yr was assumed injected into the pipe near Iceland about 6 mybp and presently nearly at the steady-state limit. Isolated diachronous ridges [11,35] might result from short "pulses" of anomalous material such as the 0.5 my long cylinder injected with V = 7.5 cm/yr to produce the flanking anomaly ridges shown in Fig. 9. The models shown in Figs. 6 to 9 may be tested by detailed basalt sampling and refraction work on and off the central block (as well as the older escarpments, more completely buried by sediment). As a first step in this direction, basalts from diachronous ridge B [11] were dredged at 60°58.3'N, 29°15,1'W and found [44] to contain 0.1 wt % P_2O_5 and 0.9 wt % TiO_2, both values rather lower than values reported for an equivalent point on the present spreading axis [4]. This tentatively suggests compositional differences transverse to the Reykjanes Ridge. Much more sampling is obviously required.

In conclusion, if the Iceland "plume" is variable in its discharge, any variations should affect not only the Reykjanes but also Kolbeinsey Ridge and Iceland itself. Perhaps the more complex geological history of Iceland has borne withess to the same mantle events that produced the central and older blocks (events A and E [11]) on the Reykjanes Ridge. Because the oldest rocks on Iceland (16 mybp [45]) were emplaced about the time the outer ("E") escarpments began to form near Iceland, it is tempting to relate both to relatively abrupt late Tertiary increases in discharge from the Iceland plume. Similarly the 8-4 mybp gap in the lava flow succession in eastern Iceland, followed by formation of the present-day north Iceland rifting zone about 4 mybp [10], probably relates in some way to the injection of new material to form the Reykjanes central block (Figs. 5 and 9) about the same time.

REFERENCES

1. G. Palmason and K. Saemundsson, <u>Ann. Rev. Earth Planet. Sci.</u>, <u>2</u>, 25, 1974.
2. G. Palmason, <u>Crustal Structure of Iceland From Explosion Seismology</u>, Soc. Sci. Islandica, <u>40</u>, 1971.
3. G. E. Sigvaldason, <u>Contr. Mineral. Petrol.</u>, <u>20</u>, 357, 1969.
4. J. G. Schilling, <u>Nature</u>, <u>242</u>, 565, 1973.
5. J. G. Schilling, D. G. Johnson and T. H. Johnston, <u>Trans. Amer. Geophys. Union</u>, <u>55</u>, 294, 1974.
6. S. P. Jakobsson, <u>Lithos</u>, <u>5</u>, 365, 1972.
7. T. J. G. Francis, <u>Earth Planet. Sci. Lett.</u>, <u>18</u>, 119, 1973.

8. P. R. Vogt and G. L. Johnson, Earth Planet. Sci. Lett., 18, 49, 1973.
9. G. L. Johnson, J. R. Southall, P. W. Young and P. R. Vogt, J. Geophys. Res., 77, 5688, 1972.
10. K. Saemundsson, Bull. Geol. Soc. Am., 85, 495, 1974.
11. P. R. Vogt, Earth Planet. Sci. Lett., 13, 153, 1971.
12. P. R. Vogt, J. G. Schilling and G. L. Johnson, Trans. Amer. Geophys. Union, 56, 1974.
13. M. Talwani, C. C. Windisch and M. G. Langseth Jr., J. Geophys. Res., 76, 473, 1971.
14. O. Meyer, D. Voppel, U. Fleischer, H. Closs and K. Gerke, Deut. Hydrogr. Z., 25, 193, 1972.
15. J. Ulrich, Kieler Meeresforschungen, 16, 155, 1960.
16. J. T. Wilson, Phil. Trans. Roy. Soc. London (Symposium on Continental Drift), 258, 145, 1965.
17. R. N. Anderson, D. McKenzie and J. G. Sclater, Earth Planet. Sci. Lett., 18, 391, 1973.
18. D. L. Turcotte and E. R. Oxburgh, Nature, 244, 337, 1973.
19. H. R. Shaw and E. D. Jackson, J. Geophys. Res., 78, 8643, 1973.
20. P. L. Ward, Bull. Geol. Soc. Am., 82, 2991, 1971.
21. W. J. Morgan, Am. Assoc. Petrol. Geol. Bull., 56, 203, 1972.
22. K. S. Deffeyes, Nature, 240, 539, 1972.
23. M. H. P. Bott, Geophys. J. R. Astr. Soc., 9, 275, 1965.
24. T. J. G. Francis, Geophys. J. R. Astr. Soc., 17, 507, 1969.
25. R. E. Long and M. G. Mitchell, Geophys. J. R. Astr. Soc., 20, 41, 1970.
26. E. Tryggvason, Bull. Seism. Soc. Am., 54, 727, 1964.
27. W. M. Kaula, Global Gravity and Mantle Convection, in: The Upper Mantle, ed. by A. R. Ritsema, Elsevier, 1972.
28. G. L. Johnson and P. R. Vogt, Marine Geology of the Atlantic Ocean North of the Arctic Circle, in: Arctic Geology, Mem. 19, Am. Assoc. Petrol. Geol., 1973.
29. P. R. Vogt and O. E. Avery, J. Geophys. Res., 79, 363, 1974.
30. P. R. Vogt and G. L. Johnson, Bull. Am. Assoc. Petrol. Geol., in press, 1974.
31. R. L. Parker and D. W. Oldenburg, Nature Phys. Sci., 242, 137, 1973.
32. B. I. R. Haigh, Geophys. J. R. Astr. Soc., 33, 405, 1973.
33. Anonymous, World Seismicity, 1961-1969, Nat. Earthquake Info. Center, U. S. Department of Commerce, Washington, D. C., 1970.
34. P. R. Vogt, E. D. Schneider and G. L. Johnson, The Crust and Upper Mantle Beneath the Sea, in: The Earth's Crust and Upper Mantle, Geophys. Monogr. Ser., ed. by P. Hart, Am. Geophys. Union, Washington, D. C., 13, 1969.
35. P. R. Vogt and G. L. Johnson, Earth Planet. Sci. Lett., 15, 248, 1972.
36. Tr. Einarsson, Jökull, 16, 157, 1966.
37. N. H. Sleep and S. Biehler, J. Geophys. Res., 75, 2748, 1970.
38. J. G. Schilling, in preparation, 1974.

39. K. F. Scheidegger, J. Geophys. Res., 78, 3340, 1973.
40. R. L. Post Jr. and D. T. Griggs, Science, 181, 1242, 1973.
41. S. C. Solomon, J. Geophys. Res., 78, 6044, 1973.
42. A. B. Metzner, Flow of Non-Newtonian Fluids, Section 7, in:
     Handbook of Fluid Dynamics, ed. by V. L. Streeter, McGraw-
     Hill, New York, 1961.
43. P. R. Vogt, Nature, 240, 338, 1972.
44. J. Campsie, J. C. Bailey, M. Rasmussen and F. Dittmer,
     Nature Phys. Sci., 244, 71, 1973.
45. S. Moorbath, H. Sigurdsson and R. Goodwin, Earth Planet. Sci.
     Lett., 4, 197, 1968.

TERTIARY BASALTS OF BAFFIN BAY: GEOCHEMICAL EVIDENCE FOR A FOSSIL HOT-SPOT

M.J. Keen and D.B. Clarke

Department of Geology, Dalhousie University,
Halifax, N.S., Canada.

ABSTRACT.   Tertiary basalts found on land in Baffin Island and western Greenland, and offshore western Greenland have been divided into several types produced at comparable evolutionary stages within the different areas. Their occurrence, seismic evidence and bathymetric evidence suggests that the general region of Davis Strait, the sill between Baffin Bay and Labrador Sea, was a hot-spot in the early Tertiary. Chemical variations within basalts at comparable evolutionary stages change geographically in the same manner as do changes in basalt chemistry from the Reykjanes Ridge to Iceland, a modern hot spot. If the analogy is correct the sense of direction of the changes suggests that the centre of the hot spot in the early Tertiary lay in the northern part of the offshore province, some distance from the present bathymetric sill of Davis Strait. The analogy may not, of course, be correct - the changes in chemistry observed may reflect other factors (such as height within the basalt pile), and the spatial variations may be simply an accidental by-product of these factors. If it is real, it suggests that geochemical search for former hot-spots, now dormant, may be profitable.

1. INTRODUCTION

It has been suggested that areas of persistent volcanic activity, termed hot spots, are related to mantle plumes [13,19]. Moreover, a group of chemical criteria for the identification of such hot spots (and plumes) has been advanced using Iceland as an example [17]. Davis Strait, which lies between Canada and Greenland (Fig. 1) is a bathymetric sill joining two Tertiary lava provinces and

Kristjansson (ed.), Geodynamics of Iceland and the North Atlantic Area. 127-137. All Rights Reserved.
Copyright © 1974 by D.Reidel Publishing Company, Dordrecht-Holland.

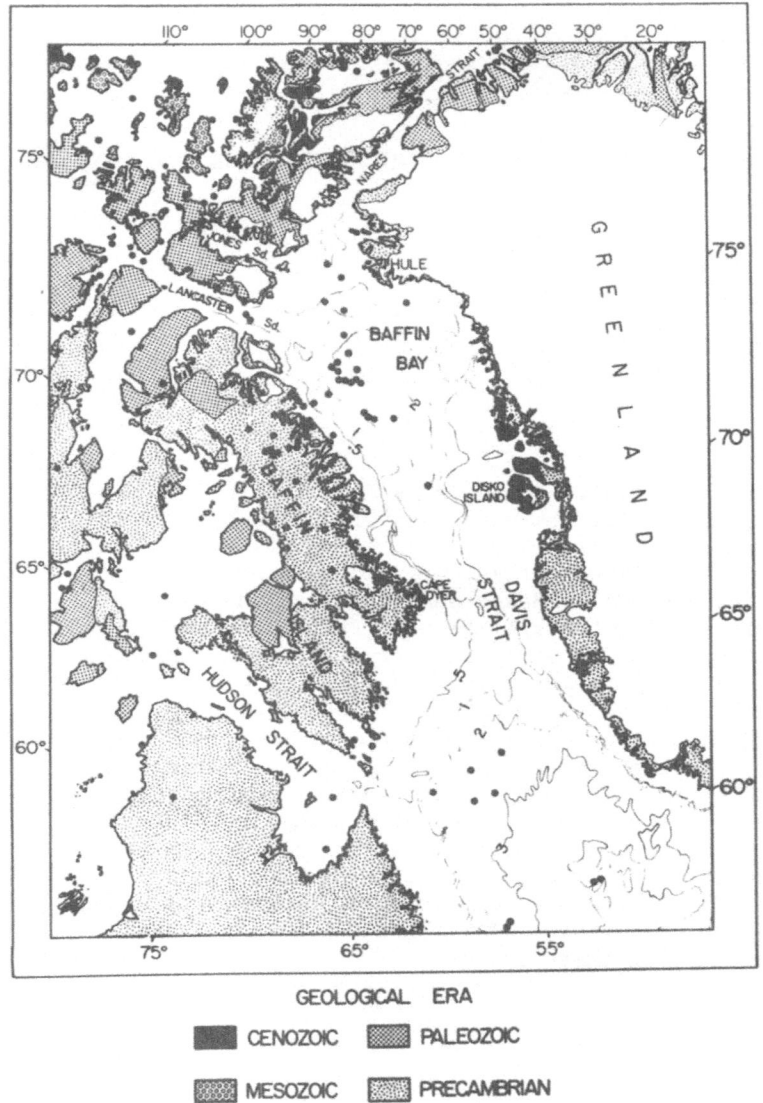

Fig. 1. A map of the area of Baffin Bay and Davis Strait showing in simplified form the geology and bathymetry. Cenozoic basalts are black. The bathymetric contours are in kilometres. The dots show earthquake epicentres. (Reproduced from [9]).

may be a fossil hot spot [6,8]. It is the purpose of this paper
to test the suitability of the Icelandic chemical criteria to the
identification of this ancient hot spot.

Igneous rocks of Paleocene age are found in western Greenland and eastern Baffin Island, and occupy a substantial part
of the shelf of western Greenland offshore, north of the Strait
[1,3,4,11,15,16]. They are important in any consideration of
spreading between Greenland and Canada, because they give the
only direct evidence of the time that Baffin Island and Greenland separated. They are important too in attempting to relate
depth of oceanic crust and the history of spreading in the
Labrador Sea, because the centre of the hot-spot has to be defined [6,18]. Hyndman [6] computed ages for the crust in the
Labrador Sea by using the age-depth relations of Sclater et al.
[18]. He obtained ages in close agreement with those deduced
from magnetic anomalies, transform fault patterns and sedimentation history [12], but to do so had to assume that a correction
was needed for the excess elevation caused by a hot spot in Davis
Strait, analogous to the excess elevation of Iceland by comparison with the Reykjanes Ridge. Seismic evidence also suggests
that Davis Strait is underlain by rocks comparable to those of
Iceland [8]. The crustal velocities beneath the Strait are similar,
the crustal thickness is comparable, and the top of oceanic basement, as inferred from reflection studies can be traced from the
deeper waters of Baffin Bay to the shallower waters of the
Strait [6].

There are systematic changes in chemical composition of
basalts along the Reykjanes Ridge from south to north to Iceland,
and the claim has been made that these changes reflect a distinctive separate source for the igneous rocks of Iceland [17].
Although this has been challenged, and the changes observed
ascribed to different mechanisms of production of igneous rocks
[14], nevertheless we should at least see if such changes can be
found in the igneous rocks of the region of the Davis Strait.
Although we cannot do this in quite the same way as was done along
the Reykjanes Ridge, we can compare the basalts of Baffin Island
with those of western Greenland, and can compare rocks dredged offshore western Greenland with rocks on land.

2. BASALTS OF BAFFIN ISLAND, SVARTENHUK, AND OFFSHORE

The Tertiary basalts of Baffin Island, Svartenhuk and the offshore
province are tholeiites, comparable in many respects to the so-
called oceanic tholeiites [1,2,3,7,15]. They can be grouped into
several types in the order of their eruptive history. These types
are picrites, olivine basalts, olivine-poor basalts and feldspar-
phyric basalts [1]. The complete range is found on Svartenhuk;

TABLE 1
COMPARISON OF GREENLAND (SVARTENHUK) ANALYSES
WITH BAFFIN ISLAND ANALYSES: MAJOR ELEMENTS
Greenland/Baffin Is. basalts

Oxide	Picritic	Olivine	Olivine-poor	Iceland/Reykjanes R.
SiO_2	0.97	0.98	0.99	Decrease
$(Fe_2O_3+FeO)/MgO$	1.06	1.14	1.15	Increase
Fe_2O_3+FeO	1.05	1.08	1.09	Increase
K_2O	3.5	1.6	2.5	Increase
Na_2O/CaO	1.3	1.3	1.3	Increase
TiO_2	1.7	1.6	1.6	Increase
P_2O_5	1.7	1.6	1.8	Increase

Analyses from Reference 1, Table 1.
Iceland/Reykjanes Ridge behaviour from Reference 17.

TABLE 2
COMPARISON OF GREENLAND (SVARTENHUK) ANALYSES
WITH BAFFIN ISLAND ANALYSES: TRACE ELEMENTS IN OLIVINE BASALTS

Element	Greenland/Baffin Is.	Iceland/Reykjanes R.
Ba	1.4	Increase
U	3.1	Increase
Rb	19	Increase
Sr^{87}/Sr^{86}	1.0006	Increase
Ce	2.81	Light REE enriches
Nd	2.24	
Sm	2.18	
Eu	1.61	
Gd	1.77	
Dy	1.39	
Er	1.39	
Yb	1.22	

Sources: Ba, Ref. 1, Table 3
           Rb, Sr^{87}/Sr^{86}, Ref. 15, Table 2
           U, Ref. 15, Table 3
           Rare Earth Elements, Ref. 15, Table 4

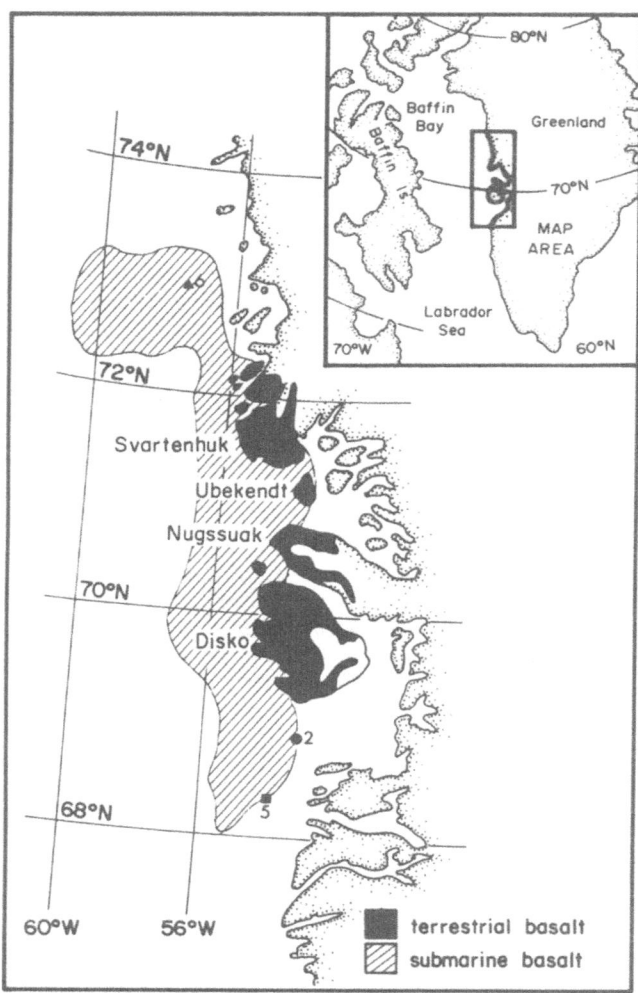

Fig. 2. Location of dredge hauls off western Greenland, in relation to the offshore basalt province described by Ross and Henderson [16], and M. J. Keen et al. [11].

feldspar-phyric basalts are the only basalts found so far in the offshore province [2], and are not found on Baffin Island [3]. Consequently, we can compare each of the olivine-bearing types from Baffin Island with the equivalent type from Svartenhuk, and can compare the feldspar-phyric basalts from the southern dredge hauls with those of Svartenhuk and the northern dredge haul (Fig. 2).

Tables 1 and 2 show comparisons between Baffin Island and Svartenhuk, and also indicate the changes found on the Reykjanes Ridge towards Iceland. We see that the major elements, and the incompatible elements all behave in the same manner as they do on the Reykjanes Ridge. Fig. 3 and Table 2 show that the light rare-earth elements are enriched in West Greenland by comparison with Baffin Island. Table 3 and Figs. 4a and b show chemical changes between similar rock types from south to north in western Greenland, with the means of analyses for the two southern hauls (2 and 5). We see that all changes are the same as found on the Reykjanes Ridge, except for total iron and the ratio Na_2O/CaO in the comparison of dredge hauls with Svartenhuk, starred in Table 3.

3. DISCUSSION AND CONCLUSIONS

Almost all comparisons suggest that we see geochemical trends from Baffin Island to western Greenland, and from south to north off western Greenland which are identical with those of the Reykjanes Ridge. It is at first sight impressive because we have been able to use assemblages thought to have been erupted at comparable evolutionary stages [1] and impressive too because although the differences are so small the geochemical trends nevertheless stand out. The data suggest that, if Iceland is a guide, the centre of any hot spot lay to the north of the offshore province (Fig. 5). This is contrary to the bathymetric evidence which naturally suggests that the centre of the hot spot lay in the region of the shallow sill of Davis Strait. If true, thoughts concerning the spreading history of the Bay will have to be revised.

However, there are several reasons to be cautious in accepting such an hypothesis. First, the dredge hauls offshore could, possibly, have been erratic. This is unlikely [2], but must not be discounted. Second, it has not been established with certainty that the changes seen on the Reykjanes Ridge are due to tapping of a separate mantle source beneath Iceland; the possibility that the changes reflect differences in fractionation caused by elevation differences, for example, still remains. While we have been careful to try to select stratigraphic equivalents for comparison in this study, the possibility does remain that, for instance, the rocks of Dredge Haul 6 are much higher in the volcanic section than the rocks of Svartenhuk or the southern Dredge Hauls 2 and 5.

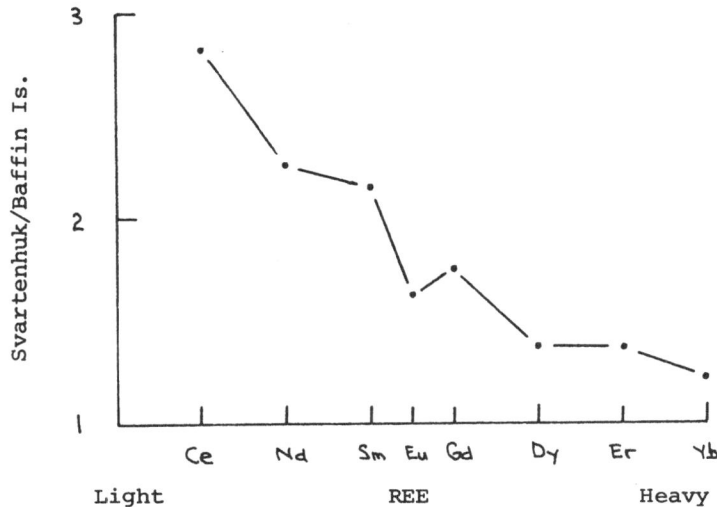

Fig. 3. Comparison of Rare-Earth Elements between Greenland (Svartenhuk) and Baffin Island for olivine-basalts. (Adapted from [15]).

If this is the case then the observed compositional variations may be a function of height in the section, not distance from a hot spot. This possibility is examined in detail by Clarke [2]. Third, the geochemical differences are small, and comparisons of individual elements from Baffin Island to Svartenhuk have been shown to be statistically not significant [15]. Fourth, many geochemical changes will follow once one is established. We should expect, for example, elements like barium to follow the behaviour established by potassium, and this naturally diminishes the apparently impressive impact of all the data.

Nevertheless, the trends we report appear to be real, and may serve as a guide in future work in the area. It is important that more dredge hauls be obtained, and that the age of the basalts be firmly established radiometrically. The absence of feldspar-phyric basalts in Baffin Island suggests that Baffin Island had drifted too far from the basalts' source by the time the feldspar-phyric basalts were being produced [3]. It is important that the discrepancies in the maps of the offshore basalts be resolved [4, 16] and their distribution in Davis Strait be established with certainty. It could be, for example, that the Strait is the site of a fracture zone.

Finally, we should look northwest of the most northerly part

TABLE 3
COMPARISON OF DREDGED MATERIAL WITH SVARTENHUK (GREENLAND)
FELDSPAR-PHYRIC BASALTS, FROM SOUTH TO NORTH

	Southern Dredge Hauls 2 & 5	Svartenhuk	Northern Dredge Haul 6	Iceland/ Reykjanes Ridge
SiO_2	49.5	48.4	47.5	Decrease
$(Fe_2O_3+FeO)/MgO$	1.96	2.12	2.56	Increase
Fe_2O_3+FeO	12.8	12.3 *	14.4	Increase
K_2O	0.24	0.38	0.68	Increase
Na_2O/CaO	0.26	0.22 *	0.31	Increase
TiO_2	2.24	2.30	3.25	Increase
P_2O_5	0.23	0.26	0.46	Increase
Ba (ppm)	83	141	170	Increase
Rb (ppm)	4.6	5.9	13.1	Increase

Analyses from Refs. 1 and 2

of the offshore province to see if any other indication of the
hot-spot can be found, within the deeper waters of the Bay.

→

Fig. 4. (opposite). Geochemical changes in the feldspar-phyric
basalts with latitude, from south to north, for dredge hauls and
Svartenhuk (see Table 3). Values for Svartenhuk and dredge 6
have been normalised to the mean of values for dredges 2 and 5.

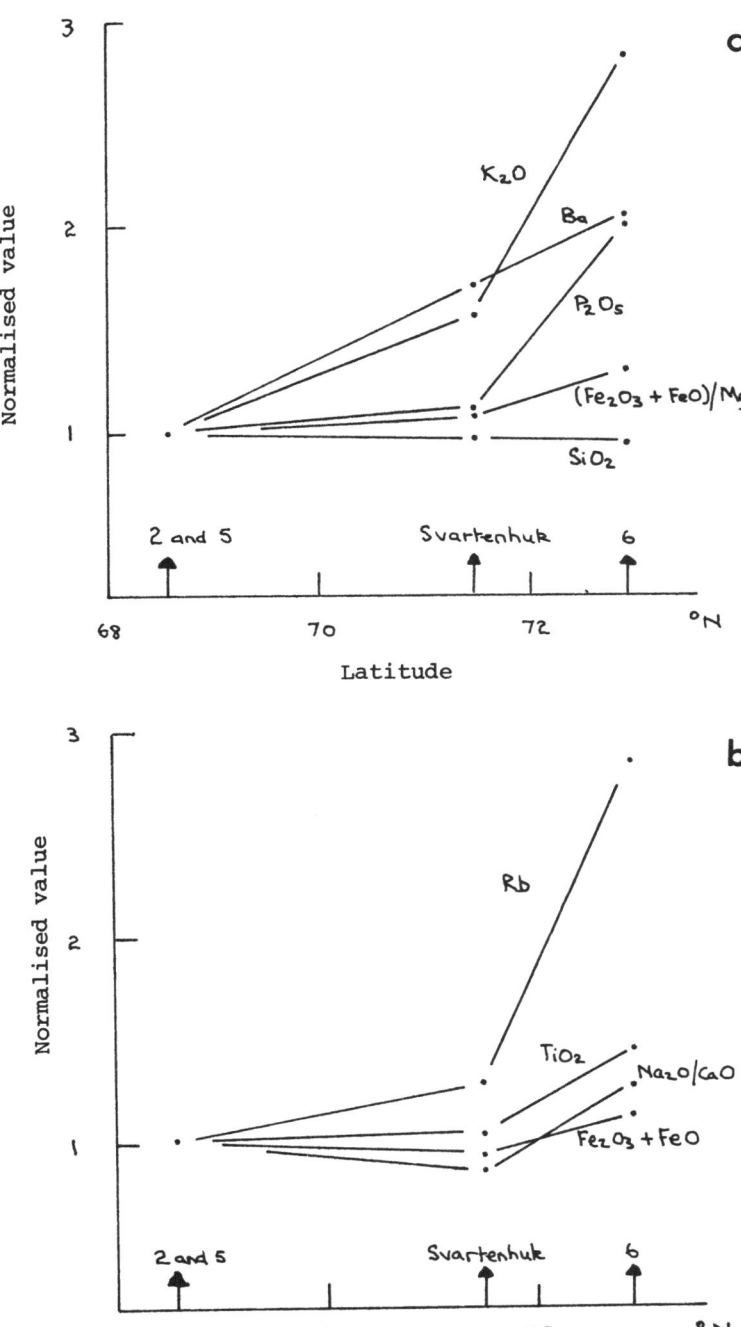

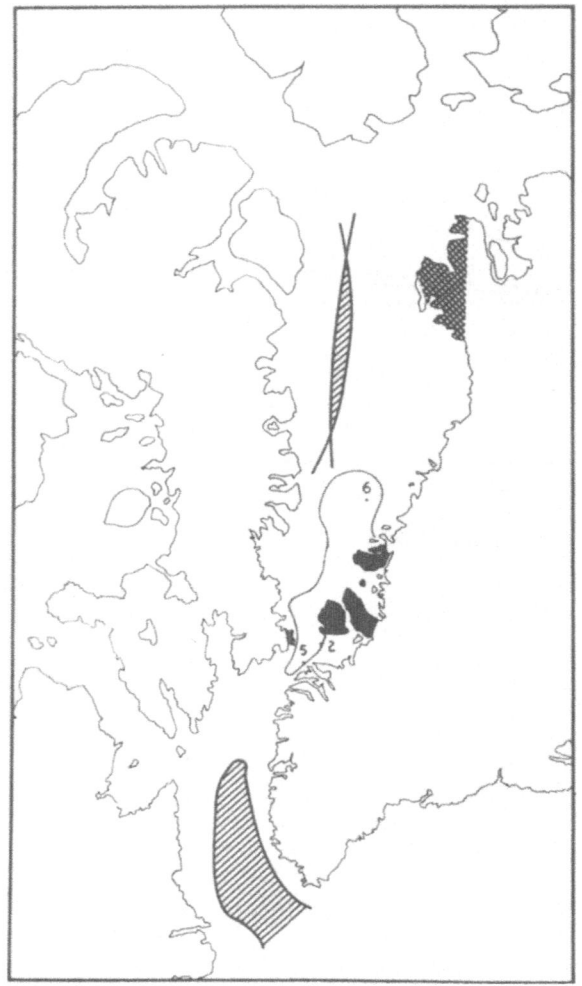

Fig. 5. The relative positions of Baffin Island and Greenland before spreading. An oblique Mercator projection about a pole at 77°N, 100°W has been used. The areas occupied by diagonal lines show the approximate extent of oceanic crust. The black areas on land indicate the location of Tertiary basalts, and their seaward extension offshore is shown by the solid black lines. The crosses show the positions of the dredge hauls of Fig. 2. (Adapted from [9]).

ACKNOWLEDGEMENTS

We appreciate the help of our colleagues in the Atlantic Geoscience Centre and Dalhousie University, and the help of the master, officers and crew of C.S.S. DAWSON, and C.S.S. HUDSON. Part of this work was supported by the National Research Council of Canada. Discussions with C. E. Keen were most helpful.

REFERENCES

1. D. B. Clarke, Contr. Mineral. and Petrol., 25, 203, 1970.
2. D. B. Clarke and M. J. Keen, in preparation, 1974.
3. D. B. Clarke and B. J. Upton, Can. J. Earth Sci., 8, 248, 1971.
4. L. R. Denham, Geol. Surv. Greenland, Report No. 63, 1974.
5. G. Henderson, Geol. Surv. Can. Paper 71-23, 521, 1971.
6. R. D. Hyndman, Can. J. Earth Sci., 10, 637, 1973.
7. B. G. Jamieson and D. B. Clarke, Jour. Petrol., 11, 183, 1970.
8. C. E. Keen and D. L. Barrett, Geophys. J. R. Astr. Soc., 30, 253, 1973.
9. C. E. Keen, D. L. Barrett, K. S. Manchester and D. I. Ross, Can. J. Earth Sci., 9, 239, 1972.
10. C. E. Keen, M. J. Keen, D. I. Ross and M. Lack, Am. Assoc. Petrol. Geol. Bull., 58 (in press, 1974).
11. M. J. Keen, J. Johnson and I. Park, Can. J. Earth Sci., 9, 689, 1972.
12. X. LePichon, R. D. Hyndman and G. Pautot, J. Geophys. Res., 76, 4724, 1971.
13. W. J. Morgan, Bull. Am. Assoc. Petrol. Geol., 56, 203, 1972.
14. M. J. O'Hara, Nature, 243, 507, 1973.
15. R. K. O'Nions and D. B. Clarke, Earth Planet. Sci. Lett., 15, 436, 1972.
16. D. I. Ross and G. Henderson, Can. J. Earth Sci., 10, 485, 1973.
17. J. G. Schilling, Nature, 242, 565, 1973.
18. J. S. Sclater, R. N. Anderson and M. L. Bell, J. Geophys. Res., 76, 7888, 1971.
19. J. T. Wilson, Nature, 198, 925, 1963.

PETROCHEMISTRY OF THE VOLCANIC ROCKS OF THE NORTH ATLANTIC RIDGE SYSTEM

C. Kent Brooks   and   S.P. Jakobsson

Geologisk Centralinstitut,   Museum of Natural History,
Copenhagen University,   Reykjavik, Iceland.
Copenhagen, Denmark.

1. INTRODUCTION

The North Atlantic has been a key area since Vine [1] demonstrated the regularity of the magnetic anomalies along the Reykjanes Ridge while Iceland and the transverse ridge extending from the Faeroes to East Greenland represent the trace of a hot spot generated throughout the development of this part of the North Atlantic. It is our intention here to review the chemistry of oceanic tholeiites such as are recovered from abyssal parts of the mid-ocean rift system, contrast them with those of hot spot areas such as Iceland, and try to trace the chemical development of these lavas throughout the 60 m.y. spreading history. In addition the alkaline lavas of the area will be briefly described.

2. MID-OCEAN RIDGE BASALTS AND THE NORTH ATLANTIC

Engel et al. [2] first recognized that the predominant basaltic type dredged from depths of >1000 m was a type poor in the large ionic radius lithophile (LIL) elements and, although significant chemical variation is now known to exist among these basalts, this has been abundantly confirmed by subsequent workers, e.g. [3-7]. These rocks are formed exclusively in the median rift valley but other types may be present on the flanking mountains [8]. This mid-ocean ridge basalt (abbreviated "MORB", typical composition shown in Table 1) is characteristically an olivine tholeiite after the classification of Yoder and Tilley [9] and has very low concentrations of the LIL elements (such as K, Ti, P, Rb, Cs, Sr, Ba, Zr, and U) relative to other basaltic types (see Table 2). It is further depleted in the light rare earths relative to chondritic

Kristjansson (ed.), Geodynamics of Iceland and the North Atlantic Area. 139-154. All Rights Reserved.
Copyright © 1974 by D. Reidel Publishing Company, Dordrecht-Holland.

meteorites [10-12], while radiogenic Sr [13,14] and Pb [15,16] are also low. Both the major element [17] and trace element (e.g. [12]) compositions indicate large percentages of partial melting (ca. 30%). Shido et al. [18] showed that, in the typical MORB from the Atlantic either olivine or plagioclase are the first minerals to crystallize and the residual liquid rapidly reaches a cotectic so that both minerals crystallize together and are subsequently joined by pyroxene. They proposed on this basis a classification into two classes, viz: OL- and PL-tholeiite.

Basalts of this type generally show rather limited ranges of differentiation. Indeed, this is a powerful argument for homo- geneity of the mantle [12]. The ratio FeO*/ MgO (FeO* = total iron as FeO) usually lies within the range 1.0 to 1.5 (the extreme values quoted by Miyashiro et al. [19] being 0.76 and 1.85) and variations appear to be related to fractionation of the principle phenocryst phases [20,12,18,21] this being parti- cularly well seen for Cr and Ni in the OL-tholeiites [22]. How- ever, a greater range is observed for the associated gabbros [19,23], which are presumably cumulates, and for the drilled samples away from the mid-ocean ridge [24]. Regional variations are also reported (see below), particularly in trace elements [25] and this was interpreted as probably reflecting variations in the upper mantle. A few rare differentiated rocks have been recorded,e.g. diorites [26] and a single aplite [19].

MORB was originally interpreted as being of a very primitive nature owing to its similarity to calcium-rich achondrites [2] but later the low content of LIL elements led to the idea that it was derived from a layer which has been depleted in LIL elements by previous partial melting events [27,28]. This idea appears to be supported by the isotopic composition of Sr [29,30] as radiogenic Sr is present in greater amounts than one would expect for the source regions of basalts with such low Rb/Sr ratios. Isotopic data for Pb is more scanty but appears to lead to the same result [15,16]. Sr isotopic compositions further show that the basalts have lower concentrations of radiogenic Sr than the ultramafic rocks dredged from the fracture zones (e.g. [31]) and this is generally taken to indicate that the ultramafics cannot be residuals after removal of the basalt as was earlier thought. However, O'Nions and Pankhurst [14] show how all these observations can be reconciled into a simple model if equilibrium between the minerals of the mantle is not attained, although the reality of such a process is perhaps doubtful.

One further characteristic of MORB may be mentioned, namely the gas contents. Owing to extrusion and rapid chilling under high pressure, the primitive gas content of the magma may be preserved. Thus Dymond [32] and others have demonstrated the presence in glassy pillow margins of excess Ar^{40}, presumably derived from

TABLE 1. Major element compositions and norms of typical basaltic rock-types of the North Atlantic.

	1.	2.	3.	4.	5.
SiO_2	50.01	50.26	47.40	45.22	38.92
Al_2O_3	15.62	13.52	15.80	13.37	4.48
Fe_2O_3	1.18	3.52	3.26	3.17	4.89
FeO	8.58	11.87	7.97	7.71	11.12
MgO	8.24	5.58	5.60	9.27	18.09
CaO	10.63	9.18	9.76	8.77	10.96
Na_2O	2.91	2.48	3.25	4.48	2.43
K_2O	0.13	0.56	2.60	2.45	1.19
MnO	0.16	0.24	0.20	0.21	0.20
TiO_2	1.69	2.47	3.25	2.77	4.10
P_2O_5	0.17	0.23	0.50	1.07	0.58
H_2O^+	0.51	0.31	0.20	0.62	1.86
H_2O^-	0.10	–	–	0.34	–
Sum	99.93	100.22	99.79	99.75	98.82
FeO^*	9.64	15.04	10.90	10.56	15.52

C.I.P.W. weight norms

	1.	2.	3.	4.	5.
Q	–	1.98	–	–	–
or	0.77	3.31	15.38	14.64	5.51
ab	24.66	20.97	17.00	10.48	8.89
an	29.22	24.09	20.87	9.24	3.65
lc	–	–	–	–	32.33
ne	–	–	5.70	15.08	–
ac	–	–	–	–	–
di	18.29	16.66	20.00	22.43	29.09
hy	14.04	24.59	–	–	–
ol	7.04	–	11.42	17.65	3.09
ln	–	–	–	–	5.26
mt	1.85	2.88	2.09	2.05	5.26
il	3.21	4.69	6.18	5.32	7.79
ap	0.39	0.53	1.16	2.51	1.34

1. Mid-ocean ridge basalt: glassy margin of pillow, V25-1-Tl, from median valley of MAR near 25°N [20] with revised value of H_2O^+ and total from Shido et al. [18].

2. FETI basalt: av. of 12 analyses of basalts from the Askja-Myvatn area, Iceland [42].

3. Alkali basalt: av. of 4 analyses of the 1970 eruption, Jan Mayen [60].

4. Basanite: av. of 3 analyses of Sverrefjeld volcano, Bockfjord, Vestspitsbergen [71].

5. Olivine nephelinite: Kangerdlugssuaq, East Greenland [85].

NOTE. norms in col. 1-4 have been calculated after adjusting Fe_2O_3/FeO ratios to 0.15, based on a rough average for ca. 30 analyses of fresh basaltic glass taken from the literature. The effect of this adjustment is negligible on the 1st, 3rd and 4th analyses, but considerable for the 2nd which probably reflects deuteric alteration. No adjustment has been made to the last analysis as the degree of oxidation of such magmas is not known.

K present in the Earth's mantle, together with traces of
primordial rare gases [33]. Similarly, Moore [34] showed that
MORB had a very low content of H_2O+ which correlates closely
with K_2O, P_2O_5, F and Cl among the different basalt types.

In the North Atlantic area, MORB-type materials have been
recovered from abyssal regions of the central rift valley both
south [8,35-37] and north [38,39] of Iceland. Kolbeinsey [40]
is the only subaerial example of this basalt type known to us.

3. HIGH Fe-Ti THOLEIITES OF THE ICELAND MANTLE PLUME

Noe-Nygaard [41] was apparently the first to draw attention to
the distinct chemical difference between the Faeroes-Iceland-
Greenland basalts and those of the mid-Atlantic ridge to the
south. In particular he noted the higher contents of *FeO, TiO_2,
P_2O_5 and K_2O and lower values of Al_2O_3 and MgO, differences
further affirmed by Jakobsson [42] for Iceland, cf. Table 1.
Caution should be exercised in the use of this table as data for
Iceland is taken from widely separated locations. These chemical
characteristics together with their higher level of LIL trace
elements (Table 2) of Icelandic basalts have normally been
ascribed to "continental tholeiites"; however, it now seems
probable that all such occurrences of continental tholeiites are
the result of hot spots such as Iceland, whose excess basalts
spilled out over the continental crust during the early stages
of continental break-up. This high iron-titanium type of basalt,
subsequently referred to as "FETI", shows, among other characteristics a greater variability than MORB, both in the critical
elements such as Fe, Ti, K and LIL trace elements, with values
ranging right down to those characteristic of MORB [28,43,44]
and in the differentiation index. Tholeiitic andesites (icelandites) are relatively common while the high abundance of acid
rocks in Iceland has frequently been remarked on [45]. Jakobsson
[42] reviewed chemical data for recent Icelandic basalts and
showed that the FETI basalts from the neovolcanic zones are
more saturated than typical MORB, although olivine tholeiites are
predominant in Reykjanes, farthest from the centre of Iceland.
He further showed that a maximum basalt discharge occurs over
Iceland. These basalts often do not belong to the OL- and PL-
tholeiite groups of Shido et al. [18] but belong to a third
PX-tholeiite group which crystallizes pyroxene as a first phase
[37].

Information on the volatile content of FETI magmas, which
have generally degassed on eruption, is much poorer than that
for MORB whose volatiles have been retained by rapid chilling
and high confining pressures. Moore [34] and Moore and Schilling
[46] have nevertheless shown that the basalts which are higher

TABLE 2. Trace elements and radioisotopes in three major basalt types of the North Atlantic.

	1.	2.	3.
percent			
CO_2	0.01	-	-
Cl	0.04	-	-
F	0.02	-	-
ppm			
S	843	-	-
Rb	1.11	13	60
Cs	0.016	-	0.64
Sr	135	182	826
Ba	10	130	783
$(\frac{La}{Sm})_{E.F.*}$	0.62	1.27	3.68
$(Yb)_{E.F.*}$	16.22	11.82	13.9
V	313	460	352
Cr	303	25	88
Co	41	47	79
Ni	135	30	36
Cu	62	91	- -
Zr	89	180	-
Hf	2.2	-	4.9
U	0.026	0.47	1.5
Th	0.2	1.6	5.2
$\frac{^{87}Sr}{^{86}Sr}$	0.70265	0.70304	0.70343
$\frac{^{206}Pb}{^{204}Pb}$	18.672	18.82	-
$\frac{^{207}Pb}{^{204}Pb}$	15.543	15.59	-
$\frac{^{208}Pb}{^{204}Pb}$	38.307	38.68	-

1. Mid-ocean ridge basalt: CO_2, Cl, F; av. for 9 basalts from the Juan de Fuca ridge [34], S; av. for 62 determinations of outermost zone of pillows from the Reykjanes ridge [46], Rb, Sr, Ba & Sr isotopes; av. for 15 samples of MORB tholeiites [21]. La/Sm & Yb; av. for 19 MORB tholeiites [11], V, Co, Cr, Ni, Cu; av. for 12 MORB samples from 25 & 30° N on MAR [88], Zr, Hf; av. for 2 samples from Carlsberg ridge and 2 from MAR [89], U; av. for 6 samples MORB glass from D.S.D.P. cores [90], Th, Pb·isotopes; drill sample from D.S.D.P., site 226 on axial trough of Red Sea [16].

2. FETI basalt: Rb, Sr, Ba, V, Cr, Co, Ni, Zr, U, Th; Askja (Iceland) 1961 eruption [91], Cu; av. for 36 samples of tholeiites from the Icelandic northern volcanic zone [44], La/Sm & Yb; av. for 9 samples from Reykjanes area, Iceland [28], Sr isotopes; av. for 16 tholeiites from the Icelandic neovolcanic zone and first 200 km of Reykjanes ridge [13], Pb isotopes; rhyolite from Kerlingafjöll, Iceland (sample no. 1-69 of Welke et al. [92]), example of Group II leads of these authors.

3. Alkali basalt: trace elements; av. for 4 samples of the 1970 eruption, Jan Mayen ([93]-Yb value interpolated), Sr isotopes; av. for 4 samples, variation within experimental error, of Jan Mayen basalts [14].

*E.F. = enrichment factor relative to chondrites.

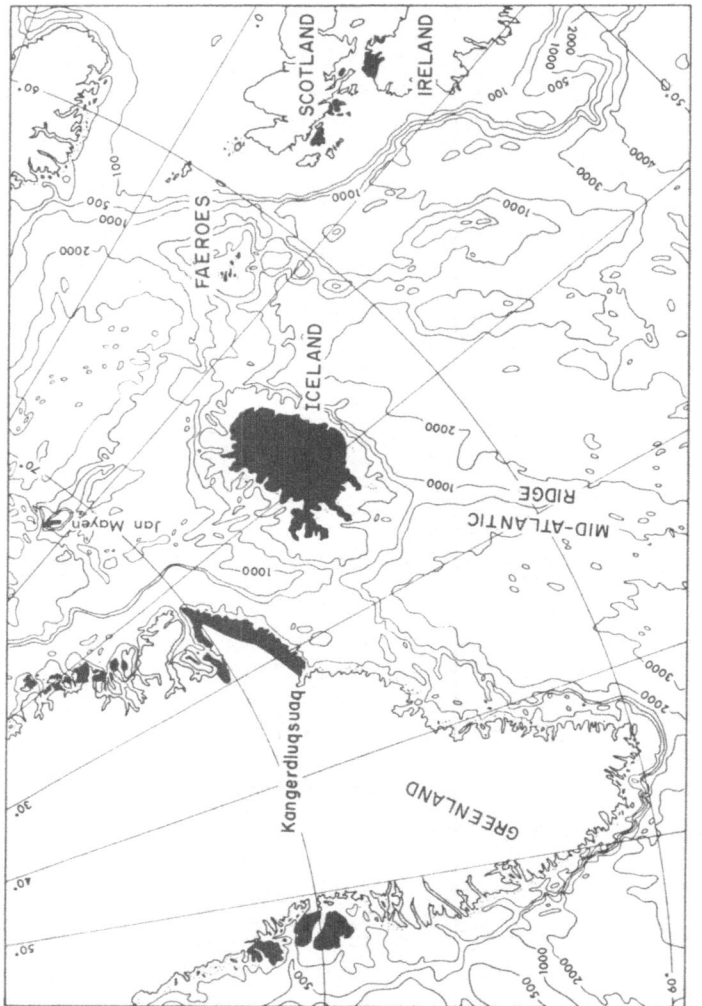

Fig. 1. Map showing locations of the principal Tertiary to recent volcanic areas of the North Atlantic. Depths in meters.

in K_2O are also higher in H_2O+, F and Cl, although the correlation with S is less clear. Moore and Fabbi [47], found no correlation between S content and basalt type. The geochemistry of Icelandic volcanics will be discussed further by Sigvaldason and Steinthórsson (this volume).

Basalt of the FETI type occur on several oceanic islands, often in large volume, as in Iceland, notably Hawaii [48] and the Galápagos Islands [49]. They also occur on continental margins such as East Greenland and the Faeroes (see below), the Deccan of India and the Columbia River province of western U.S.A. These areas are closely associated with active lithospheric spreading (except Hawaii) and regions of excess basalt discharge or hot spots.

These factors have led to the idea of mantle plumes, which has been actively applied to the Icelandic region by Schilling [28,50]. FETI basalts are regarded as products of a primary hot mantle plume, the MORB-type of the depleted low velocity layer. The idea of the two separate sources in this area is an attractive one and the same ideas can readily be applied to, for example, Hawaii and Galápagos and the Afar triangle [51].

4. TRANSITION BETWEEN MORB AND FETI NORTH AND SOUTH OF ICELAND

This transition has been studied by Schilling [28] (rare earths and other LIL elements); Hart et al. [13] (Sr isotopes) and O'Nions and Pankhurst [14] (Sr isotopes), while specialized studies have been made by Moore and Schilling [46] (vesicles, water and sulphur), Unni and Schilling [52] (chlorine) and Hermes and Schilling [53] (olivines). Additional data are presented by Brooks et al. [43] and Jakobsson and Sigvaldason [39], from which Fig. 1 is reproduced. As shown in this figure, the chemical transitions are gradual, LIL elements increasing in concentration and scatter from typical MORB values south of about 400 km from Reykjanes, to typical FETI values over Iceland and this correlates with increased discharge rate [42]. Water and chlorine, but not sulphur, apparently show similar patterns which have been complicated by degassing at shallow depths. Sr isotopes however, show an abrupt transition about 200 km south of Reykjanes and a satisfactory reconcilation of these apparently conflicting observations has not yet appeared. North Iceland data are at present sparse, but what there is [38,39], shows clearly that there is an abrupt transition northwards across the Tjörnes fracture zone to typical MORB material. This is also supported by the petrochemistry [40] and Sr isotopic composition [14] of Kolbeinsey.

Campsie et al. [37] presented evidence to show that the Charlie Gibbs fracture zone (56°N) represents a petrological boundary between low Al_2O_3, pyroxene tholeiites to the north and high Al_2O_3 plagioclase and olivine tholeiites to the south. This is in agreement with the observation of Vogt ([55] and this volume) that there is a major change in crustal thickness across such fracture zones, which could lead to a damming of the radial flow in the asthenosphere which is postulated to occur along the ridges around plumes. A similar explanation could be applied to the abrupt geochemical change across the Tjörnes fracture zone, noted above. Similarly Shido and Miyashiro [56] showed an abrupt fall in CaO/Na_2O for basalts from the ridge north of the Azores to that south of the Azores. In neither case were the variations simply related to varying degrees of fractionation. However, Shido and Miyashiro [56] point out that the number of dredge hauls from this region is probably too small to allow definite conclusions to be drawn at this point.

5. THE ALKALINE ROCKS OF ICELAND AND JAN MAYEN

Alkali basalts are more variable mineralogically, chemically and isotopically than tholeiites. Major element compositions for a typical example are shown in Table 1 (col. 3) and trace elements in Table 2. In the North Atlantic region they occur mainly in the Vestmannaeyjar and Snaefellsnes zones of Iceland [42], Jan Mayen, Kong Oscars Fjord and Kangerdlugssuaq (East Greenland) and the Hebridean Province. All these areas appear to be charac- terized by a thick crust, in the three last cases by typical continental crust, in the case of Iceland, a basaltic crust thickening from the tholeiitic neovolcanic zone to the flanking alkalic zones [42]. The situation in Iceland appears to be similar to that described for the MAR at 45°N by Aumento [8]. It is probable that Jan Mayen is also situated on continental material which has been split off Greenland by a jump in the position of the spreading axis [57,58] and it is tempting to relate the special character of the Jan Mayen rocks to either this continental material or to the major Jan Mayen fracture zone. The lavas of Jan Mayen consist of a differentiated alkaline series with abundant ankaramites bearing mantle-derived chromian diopside and nickeli- ferous olivine [59,60]. The sequence is notably potassic (leucite is recorded [61]) and resembles Tristan da Cunha [62] and Gough Island [63]. It has several times been suggested as the site of a

$\longrightarrow$

Fig. 2 (opposite). Distribution of TiO_2 and K_2O in tholeiites with distance along the ridge axis in the Iceland area. The position of the profile (solid line on the inset) follows that part of the neovolcanic zone from which tholeiitic lavas are being erupted at the present time (stippled on inset). Solid points are those of Brooks et al. [43] while circles are values of Jakobsson and Sigvaldason [39] on samples recovered by USNS Lynch, 1971 & 1973.

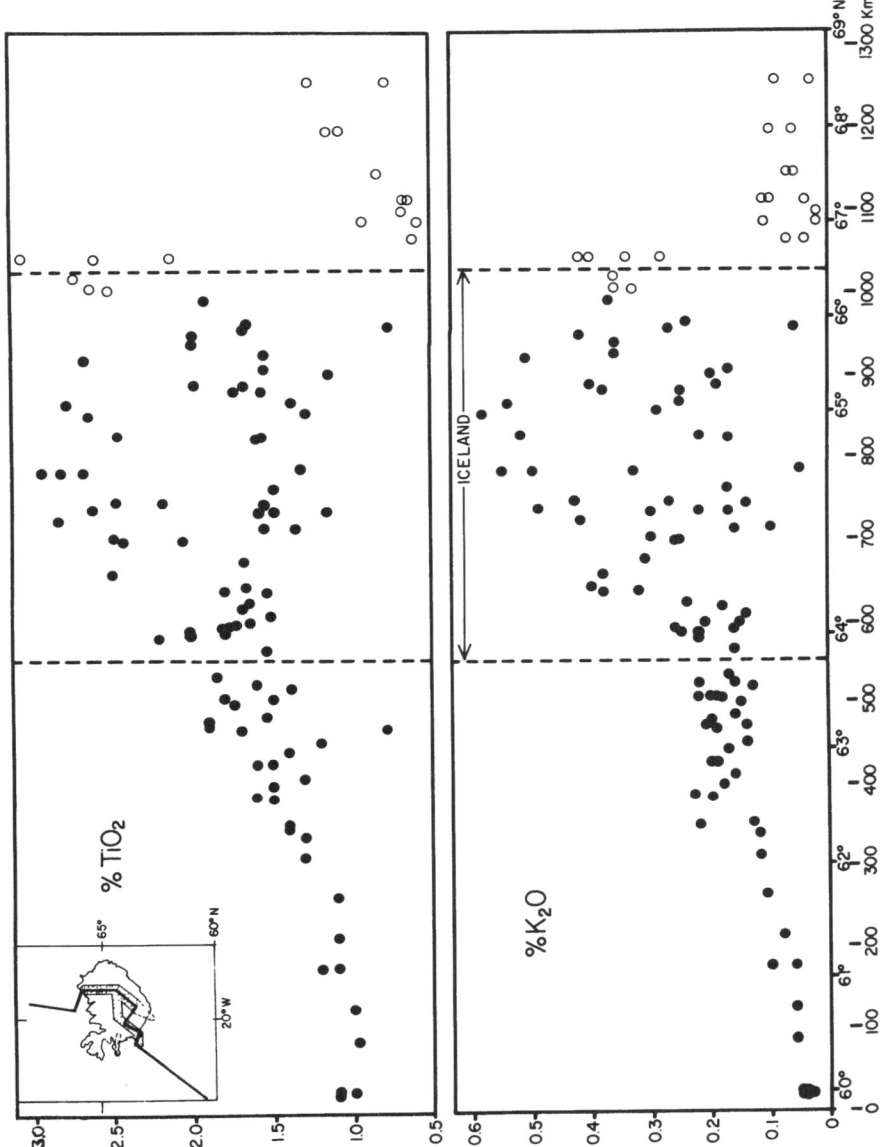

plume, most recently by Schilling et al. [38], but it is remarkable that one plume in Iceland produces FETI basalts while another
so close has such different products. The light rare earth element
enrichment seen in alkaline rocks of this type is probably only a
reflection of melting in small amounts at greater depth than seen
in the case of tholeiites, as shown by Kay and Gast [6]. The
$^{87}Sr/^{86}Sr$ ratio is similar to that of the Snaefellsnes zone of
Iceland but higher than for either alkali basalts of the eastern
zone or tholeiites from Iceland [14].

It is possible that earlier activity at the position of Jan
Mayen is shown by the alkaline rocks of the Kong Oscars Fjord
region of East Greenland. The age of these rocks is uncertain
although a K-Ar value of 29 m.y. was reported by Beckinsale et al.
[64] and Miocene faulting was mapped by Birkenmajer [65]. They
include nepheline syenites, nordmarkites and granites [66]. Of
the basic dike swarms many are alkaline and potassic (cf. p. 149
and 158 of [67]). According to the sources cited above, the Jan
Mayen ridge was detatched from Greenland between 10 and 30 m.y.
ago so this magmatic activity may be related to this event.

Of the other areas of alkaline rocks in the region, Kangerdlugssuaq is discussed below but the Hebridean province lies outside the scope of this review. It extends well into continental
areas and contains both tholeiitic and mildly alkaline rocks [68].
The origin of these magmas is widely regarded as being related
to continental break-up although the precise relationship remains
obscure; however, the proposal put forward by Duncan et al. [69]
must be discounted for the reasons set out by Meighan and Gamble
[70]. The Bockfjord volcanics of NW Spitsbergen [71] appear to
resemble those of Jan Mayen and contain many upper mantle xenoliths [72]. However, no modern chemical data are available.

6. GEOCHEMICAL HISTORY OF THE ICELAND MANTLE PLUME

Submarine ridges extending from Iceland to areas of volcanic rocks
in the Faeroes and East Greenland apparently record earlier positions of the Icelandic hot spot [66,73]. The Paleocene ages of
basalts in these areas [64,74], the magnetic anomalies in intervening areas [75] and extrapolations of present spreading rates
[1] all lend support to this model. A study of these areas might
therefore be expected to throw light on secular aspects of the
geochemistry and rate of basalt discharge of the Icelandic hot
spot.

Vogt [76] assessed topographic evidence along the Faeroes-
Iceland-Greenland ridge and showed that discharge rate was high
50-60 m.y. ago, decreased to a minimum in the mid-Tertiary and
increased again in the late Tertiary, a pattern perhaps worldwide.

At present, it is only possible to compare the earliest and latest volcanics, those of the mid-Tertiary being submerged, and detailed correlation of composition with age must await drilling of these areas.

Within Iceland, petrochemical studies of Tertiary rocks [42; Bailey and Noe-Nygaard, unpubl. data] have not revealed any systematic difference from neovolcanic tholeiites, but Sr isotope studies show a progressive fall in $^{87}Sr/^{86}Sr$ [77] with time. Several possibilities to explain this were discussed but no definite conclusion reached.

On the Faeroe Islands a distinct secular (stratigraphic) variation was shown by the data of Rasmussen and Noe-Nygaard [78] and discussed by Noe-Nygaard and Rasmussen [79]. Here, the lowermost basalts have values for total iron and TiO_2 typical of FETI basalts, while the uppermost basalts resemble MORB. This impression is strengthened by the bulk chemistry and the trace elements [Bollingberg and Noe-Nygaard, unpubl. data], the upper basalts being distinctly poorer in Sr, Cu, Sr and Ba than the lower series.

Schilling and Noe-Nygaard (in prep.) have found the expected rare earth patterns. The middle basalts occupy an intermediate position chemically. It is tempting to speculate that this transition from FETI basalt to MORB type is correlated with the decrease in basalt discharge round about this time described by Vogt [76].

Brooks [76] has reviewed the Tertiary geology of the Kangerdlugssuaq district of East Greenland and described the early effects of the Icelandic hot spot here. Volcanism was associated with the development of a considerable dome some 200 km across and 3.5 km high, across whose top a three-armed rift developed in the manner depicted for Hawaiian volcanoes (but on a smaller scale) by Macdonald and Abbott ([80], p. 36) and elaborated for many examples by Burke and Dewey [81], who attributed their formation to underlying mantle plumes. Subsequent crustal spreading occurred along the arms parallel to the present coastline accompanied by the eruption of massive quantities of tholeiites of FETI type [82], formation of many layered gabbros (e.g. Skaergaard) and injection of an intense coast-parallel dikeswarm [83]. It was also proposed that the Faeroe plateau represents the eastern half of the original dome which was detached by seafloor spreading. However, no MORB-type volcanics similar to the uppermost Faeroe lavas are yet reported from East Greenland. Kangerdlugssuaq, a major fjord extending inland from the point at which the coast-line is abruptly inflected, is apparently tectonically controlled and exhibits volcanism similar to that of the African Rift, becoming progressively more alkaline inland, just

as is seen in Africa with increasing distance from rift inter-
sections [84]. The following progression is observed along the
length of the fjord: comendites and nordmarkites, basanites with
kaersutite gabbro inclusions [85] and in the innermost parts of
the fjord, olivine nephelinites [94], uncompahgrites (melilite-
rich intrusives) and leucite-bearing rocks of the kamafugitic
suite. This fjord apparently represents a non-spreading rift
or "failed arm" [81] with increasing depth of magma generation
away from the rift intersection. Dikes with mantle-derived
xenoliths described from the coastal areas to the north, [86], show
however, that this is not the only place where deep-seated
magmas rose rapidly to the surface.

In summary, it appears that the bulk of the magma erupted by
the Icelandic hot spot during its initial stages was very similar
to that appearing at the present time, although it may well be
that during the mid-Tertiary it had a much greater MORB character.
Alkaline types occupied a subsidiary role just as they do today
but were of a much more extreme character.

7. CONCLUSIONS

Quantitatively the most important rock types of the North Atlantic
basin are the MORB and FETI tholeiites, whose chemical characteristics were described above. MORB is the predominant volcanic
of the abyssal part of the ridge and presumably also the older
areas away from the ridge, although the recovery of fresh materials from these areas must await drilling. FETI basalts are
found in regions of high topography, heat flow and basalt discharge, such as Iceland and the earlier positions of the Icelandic hot spot.

Discussion essentially revolves around whether or not FETI
can be derived from MORB by fractional crystallization as maintained by O'Hara [87] or is itself primary material, possibly of
a more primitive nature from the deep mantle. O'Hara observed
that FETI generally has a higher FeO*/MgO ratio than MORB and
used this as an argument for advanced differentiation. However,
it can readily be shown that on variation diagrams FETI has
significantly higher concentrations of LIL trace elements than
MORB for comparable values of FeO/MgO. Schilling [28] showed how
O'Hara's arguments regarding enrichment of LIL elements without
significant variation in major element chemistry could not be
applied to the rare earths and Schilling's conclusions appear
to be supported by isotopic evidence. The increased discharge
rate over Iceland of LIL enriched basalts is also powerful
evidence against relation by differentiation or varying partial
melting. We therefore believe that, at the present time, the
weight of geochemical evidence is in favour of a multiple source

hypothesis (i.e. the mantle plumes) for the mutual relationships of these two important magma types.

Alkali magmas, which are quantitatively of little importance appear to form where deep-seated fractures cut thick crustal piles, as in parts of Iceland, Jan Mayen and especially Kangerdlugssuaq, where the structure is that of a failed arm and the crust is Precambrian shield. All evidence favours formation of these magmas at great depth in regions of relatively low heat flow and rapid transport to the surface along the fractures.

REFERENCES

1. F.J. Vine, Science, 154, 1405, 1966.
2. A.E.J. Engel, C. Engel and R.G. Haven, Bull. Geol. Soc. Am., 76, 719, 1965.
3. F. Aumento, Can. Jour. Earth Sci., 5, 1, 1968.
4. W.G. Melson, G. Thompson and T. van Andel, J. Geophys. Res., 73, 5925, 1968.
5. J.R. Cann, Jour. Petrol., 10, 1, 1969.
6. R. Kay and P.W. Gast, Jour. Geol., 81, 653, 1974.
7. W.G. Melson and G. Thompson, Phil. Trans. Roy. Soc. Lond., A268, 423, 1972.
8. F. Aumento, Earth Planet. Sci. Lett., 2, 225, 1967.
9. H.S. Yoder and C.E. Tilley, Jour. Petrol., 3, 342, 1962.
10. F.A. Frey and L.A. Haskin, J. Geophys. Res., 69, 775, 1964.
11. J.G. Schilling, Phil. Trans. Roy. Soc. Lond., A268, 663, 1971.
12. R. Kay, N.J. Hubbard and P.W. Gast, J. Geophys. Res., 75, 1585, 1970.
13. S.R. Hart, J.G. Schilling and J.L. Powell, Nature Phys. Sci., 246, 104, 1973.
14. R.K. O'Nions and R.J. Pankhurst, Jour. Petrol. (in press, 1974).
15. M. Tatsumoto, Science, 153, 1094, 1966.
16. R.G. Coleman, M. Tatsumoto, D.G. Coles, C.E. Hedge and R.E. Mays, Trans. Am. Geophys. Union (EOS), 54, 1001, 1973.
17. D.H. Green and A.E. Ringwood, Contr. Mineral. Petrol., 15, 103, 1967.
18. F. Shido, A. Miyashiro and M. Ewing, Contr. Mineral. Petrol., 31, 251, 1971.
19. A. Miyashiro, F. Shido and M. Ewing, Earth Planet. Sci. Lett., 7, 361, 1970.
20. A. Miyashiro, F. Shido and M. Ewing, Contr. Mineral. Petrol., 23, 38, 1969.
21. S.R. Hart, Phil. Trans. Roy. Soc. Lond., A268, 573, 1971.
22. R.N. Thompson, J. Esson and A.C. Dunham, Jour. Petrol., 13, 219, 1972.
23. G. Thompson, Chem. Geology, 12, 99, 1973.

24. Anonymous, Trans. Am. Geophys. Union (EOS), 54, 972, 1973.
25. J.B. Corliss, Mid-ocean ridge basalts: II. Regional diversity
     along the mid-Atlantic ridge. Ph.D. Dissertation, Univ. Cali-
     fornia, San Diego, p. 45-108, 1970.
26. F. Aumento, Science, 165, 1112, 1969.
27. P.W. Gast, Geochim. Cosmochim. Acta, 32, 1057, 1968.
28. J.G. Schilling, Nature, 242, 565, 1973.
29. M. Tatsumoto, C.E. Hedge and A.E.J. Engel, Science, 180, 886,
     1965.
30. Z.E. Peterman and C.E. Hedge, Bull. Geol. Soc. Am., 82, 493,
     1971.
31. E. Bonatti, J. Honnorez and G. Ferrara, Phil. Trans. Roy.
     Soc. Lond., A268, 385, 1971.
32. J. Dymond, Bull. Geol. Soc. Am., 81, 1229, 1970.
33. J. Dymond, Trans. Am. Geophys. Union (EOS), 54, 485, 1973.
34. J.G. Moore, Contr. Mineral. Petrol., 28, 272, 1970.
35. G.D. Nicholls, Miner. Mag., 34, 373, 1965.
36. R. Hekinian and F. Aumento, Marine Geol., 14, 47, 1973.
37. J. Campsie, J.C. Bailey, M. Rasmussen and F. Dittmer, Nature
     Phys. Sci., 244, 71, 1973.
38. J.G. Schilling, D.G. Johnson and T.H. Johnston, Trans. Am.
     Geophys. Union (EOS), 55, 294, 1974.
39. S.P. Jakobsson and G.E. Sigvaldason, (in preparation, 1974).
40. H. Sigurdsson and G.M. Brown, Jour. Petrol., 11, 205, 1970.
41. A. Noe-Nygaard, Nature, 212, 272, 1966.
42. S.P. Jakobsson, Lithos, 5, 365, 1972.
43. C.K. Brooks, S.P. Jakobsson and J. Campsie, Earth Planet.
     Sci. Lett., 22, 320, 1974.
44. G.E. Sigvaldason, Jour. Petrol., (in press, 1974).
45. H. Sigurdsson, The Icelandic basalt plateau and the question
     of sial, in: Iceland and Mid-Ocean Ridges (ed. S. Björnsson)
     Soc. Sci. Islandica, Rit 38, 32, 1967.
46. J.G. Moore and J.G. Schilling, Contr. Mineral. Petrol., 41,
     105, 1973.
47. J.G. Moore and B.P. Fabbi, Contr. Mineral. Petrol., 33, 118,
     1971.
48. G.A. Macdonald and T. Katsura, Jour. Petrol., 5, 82, 1964.
49. A.R. McBirney and H. Williams, Mem. Geol. Soc. Am., 118,
     197 pp, 1969.
50. J.G. Schilling, Nature, 246, 141, 1973.
51. J.G. Schilling, Nature, 242, 2, 1973.
52. C.K. Unni and J.G. Schilling, Trans. Am. Geophys. Union (EOS),
     55, 454, 1974.
53. O.D. Hermes and J.G. Schilling, Trans. Am. Geophys. Union
     (EOS), 55, 454, 1974.
54. I.S.E. Carmichael, Jour. Petrol., 5, 435, 1964.
55. P.R. Vogt, Trans. Am. Geophys. Union (EOS), 54, 239, 1973.
56. F. Shido and A. Miyashiro, Nature Phys. Sci., 245, 59, 1973.
57. G.L. Johnson and B.C. Heezen, Deep-Sea Res., 14, 755, 1967.

58. O. Eldholm and M. Talwani, Trans. Am. Geophys. Union (EOS), 54, 324, 1973.
59. T.R.W. Hawkins and R.W. Roberts, Norsk Polarinst. Årbok 1970, 19, 1972.
60. P.W. Weigand, Norsk Polarinst., Årbok 1970, 42, 1972.
61. H. Carstens, Norsk Polarinst., Årbok 1962, 185, 1963.
62. P.E. Baker, I.G. Gass, P.G. Harris and R.W. LeMaitre, Phil. Trans. Roy. Soc. Lond., A256, 439, 1964.
63. R.W. LeMaitre, Bull. Geol. Soc. Am., 73, 1309, 1962.
64. R.D. Beckinsale, C.K. Brooks and D.C. Rex, Bull. Geol. Soc. Denmark, 20, 27, 1970.
65. K. Birkenmajer, Geol. Surv. Greenland, Rept. 48, 85, 1972.
66. C.K. Brooks, The Tertiary of Greenland: a volcanic and plutonic record of continental break-up, in: Arctic Geology (ed. M.G. Pitcher), Am. Assoc. Petrol. Geologists, Tulsa, Oklahoma, 1973.
67. H. Kapp, Zur Petrologie der Subvulkane zwischen Mesters Vig und Antarctic Havn (Ost-Grönland). Medd. Grönland, 153(2), 203 pp, 1960.
68. G. Thompson and W.G. Melson, Jour. Geology, 80, 526, 1972.
69. R.A. Duncan, N. Petersen and R.B. Hargraves, Nature, 239, 82, 1972.
70. I.G. Meighan and J.A. Gamble, Nature Phys. Sci., 240, 183, 1972.
71. V.M. Goldschmidt, Vid. Selsk. Skr. (Kristiania), I. Mat-naturv. kl. 9, 17 pp, 1911.
72. T. Gjelsvik, Norsk Polarinst., Årbok 1962, 50, 1963.
73. C.K. Brooks, Nature Phys. Sci., 244, 23, 1973.
74. D.H. Tarling and N.H. Gale, Nature, 218, 1043, 1968.
75. P.R. Vogt, N.A. Ostenso and G.L. Johnson, J. Geophys. Res., 75, 903, 1970.
76. P.R. Vogt, Nature, 240, 338, 1972.
77. R.K. O'Nions and R.J. Pankhurst, Earth Planet. Sci. Lett., 21, 13, 1973.
78. J. Rasmussen and A. Noe-Nygaard, Beskrivelse til geologisk kort over Faerøerne. Geol. Surv. Denmark, I Ser. 24, 370 pp, 1969.
79. A. Noe-Nygaard and J. Rasmussen, Lithos, 1, 286, 1968.
80. G.A. Macdonald and A.T. Abbott, Volcanoes in the Sea. The geology of Hawaii. Honolulu, Univ. Hawaii Press. 441 pp, 1970.
81. K. Burke and J.F. Dewey, Jour. Geol., 81, 406, 1973.
82. J.J. Fawcett, C.K. Brooks and J.C. Rucklidge, Medd. Grønland 195(6), 54 pp, 1973.
83. L.R. Wager and W.A. Deer, Geol. Mag., 75, 39, 1938.
84. P.G. Harris, Tectonophysics, 8, 427, 1969.
85. C.K. Brooks and R.G. Platt, Miner. Mag., (in press, 1974).
86. C.K. Brooks and J.C. Rucklidge, Contr. Mineral. Petrol., 42, 197, 1973.

87. M.J. O'Hara, Nature, 243, 507, 1973.
88. G. Thompson and W.G. Melson, Jour. Geology, 80, 526, 1972.
89. C.K. Brooks, Geochim. Cosmochim. Acta, 34, 411, 1970.
90. W.G. Melson, Trans. Am. Geophys. Union (EOS), 54, 1011, 1973.
91. K.S. Heier, B.W. Chappell, P.A. Arriens and W.J. Morgan,
     Norsk Geol. Tidsskr., 46, 427, 1966.
92. H. Welke, S. Moorbath, G.L. Cumming and H. Sigurdsson, Earth
     Planet. Sci. Letters, 4, 221, 1968.
93. P.W. Weigand, A.O. Brunfeld, K.S. Heier, B. Sundvoll and
     E. Steinnes, Nature Phys. Sci., 235, 31, 1972.
94. C.K. Brooks and J.D. Rucklidge, Lithos, (in press, 1974).

CHEMISTRY OF THOLEIITIC BASALTS FROM ICELAND AND THEIR RELATION TO THE KVERKFJÖLL HOT SPOT

G.E. Sigvaldason   and   S. Steinthórsson

Nordic Volcanological   Science Institute,
Institute, Reykjavik,   University of Iceland,
Iceland.   Reykjavik, Iceland.

1. INTRODUCTION

Ever since the universal occurrence and unique composition of "oceanic tholeiites" was defined by Engel et al. [1,2], the "differentiated" or "undepleted" nature of Icelandic basalts has become increasingly apparent [3-6]. This, in conjunction with the high elevation of Iceland relative to the Mid-Atlantic Ridge, and its anomalous volcanic productivity throughout 70 m.y. (as witnessed by the aseismic Wyville-Thompson ridge) has clearly defined Iceland the hot spot of the North Atlantic [7-9]. Adjacent to the hot spot a gradient has been established along the mid-ocean ridges next to Iceland, both in the elevation of the sea floor [several authors in this volume] and in the chemical and isotopic composition of the volcanics [10,11].

Morgan [8] suggested, contrary to Wilson [7], that the hot spots form an integral part of the mechanism driving the crustal plates. He visualized mantle material ascending along deep pipes to spread radially into the asthenosphere, a model subsequently used by Vogt [19] to explain the V-shaped ridges south of Iceland. In such a system partial melting would ensue at some stage as a result of the adiabatic decompression. Depending on the path travelled in vertical and horizontal direction by each volume of rock undergoing partial melting within the plume and its overflow, a geographical distribution of various types of basalt, representing various degrees and depths of partial melting, should obtain at the surface. Such geographic distribution was demonstrated by Schilling [10] and Hart, Schilling and Powell [11] for the Reykjanes Ridge. However, these authors elected to explain the chemical and isotopic evidence in terms of a model involving a number of

Kristjansson (ed.), Geodynamics of Iceland and the North Atlantic Area. 155-164. *All Rights Reserved.*
Copyright © 1974 by D. Reidel Publishing Company, Dordrecht-Holland.

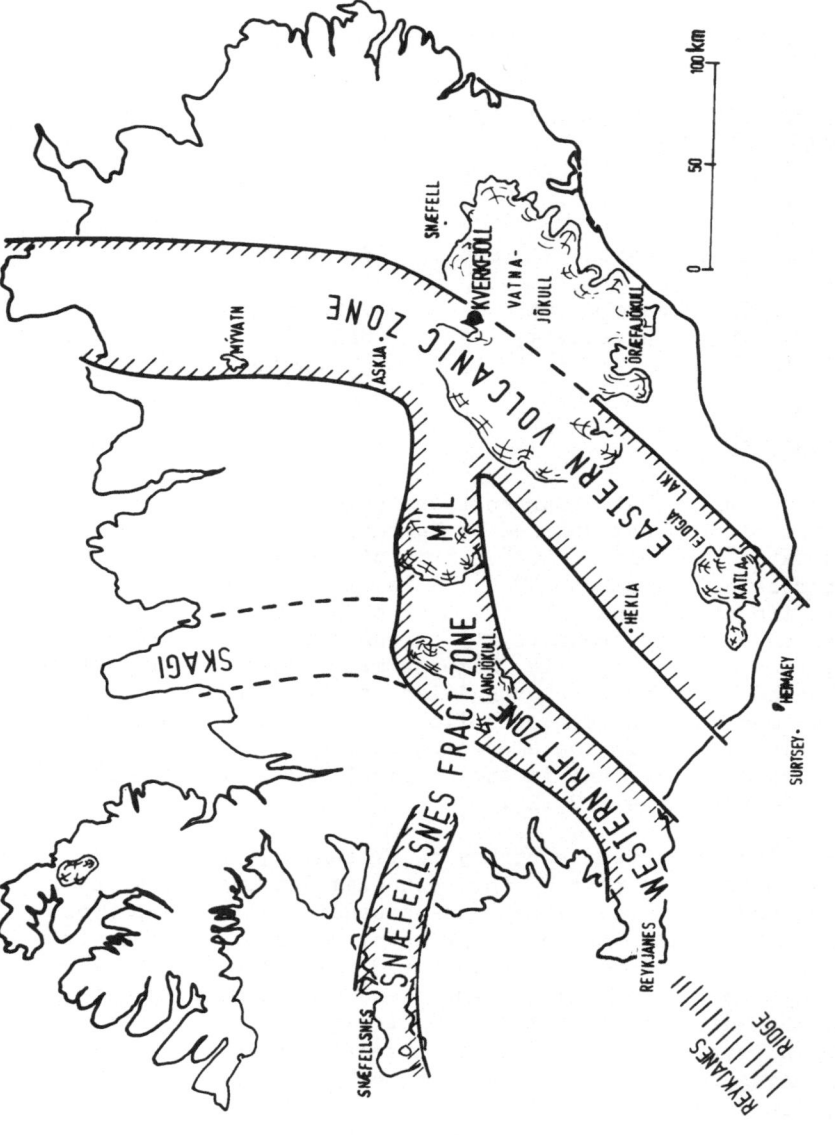

Fig. 1. An index map of Iceland, showing the neovolcanic zones and place names used in the article.

discrete mantle sources with some magma mixing at the source boundaries. We propose to show that the same variation continues across Iceland, and that it can be explained by a one-source model compatible with the mantle plume hypothesis.

2. EXTENT OF THE PETROCHEMICAL SURVEY

Active volcanism in Iceland is restricted to well-defined zones, termed the neovolcanic zones. They exhibit distinct petrological characteristics compatible with their respective tectonic natures. Thus, the volcanics in the western and north-eastern zones, which form the subaereal continuation of the Reykjanes and Kolbeinsey Ridges, respectively, are tholeiitic in composition whereas the volcanics of the Snaefellsnes and south-eastern zones are pre-dominantly alkalic [12].

In the process of the present petrochemical survey over 500 samples from all the zones have been collected and analyzed. This suite is considered to cover volcanic activity in Iceland during the Brunhes palaeomagnetic period (700,000 years). The sampling was carried out so as to represent (a) the areal extent, (b) chronological formations, (c) morphological formations, and (d) petrochemical types. The morphological formations are subglacial pillow lavas and móberg (palagonite breccia), interglacial lavas, and postglacial volcanics. The latter may be divided according to volcano types into fissure and minor eruptions on the one hand, and shield eruptions on the other. The shield volcanoes are the postglacial counterparts of the subglacial stapis (table mountains). The volumes of the two types vary greatly: the shield volcanoes and stapis range from one to 15 km^3 whereas the others, with a few very notable exceptions, are generally less than 1 km^3.

3. CHEMICAL VARIATION IN THE THOLEIITES

The chemical variation observed in the North Atlantic centers on the Kverkfjöll area in Iceland (Fig. 1). The maximum values for K_2O, P_2O_5, and TiO_2 decrease continuously with distance from Kverkfjöll in both directions along the tholeiitic zones to the southwest and north (Fig. 2). Conversely the minimum values stay about the same throughout the zones. As seen from Fig. 3, K_2O and TiO_2 vary sympathetically. Fig. 4 shows a plot of Al_2O_3 versus TiO_2. Titanium is sensitive to pressure [13], whereas Al_2O_3 has been claimed to be a function of degree of melting [14]. Thus, taking Figs. 3 and 4 at face value it is indicated that the undepleted (low Al, high K) basalts formed at relatively high pressures whereas the depleted ones (low K, high Al) formed at shallower depths.

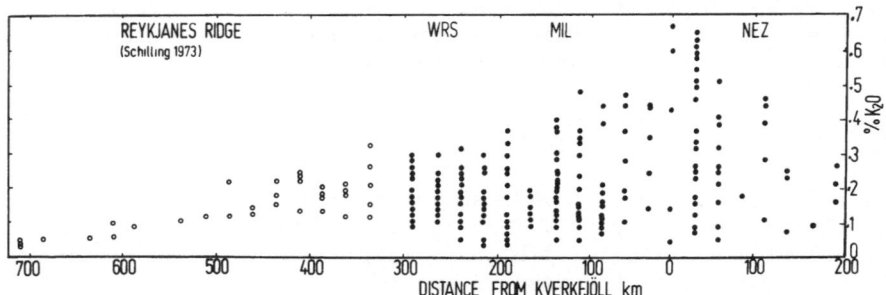

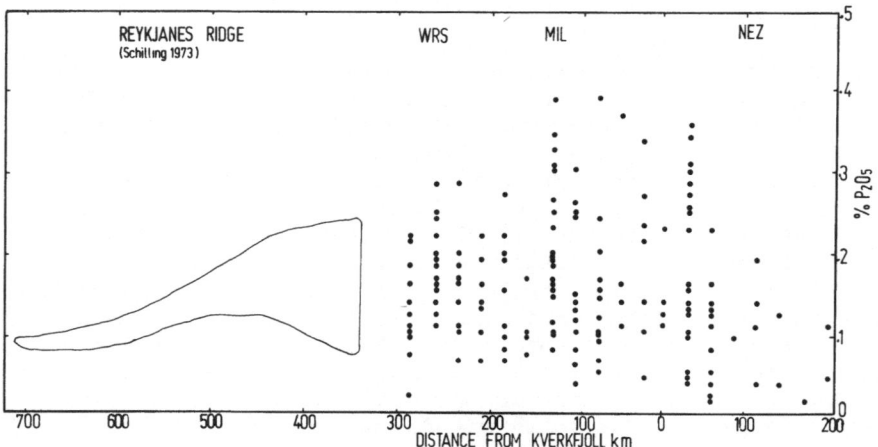

Fig. 2.   Plot of K_2O and P_2O_5 versus distance from the Kverkfjöll hot spot.   WRZ - Western Rift Zone;   MIL - Middle Iceland;
NEZ - Northeastern Iceland.

       Many other chemical parameters show a positive correlation
with K_2O, such as total iron, Zr, and Sr. Furthermore, similar
lateral variation has been demonstrated on the Reykjanes Ridge
for water and sulfur [15], rare earths [10], and Sr^{87}/Sr^{86} -iso-
tope ratios [11] and for Cl along the volcanic zones in Iceland
[Sigvaldason et al., in preparation]. The iron enrichment of the
Icelandic tholeiites relative to samples from the ocean floor
was previously noted by Sigvaldason [4]. Samples from the Juan de
Fuca Ridge resemble the Icelandic ones in this respect, in accor-
dance with the fact that that area is reputed to be the site of
a mantle plume as well [25].

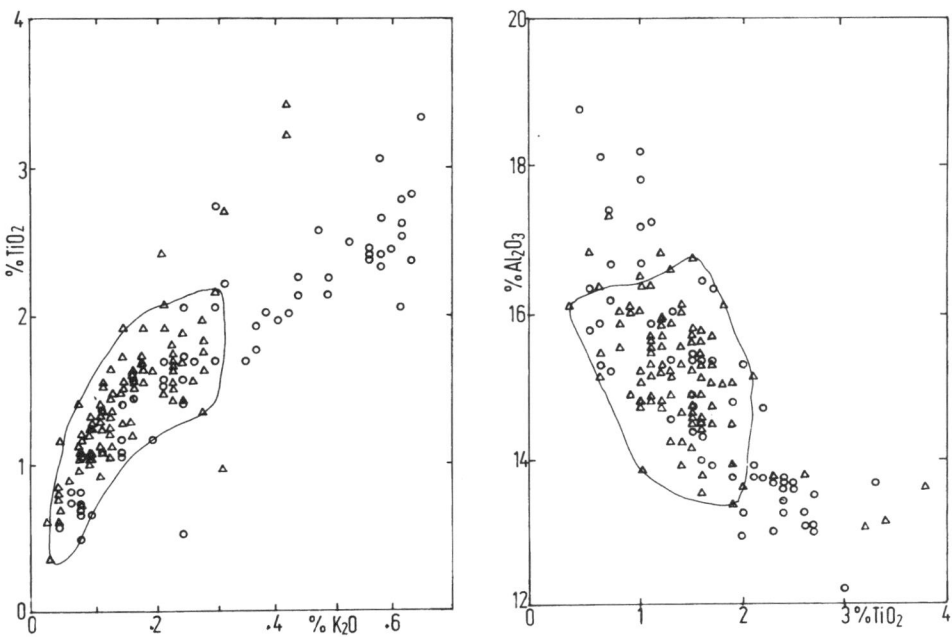

Fig. 3 (left). A plot of TiO$_2$ versus K$_2$O.

Fig. 4 (right). A plot of Al$_2$O$_3$ versus TiO$_2$.

Circles: samples from Kverkfjöll and the north-eastern zone. Triangles: samples from Middle Iceland and the Reykjanes peninsula, the latter enclosed by the solid line.

A pronounced correlation exists between the chemistry and volume of the tholeiites under consideration (Fig. 5, from [16]). This is especially notable in Central Iceland, where the composition range is large. The shield eruptions tend to be voluminous; chemically they are the most depleted rocks in Iceland and thus resemble the oceanic tholeiites most. They belong to the specific volcano-tectonic environment realized during the isostatic rebound of the country during the waning stages of the pleistocene glaciation. The upper mantle was suddenly disturbed toward lower pressure resulting in the melting of large batches of mantle material. Thus, a local decompression can produce a chemical variation in one area that is similar to the horizontal variation otherwise observed.

As previously noted, Schilling and his associates have demonstrated a gradient in Sr^{87}/Sr^{86} isotopic ratios and REE patterns
away from Iceland along the Reykjanes Ridge. Unfortunately, at
this writing no such systematic analysis of these parameters has
been undertaken on the Icelandic tholeiite suite. However, the
data of Shimokawa and Masuda [6] and O'Nions and Grönvold [17]
indicates that lavas of shield volcanoes have low La/Sm ratios
(i.e. are relatively depleted in the light REE), olivine tholeiites,
such as predominate in the Reykjanes peninsula, have higher La/Sm
ratios, whereas tholeiites, such as the 1961 Askja lava, are the
least depleted in light REE. Therefore, this chemical parameter
seems to follow the others in Iceland as well as on the Reykjanes
Ridge. Likewise the available Sr-isotope evidence [11,17] might be
construed to indicate that the Sr^{87}/Sr^{86} -ratios in the Icelandic
rocks are chemistry-bound as well, for basalts from the alkalic
Snaefellsnes peninsula possess slightly higher Sr-isotopic ratios
than the tholeiites. This will be tested by further work.

To summarize the chemical evidence enumerated above, a continuous petrochemical gradation exists in the North Atlantic along
the mid-ocean ridge that centers on the Kverkfjöll area in Iceland.
The variation includes major, minor and rare earth elements, as
well as Sr-isotopic ratios. Geochemical arguments, the geographic
variation, and the circumstantial evidence of the shield volcanoes
suggest that the compositional gradient results from the combination
of changing P-T environment in an upwelling and laterally spreading
mantle plume system, and a progressive depletion in the low-temperature melting fraction at the source.

4. A SINGLE SOURCE MODEL FOR THE NORTH ATLANTIC

The chemical evidence at hand for the North Atlantic has been explained both in terms of two or more primary magma sources [10,11],
and in terms of one source [18]. According to the two-source model,
which receives its chief support from the conventional doctrine of
the "immutability of Sr-isotopes", magmas derived from the Icelandic plume source possess Sr^{87}/Sr^{86} -ratios of the order of 0.7032,
show undepleted REE-patterns, have high Fe/Mg-ratios and are
relatively rich in K, Ti, P, and volatiles. Magmas from the oceanic
source, on the other hand, have much more depleted chemistry and
Sr^{87}/Sr^{86} -ratios about 0.7026. The compositional gradient along
the Reykjanes Ridge results, according to this theory, from the
mixing of the two magmas.

O'Nions and Pankhurst [19] and Sigvaldason et al. [18] have
suggested that Sr-isotopic disequilibrium may exist in the mantle
between mineral phases. Assuming this, the analytical evidence from
Iceland [17] and the Reykjanes Ridge [11] becomes intelligible in
terms of a single-source model: The plume is richer in volatiles

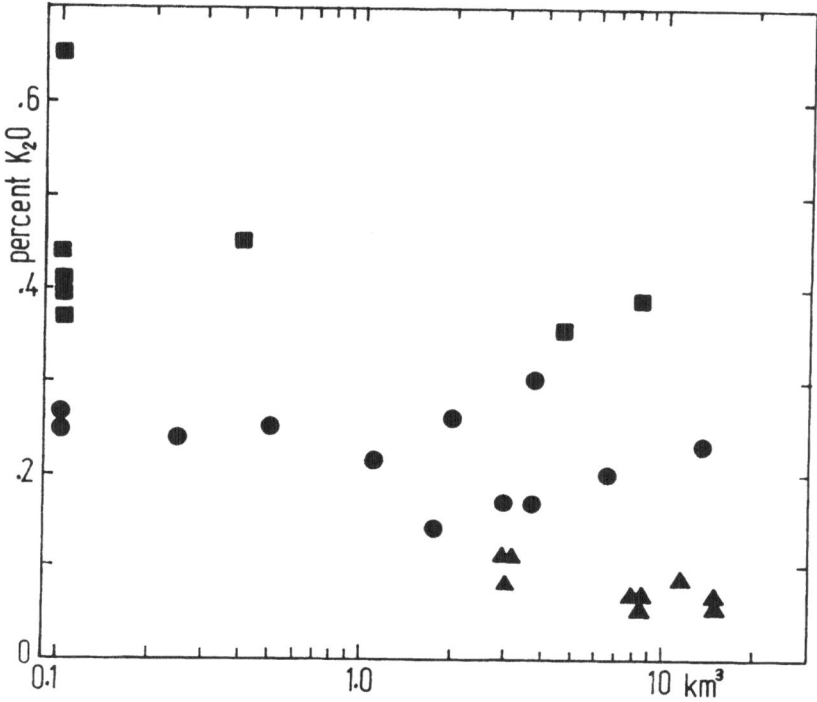

Fig. 5. A plot of lava volume vs. K$_2$O for the north-eastern volcanic zone (from [16]).

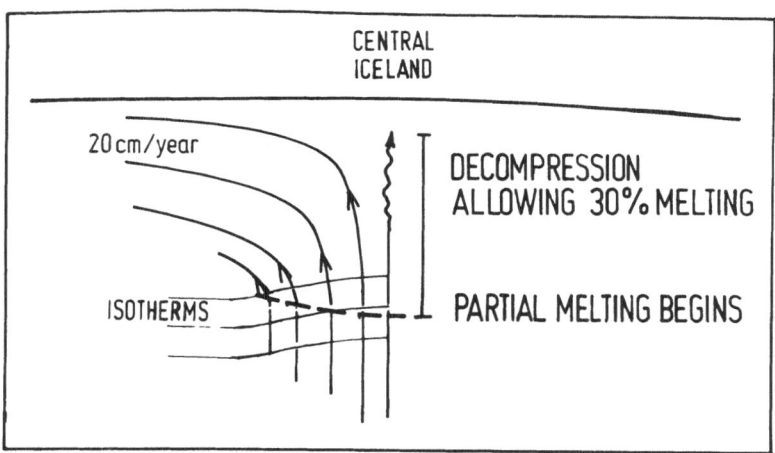

Fig. 6. A schematic section showing hypothetical flow lines within a mantle plume system.

and radioactivities than the surrounding mantle. As a result it is
hotter and lighter and ascends diapirically through the mantle,
providing the required buoyancy beneath the high-rising hot spots.
The plume material appears to spread out laterally along preferred
channels [9] as evidenced by the seismic work of Palmason [20].
Sigvaldason [16] predicted a temperature regime in and around the
plume comparable to that of a hydrothermal system. Within the plume
the thermal gradient is positive throughout, providing a thermal
environment suitable for partial melting over an extended depth
range. Conversely, in the plume overflows a temperature maximum
occurs at the depth of partial fusion.

The plume is a closed thermal system -- partial melting only
occurs as a result of decompression (Fig. 6). In the center of the
plume the entire decompression projects on a small area at the
surface, the center of the hot spot, where up to 30% partial melting can take place in a vertical column. Towards the margin of
the plume the vertical component of the movement is more restricted,
and the various products of the partial melting are spread horizontally, as evidenced by the chemical variation at hand.

The upper mantle, where magma segregation is supposed to take
place, probably consists of olivine, two pyroxenes, plagioclase,
and minor phases, including amphibole and phlogopite [14,21,22].
When such an assemblage melts, the phlogopite would tend to melt
completely before the other more refractory phases started breaking down [23]. The K- and Rb-enriched composition of the basalts
in Central Iceland result from the important contribution of phlogopite to the early partial melts of the plume. Recent experimental evidence by Mysen [26] indicates that some such "extraneous"
source is necessary to obtain all K-values other than the low ones
observed in oceanic tholeiites. Phlogopite contains of the order
of 7% K, 250 ppm Rb, and 18 ppm Sr [23]. The other components in
the upper mantle contain between 1 and 0.1 ppm Rb. The high Rb
content of the phlogopite results in an increased Sr^{87}/Sr^{86} ratio
relative to the rest of the mantle. Using the above cited values
for Rb and Sr in phlogopite the Sr^{87}/Sr^{86} -ratio would increase by
7×10^{-5} every million years. Conversely, the highest Rb/Sr ratios
observed in Icelandic tholeiites yield an increase of 2×10^{-6} every
million years, whereas the Sr-isotopic ratio of the depleted ocean
tholeiites has very little capacity for change with time, and probably reflects that of the bulk of the upper mantle [18]. Considering that phlogopite crystals must form but a small fraction of the
upper mantle (Griffin and Murthy [23] use 0.5 wt. %) they must be
relatively few and far between. Thus it is difficult, barring complete melting or recrystallization, to visualize but partial isotope equilibration. Even so only the immediate vicinity of each
phlogopite crystal should be affected - which probably will eventually be the first to melt subsequent to the phlogopite as a
result of the fluxing effect of the water released.

The chief virtue of the two-source theory is its conceptual simplicity and its adherence to well-worn notions about a pervasive equilibration of Sr-isotopes at elevated temperatures. However, the growing body of evidence on Sr-isotopes in Iceland threatens to call for an ever-increasing number of primary sources -- the Snaefellsnes province is already being suggested as a third one.

The chemical gradient observed on the Reykjahes Ridge does not terminate at Iceland, but continues uninterrupted to center at Kverkfjöll. Were this to be explained by the mixing of two homogeneous magmas the plume would have to be very thin indeed.

Finally, Schilling et al. [24] have analyzed a sequence of samples collected on a line perpendicular to the Mid-Atlantic Ridge, in which no chemical variation was found. Such uniformity through time can only be accomplished by constantly replenishing the source of partial fusion. This could be done in three ways, (a) by a convective system below the ridges, (b) by asthenospheric flow from Iceland, and (c) by drifting the plate system relative to the upper mantle to constantly tap fresh portions of the low-velocity zone.

The universal symmetry of magnetic anomaly patterns about the mid-ocean ridges shows that the plate systems cannot be rooted to subcrustal convection cells – on the contrary they must drift passively about the surface of the Earth relative to the mantle plumes. The last alternative (above) seems a doubtful one as a general process, whereas an asthenosphere flow along the ridges away from the plumes, as described above, would account for all the facts.

REFERENCES

1. A.E.J. Engel and C.G. Engel, Science, 144, 1330, 1964.
2. A.E.J. Engel and C.G. Engel, Science, 146, 477, 1964.
3. K.S. Heier, B.W. Chappell, P.A. Arriens and J.W. Morgan, Norsk Geol. Tidsskr., 46, 427, 1966.
4. G.E. Sigvaldason, Contr. Mineral. and Petrol., 20, 357, 1969.
5. N.J. Hubbard, Trans. Am. Geophys. Union, 52, 376, 1971.
6. T. Shimokawa and A. Masuda, Contr. Mineral. and Petrol., 37, 39, 1972.
7. J.T. Wilson, Nature, 207, 907, 1965.
8. W.J. Morgan, Nature, 230, 42, 1971.
9. P.R. Vogt, Earth Planet. Sci. Lett., 13, 153, 1971.
10. J.G. Schilling, Nature, 242, 565, 1973.
11. S.R. Hart, J.-G. Schilling and J.L. Powell, Nature Phys. Sci., 240, 104, 1973.
12. S.P. Jakobsson, Lithos, 5, 365, 1972.

13. I.D. MacGregor, The effect of pressure on the minimum melting composition in the system $MgO-SiO_2-TiO_2$, in: Ann. Rep. Dir. Geophys. Lab., Carnegie Inst. Year Book (1964-1965), 135, 1965.

14. D.H. Green and A.E. Ringwood, Contr. Mineral. and Petrol., 15, 103, 1967.

15. J.G. Moore and J.-G. Schilling, Contr. Mineral. and Petrol., 41, 105, 1973.

16. G.E. Sigvaldason, Jour. Petrol. (in the press, 1974)

17. R.K. O'Nions and K. Grönvold, Earth Planet. Sci. Lett., 19, 397, 1973.

18. G.E. Sigvaldason, S. Steinthorsson, N. Oskarsson and P. Imsland, Nature (in the press, 1974)

19. R.K. O'Nions and R.J. Pankhurst, Earth Planet. Sci. Lett., 21, 13, 1973.

20. G. Palmason, Crustal structure of Iceland from explosion seismology, Soc. Sci. Islandica, 40, 187 pp, 1971.

21. E.R. Oxburgh, Geophys. J. R. Astr. Soc., 8, 456, 1964.

22. R.E.T. Hill and A.L. Boettcher, Science, 167, 980, 1970.

23. W.L. Griffin and V. Rama Murthy, Geochim. Cosmochim. Acta, 33, 1389, 1969.

24. J.G. Schilling, Y. Oji, K. Oji and E. Sekator, Trans.Am. Geophys. Union, 52, 376, 1971.

25. W.J. Morgan, in: R. Shagan et al. (eds.): Studies in Earth and Space Sciences; A Memoir in Honor of Harry Hammond Hess, 1906-1969, 1972.

26. B. Mysen, Melting in a hydrous mantle: Phase relations of mantle peridotite with controlled water and oxygen fugacities. Carnegie Inst. Yearb., 72, 467, 1974.

THE PETROLOGIC NATURE OF THE LOWER OCEANIC CRUST AND UPPER MANTLE*

Nikolas I. Christensen

Department of Geological Sciences and Graduate Program
in Geophysics, University of Washington, Seattle,
Washington, U.S.A.

ABSTRACT. The compositions of the lower oceanic crust and upper
mantle are investigated using data from high pressure experiments
of compressional and shear wave velocities in rocks. Four compositional models for the lower oceanic crust are considered: 1)
partially serpentinized peridotite, 2) gabbro, 3) metabasalt and
metagabbro and 4) an ophiolite model consisting of metamorphosed
sheeted dikes overlying late differentiates and cumulate gabbros.
Comparisons of compressional wave velocities (V_p) from dredged
oceanic rocks with layer 3 refraction velocities show that peridotites 30 to 40% serpentinized, unaltered gabbro, metagabbro and
metabasalt all have velocities similar to observed lower crustal
velocities. Thus compressional wave velocity measurements alone
will not distinguish between the various crustal models. Although
only a limited amount of refraction data on shear wave velocities
(V_s) are available, it appears that lower crustal Poisson's ratios, calculated from V_p and V_s, are significantly lower than
measured values in partially serpentinized peridotite and unaltered
gabbro. In support of models 3 and 4 it is shown that Poisson's
ratios of metabasalt and metagabbro, on the other hand, agree well
with seismic data from the upper portion layer 3. The low Poisson's
ratios reported for the lower crust of Iceland ($\sim$ 0.27) suggest
that metabasalt and metagabbro are abundant constituents of layer 3.
The 7.4 km/sec layer, often found between layer 3 and oceanic upper
mantle in the Pacific Ocean, is interpreted as most likely being
composed of peridotite, 10 to 20% serpentinized, or feldspathic

* This work has been supported in part by the Office of Naval Research and the National Science Foundation.

peridotite. Compressional and shear wave velocities in eclogite and peridotite at appropriate pressures are similar to oceanic upper mantle velocities. An upper mantle of peridotite or dunite composition is favored, however, in regions where strong upper mantle anisotropy is observed.

1. INTRODUCTION

During the past few years several papers have been directly concerned with the petrologic nature of the oceanic crust and upper mantle. These papers have attracted much scientific attention because of the current interests in plate tectonics and the geological processes operating at ridge crests. Even though abundant geological and geophysical information is available on the crust and upper mantle, the relationship of this information to an understanding of composition is vague. Many rocks have been dredged from oceanic regions which offer clues to the petrologic nature of the lower levels of the crust and the upper mantle. However, it is impossible from geological criteria to accurately assess whether or not various rock types are abundant at different depth intervals. Similarly, ophiolite suites, which are believed by many to be on land exposures of oceanic crust, present complications because they are often dismembered and when complete differ significantly from one another in thickness, petrology and internal structure.

Much geophysical information on the crust and upper mantle, including heat flow, gravity and magnetic data, is also abstract and requires deciphering in order to interpret the physical significance of the measurements. Since the most direct information on the nature of the rocks beneath the oceans comes from seismology, it is appropriate to concentrate on seismic velocities of the lower oceanic crust and upper mantle. In the following discussion models of the mineralogical composition of the lower oceanic crust and upper mantle will be evaluated with the aid of experimental data on velocities in rocks.

2. SEISMIC STRUCTURE

Seismologists now recognize several layers within the oceanic crust. Average oceanic crustal structure [1] is often referred to in terms of a three layer crust in which the uppermost layer (layer 1) consists of a thin veneer of unconsolidated to semiconsolidated sediments, usually less than 1 km thick. Layer 2, often referred to as basement, is generally between 1.0 and 2.5 km in thickness and is believed to consist of basalt which is metamorphosed at its lower levels. Layer 3, immediately overlying the mantle, is quite variable in thickness (usually between 3 and 6 km) and at present is of unknown composition. Although many recent seismic studies have found a three layered crust, it now has become

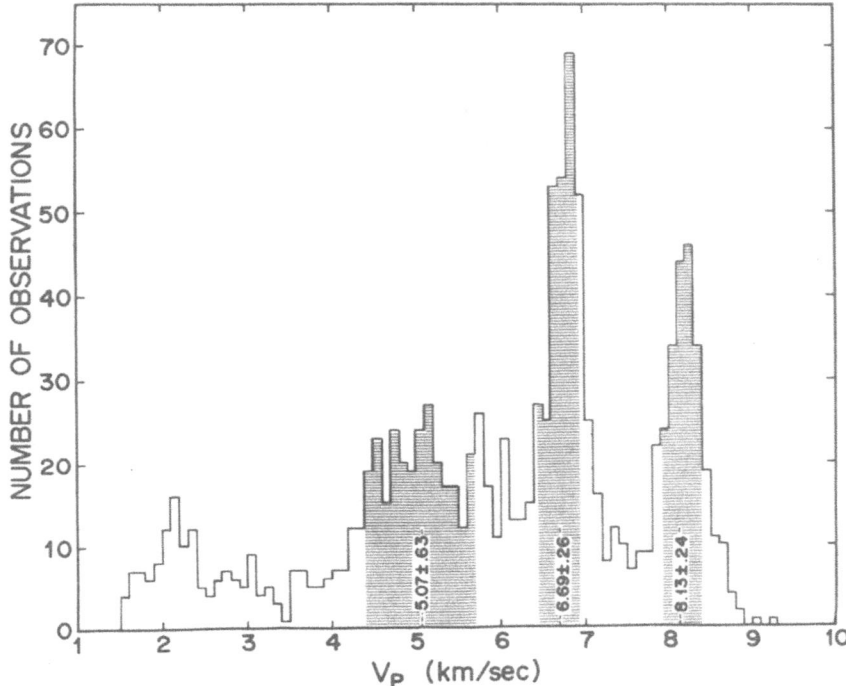

Fig. 1. Crustal and upper mantle compressional wave velocities.

increasingly apparent that in many regions crustal structure is much more complicated than this simple model. Examples include crust under ridge crests [2,3] and the high velocity basal crustal layer common in some oceanic regions [4].

In Fig. 1 crustal and upper mantle compressional wave velocities are shown in histogram form for 415 sites located in main oceanic basins. In addition, the average velocities of Raitt [1] are shown for layers 2 and 3 and the upper mantle. As can be seen from the diagram, the subdivisions of Raitt still hold, however many observed velocities fall outside the reported standard deviations. Thus even though it may be appropriate to envision the oceanic crust as containing three more or less well defined layers, such a simple model must be treated cautiously.

Much less information is available on shear velocities in the oceanic crust and the available data to 1970 have been summarized by Christensen [5]. In general for lower crustal velocities between 6.60 and 7.00 km/sec, shear wave velocities vary from 3.61 to 3.89 km/sec and calculated Poisson's ratios are between 0.25 and 0.29. It will be shown later that these values are extremely

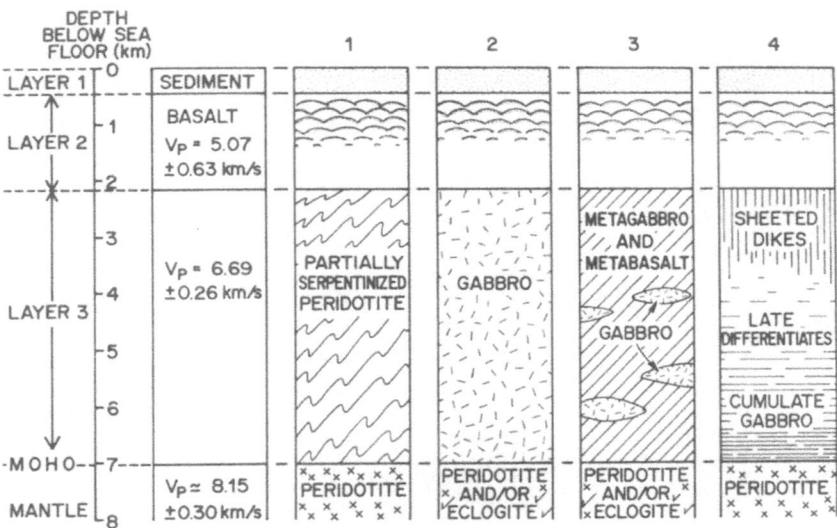

Fig. 2. Petrologic models for the lower crust and upper mantle.

important in interpreting composition from laboratory measurements of seismic velocities.

3. PETROLOGIC MODELS OF THE LOWER OCEANIC CRUST

Although the composition of the oceanic crust has been of consider-able interest for the past decade, surprisingly few models have been seriously considered as viable representations of the petro-logic nature of the crust. These are illustrated in Fig. 2 along with the average seismic structure of Raitt [1]. It should be em-phasized that all four models are probably correct at least locally. Also it seems probable that all of these models are greatly over-simplified. However, the relative abundance of the rocks given in these simple sections is still a matter of speculation and a sub-ject which can be clarified by comparisons of seismic refraction velocities with laboratory measurements.

Through the success of the Deep Sea Drilling Project, the upper portion of layer 2 has been found to consist of basalt at several localities. The laboratory measured velocities in these basalts agree well with refraction velocities of layer 2 [6]; be-cause of this it is probable that basalt is the abundant rock type below the sea floor sediments. It also appears that the lower portion of layer 2 has undergone metamorphic recrystallization [7], since the magnetic anomalies of the oceanic crust probably originate in only the upper half kilometer of layer 2.

The serpentinite lower crustal model (model 1, Fig. 3), originally proposed by Hess [8], assumes that the lower oceanic crust is generated under oceanic ridges by hydration of mantle peridotite. This model is attractive in several ways; serpentinite is simple to generate at ridge crests from an upper mantle composed of peridotite and as emphasized by Hess [8] serpentinite is easily disposed of at subduction zones by dehydration. Nevertheless this model has been strongly criticized by proponents of a mafic lower crust, three different models of which are shown in Fig. 3.

Model 2, which assumes that the lower crust consists of gabbro, has been proposed in several papers [1,9,10,11]. A slight variation of this model has hornblende gabbro as an abundant lower crustal rock, if sufficient water is present during the crystallization of basic magma within the lower crust [12].

As has been suggested as an alternative to the serpentinite and gabbro models, [12,13,14], metabasalts and metagabbros may be abundant constituents of the lower crust. For this model (model 3, Fig. 2) gabbros are assumed to occur within the lower crust as dikes and sills which have intruded the metamorphics. It is usually assumed that the metamorphism (greenschist to amphibolite facies) originates at ridge crests where high thermal gradients prevail, and the metabasalts and metagabbros are transported laterally by sea floor spreading.

Model 4 of Fig. 2 is based on rocks from ophiolites, believed by many to represent oceanic crust [15,16,17]. The hard rock portion of these sequences usually begin from the top with pillow basalts overlying sheeted dikes. The sheeted dikes, which are usually equated with the upper portion of layer 3, are vertical or near vertical and often metamorphosed to greenschist and amphibolite facies grade. Beneath the sheeted dikes are cumulate gabbros often containing abundant hornblende. Diorites, granophyres and trondhjemites, representing late stage products of differentiation, are common between the sheeted dikes and the gabbros. The lower portion of the ophiolite column (interpreted to represent mantle) consists of an uppermost ultramafic cumulate phase, containing abundant dunite and harzburgite, which overlies ultramafic tectonite composed of harzburgite.

4. LOWER CRUSTAL COMPOSITION

Since the dominant physical variable within the lower crust and upper mantle is pressure, experimental studies of seismic velocities in rocks at high pressure are the most significant for the analysis of seismic refraction velocities. Pressures within the lower oceanic crust and immediately beneath the Mohorovicic discontinuity are usually between 1 and 3 kbars. Compressional wave velo-

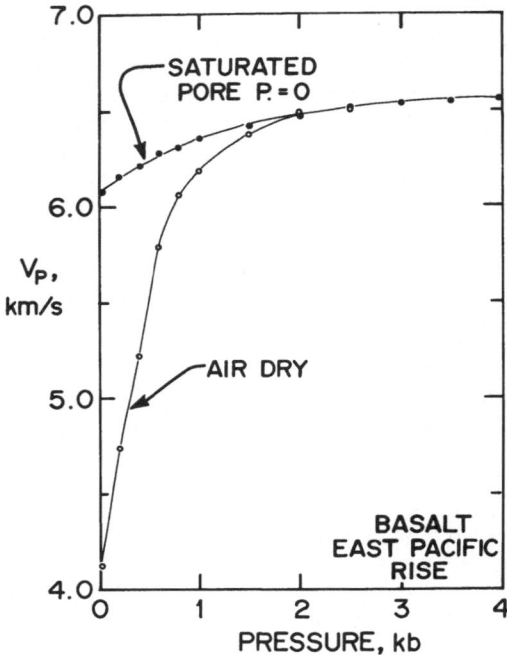

Fig. 3.   Compressional wave velocities for saturated and air-dry
basalt.

cities for a sample of oceanic basalt to 4 kbars, illustrated in
Fig. 3, show two important features of rock velocities.   First,
velocities increase rapidly over the initial few kilobars increase
in pressure, a finding ascribed to closure of grain boundary cracks
[18].   At higher pressures grain boundary porosity is essentially
eliminated, and the velocities are primarily related to the elastic
properties of the minerals within the rocks.   Second, it is significant that low pressure compressional wave velocities are higher
if the rock pore spaces are water saturated.   Since for most rocks
the effect of pore water on velocities is usually minimal above 1
to 2 kbars, water saturation is an important variable only in velocity measurements of possible oceanic crustal rocks and has little
importance in the interpretation of mantle velocities.
        Since the classic works of Birch [18,19], in which compres-
sional wave velocities were reported to 10 kbars for a wide variety
of igneous and metamorphic rocks, several papers have presented data
on velocities in rocks which are pertinent to a discussion of crustal
and upper mantle composition.   An extensive listing will not be at-

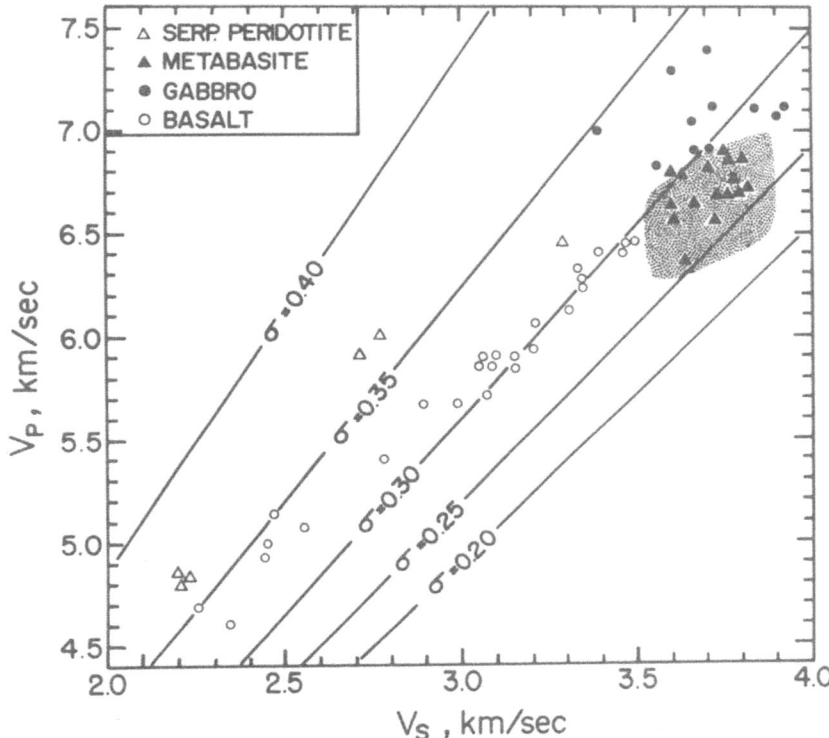

Fig. 4.   The range of observed refraction velocities for layer 3
(stippled area) and rock velocities at 1 kbar.

tempted here; however, in addition to Birch's papers, much use will
be made of the studies of Simmons [20], Christensen [5,12,21,22],
Christensen and Shaw [23], and Christensen and Ramananantoandro
[24].   Many of these papers also include data on shear wave velocities at elevated pressures, which provide important constraints on
crustal and upper mantle composition when combined with compressional wave velocities.

Compressional wave velocities reported for layer 3 are similar
to laboratory measurements of velocities in gabbro [12,18,21], amphibolite [12,18,21] and greenstone [22].   In addition, lower crustal velocities in the range of 6.7 to 6.9 km/sec agree with laboratory measurements in partially serpentinized peridotite containing

between 30 and 40% serpentine [5,12]. Thus if only oceanic compressional wave velocity refraction data are compared with rock
velocities, all of the models illustrated in Fig. 3 are probable.

Recently it has been shown that the Poisson's ratios (σ) of
partially serpentinized peridotites calculated from compressional
(V_p) and shear (V_s) wave velocities from the relation

$$2\sigma = 1 - \frac{1}{(V_p/V_s)^2 - 1},$$

are usually much higher than values calculated from lower oceanic
crustal refraction data [5]. This is illustrated in Fig. 4, in
which velocities are plotted at 1 kbar for a variety of water saturated rocks from the ocean floors. Also shown are lines of constant
Poisson's ratio and the range of lower crustal refraction data for
which both compressional and shear wave velocities have been reported [5]. Of importance is the observation that Poisson's ratios
for oceanic gabbros are generally much higher than observed values
for layer 3, whereas laboratory measurements of σ for metabasalts
and metagabbros agree well with seismic refraction measurements.
Thus models 1 and 2 of Fig. 3 are unlikely to be abundant within
the oceanic crust. It should be emphasized that refraction theory
assumes that layer velocities are for waves traveling near the upper boundary of the layer. Thus the conclusion that the upper portion of layer 3 is metamorphic agrees with models 3 and 4, since
the sheeted dikes of model 4 are usually metamorphosed.

At present it is impossible to distinguish between models 3
and 4 from seismic refraction data; however the velocity distri-
butions within layer 3 of the two models should differ significantly.
The late differentiates of model 4, which occur beneath the sheeted
dikes, have very low velocities [18] and, if abundant, should form
a low velocity region within layer 3. Also Poisson's ratio in the
lower regions of layer 3 should be high if cumulate gabbros are
abundant.

Crustal studies by seismic refraction techniques in Iceland
provide abundant data on crustal values of Poisson's ratios for the
crust of this region. Pálmason [3] reports average lower compressional and shear velocities of 6.35 km/sec and 3.53 km/sec, respectively, with Poisson's ratios near 0.27. The low velocities,
which are also common in the upper mantle, are most likely related
to the high temperature gradient in the region. Poisson's ratio,
on the other hand, shows little change with temperature; thus from
Fig. 4 it appears that metabasalt and metagabbro are abundant
within the lower crust of Iceland.

A high velocity basal crustal layer with compressional wave
velocities averaging 7.4 km/sec and a thickness of about 3 km has
been found in several locations in the Pacific Ocean basin [4]. At
present no shear velocity data have been published for this layer,
so one can only speculate as to its nature using compressional wave
velocity measurements. The velocities for this layer are in general

much higher than those observed for amphibolite and gabbro. Nevertheless, appropriate velocities have been reported for specific directions in anisotropic varieties of these rocks [21]. The most probable composition for this layer, not to be confused with "anomalous mantle" under ridge crests, is that of peridotite 10 to 20% serpentinized or feldspathic peridotite, both of which have appropriate velocities. Either of these compositions is consistent with model 3 of Fig. 3. Feldspathic peridotite does not occur in sufficient amounts within ophiolites to form a 3 km thick basal crustal layer. Thus a serpentinized peridotite origin for the 7.4 km/sec layer is the most favored for model 4.

5. UPPER MANTLE COMPOSITION

Compressional wave velocities in the oceanic upper mantle are generally between 7.8 and 8.6 km/sec, with a mean of approximately 8.2 km/sec (Fig. 1). The velocities severely limit permissible mineralogies, such that the most likely upper mantle rocks are peridotite and eclogite. The relative abundances of these two rock types, however, are still highly debated.

Recently Hart and Press [25] reported lithospheric S_n velocities ranging from 4.58 to 4.71 km/sec. Based on calculated velocities for various upper mantle petrologic models [26], they concluded that the upper mantle is composed of an eclogite-peridotite mix, because shear velocities of 4.71 km/sec are higher than that of peridotite. This, however, does not agree with laboratory measurements of shear wave velocities in fresh peridotites and dunites [24,27], which show that at pressures of 6 to 10 kbars unaltered peridotites have shear velocities between 4.6 and 4.8 km/sec. Similar shear velocities have been reported for eclogites [27]. Thus, in contrast with lower crustal studies, shear velocities are of

TABLE 1. Compressional wave velocities (km/sec) and anisotropies at 2 kbars

Rock	Highest Velocity	Lowest Velocity	Difference, % of mean	Reference
Dunite (North Carolina)	8.35	7.61	9.3	[18]
Dunite (Washington)	8.85	7.92	11.1	[18]
Dunite (New Zealand)	8.25	7.48	9.8	[18]
Dunite (Washington)	8.96	8.22	8.6	[28]
Dunite (Washington)	8.62	7.81	9.9	[24]
Dunite (Washington)	9.00	7.70	15.6	[24]
Eclogite (Colorado)	8.34	8.23	1.3	[29]
Eclogite (Colorado)	8.04	7.96	1.0	[29]
Eclogite (Japan)	8.54	8.40	1.8	[29]
Eclogite (Japan)	8.12	7.88	3.0	[29]

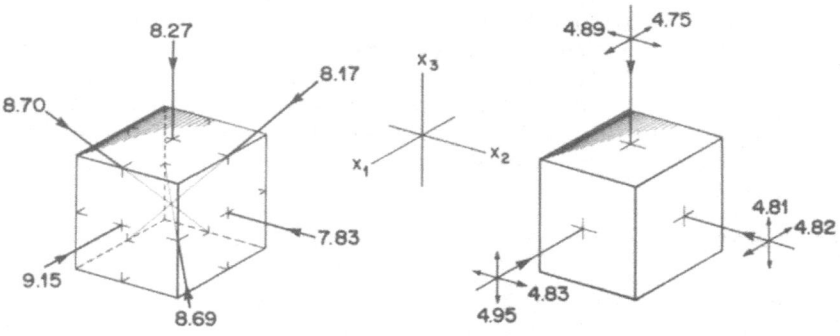

Fig. 5. Seismic velocities for an anisotropic dunite [24].

little value in deciphering upper mantle composition.

The elastic properties of rocks are usually considered to be isotropic, in which the elastic responses such as seismic wave velocities do not vary with direction. Under this simplifying assumption of isotropy, only two independent constants are required to completely describe elastic behavior. However, several recent investigations have shown that most ultramafic rocks are not adequately described by isotropic models, but require anisotropic elastic descriptions.

Compressional wave velocities and associated anisotropies at high pressure for several ultramafics and eclogites are given in Table 1. As was originally shown by Birch [18,19], the strong variation of compressional wave velocity with propagation direction in ultramafic rocks is related to preferred olivine orientation. Single crystal olivine is highly anisotropic with a maximum compressional wave velocity of 9.9 km/sec for propagation parallel to the a crystallographic axis and a minimum velocity of 7.7 km/sec for propagation parallel to the b axis. More recent studies of seismic anisotropy [24,28,29] have shown that the degree of preferred mineral orientation controls the magnitude of the anisotropy. Also, studies of shear wave propagation in olivine bearing ultramafic rocks have found that shear wave velocities within rocks with strong fabrics vary with displacement direction as well as propagation direction [24,30]. These relationships and the magnitude of anisotropy for a specimen of dunite with strong olivine orientation are illustrated in Fig. 5.

Detailed seismic refraction studies of the upper mantle in the northeast Pacific [31,32,33] have clearly shown that compressional wave velocities vary significantly with azimuth, the highest velocities being normal and the lowest parallel to ridge crests. The magnitude of the anisotropies are usually between 0.3 and 1.0 km/sec, in agreement with many laboratory measurements in olivine-rich ultramafics.

It is significant that eclogites, in general, do not show the strong anisotropy characteristic of many ultramafics [18,27,34]. Thus the observed anisotropies in the Pacific are most likely accounted for by dunite and peridotite with preferred olivine orientation. Upper mantle seismic velocities in other oceanic regions, however, which do show anisotropy may very well be accounted for by an eclogite model.

REFERENCES

1. R. Raitt, The Crustal Rocks, in: The Sea, ed. by M. N. Hill, New York, 1963.
2. X. LePichon, R. E. Houtz, C. L. Drake and J. E. Nafe, J. Geophys. Res., 70, 319, 1965.
3. G. Palmason, Crustal Structure of Iceland from Explosion Seismology, Soc. Sci. Islandica, 40, Reykjavik, 1971.
4. G. H. Sutton, G. L. Maynard and D. M. Hussong, Widespread Occurrence of a High-Velocity Basal Layer in the Pacific Crust Found with Repetitive Sources and Sonobuoys, in: The Structure and Physical Properties of the Earth's Crust, ed. by J. G. Heacock, AGU, Washington D.C., 1971.
5. N. I. Christensen, J. Geol., 80, 709, 1972.
6. N. I. Christensen and M. H. Salisbury, Earth Planet. Sci. Lett., 19, 461, 1973.
7. Tj.H. van Andel, J. Mar. Res., 26, 144, 1968.
8. H. H. Hess, History of the Ocean Basins, in: Petrologic Studies. Buddington Vol., ed. by A. E. J. Engel, H. L. James and B. F. Leonard, Geol. Soc. Am., New York, 1962.
9. J. Ewing and M. Ewing, Bull. Geol. Soc. Am., 70, 291, 1959.
10. B. Gutenberg, Physics of the Earth's Interior, Academic Press, New York, 1959.
11. P. J. Fox, E. Schreiber and J. J. Peterson, J. Geophys. Res., 78, 5155, 1973.
12. N. I. Christensen, Marine Geol., 8, 139, 1970.
13. J. R. Cann, Geophys. J. R. Astron. Soc., 15, 331, 1968.
14. A. Miyashiro, F. Shido and M. Ewing, Deep-Sea Res., 17, 109, 1970.
15. I. G. Gass, Nature, 220, 39, 1968.
16. A. Gansser, Econ. Geol. Helv., 52, 659, 1959.
17. J. F. Dewey and J. M. Bird, J. Geophys. Res., 76, 3179, 1971.
18. F. Birch, J. Geophys. Res., 65, 1083, 1960.
19. F. Birch, J. Geophys. Res., 66, 2199, 1961.

20. G. Simmons, J. Geophys. Res., 69, 1123, 1964.
21. N. I. Christensen, J. Geophys. Res., 70, 6147, 1965.
22. N. I. Christensen, Bull. Geol. Soc. Am., 81, 905, 1970.
23. N. I. Christensen and G. H. Shaw, Geophys. J. R. Astron. Soc., 20, 271, 1970.
24. N. I. Christensen and R. Ramananantoandro, J. Geophys. Res., 76, 4003, 1971.
25. R. S. Hart and F. Press, J. Geophys. Res., 78, 407, 1973.
26. D. W. Forsyth and F. Press, J. Geophys. Res., 76, 7963, 1971.
27. N. I. Christensen, J. Geophys. Res., 79, 407, 1974.
28. N. I. Christensen, J. Geophys. Res., 71, 5921, 1966.
29. M. Kumazawa, H. Helmstaedt and K. Masaki, J. Geophys. Res., 76, 1231, 1971.
30. N. I. Christensen, J. Geophys. Res., 71, 3549, 1966.
31. R. W. Raitt, G. G. Shor and T. J. G. Francis, J. Geophys. Res., 74, 3095, 1969.
32. G. B. Morris, R. W. Raitt and G. G. Shor, J. Geophys. Res., 74, 4300, 1969.
33. C. E. Keen and D. L. Barrett, Can. J. Earth Sci., 8, 1056, 1971.
34. N. L. Carter, D. W. Baker and R. P. George Jr., Seismic Anisotropy, Flow, and Constitution of the Upper Mantle, in: Flow and Fracture of Rocks, ed. by H. C. Heard, I. Y. Borg, N. L. Carter and C. B. Raleigh, AGU, Washington D. C., 1972.

THE STRUCTURE OF EASTERN ICELAND

George P.L. Walker

Department of Geology,
The Imperial College of Science and Technology,
London SW7 2BP, England.

ABSTRACT. This review is based on compilations of dips, the distribution of amygdale minerals, and the intensity of the dyke swarms. These compilations enable the position of the original top of the crust to be deduced. The top lies generally well above the present summit levels, but is highest in the belt of country within which the highest summits occur. There is also a general rise south-wards along the length of the belt to a maximum in the Quaternary volcanic district along the edge of the Vatnajokull. The non-parallelism of lava isochrons and dyke isochrons implies that there was a progressive southward shift in the zone of maximum spreading and probably a progressive intensification of activity as well.

1. INTRODUCTION

The Tertiary/Quaternary lava pile of eastern Iceland is inclined in a general westerly direction to expose a 10 km thickness of lava flows covering the period from approximately 15 m.y. ago [1,2,3]. Practically the entire Tertiary sequence was erupted sub-aerially: occasional hyaloclastites and pillow lavas found for example at 14°02'W, 64°50'N, and 14°39'W, 64°33½'N, were probably formed in temporary ponds, and lignite beds with overlying columnar basalt were formed in temporary depressions [4]. Pyroclastic rocks and terrigenous sediments interstratified with the lavas constitute approximately 6% of the total thickness; the former include acid ignimbrites, some of them welded [5,6].

Silicic volcanic rocks are more or less restricted to several discrete areas in which locally great thicknesses of intermediate and acid rocks occur. These discrete areas are called central

Kristjansson (ed.), Geodynamics of Iceland and the North Atlantic Area. 177-188. *All Rights Reserved.*
Copyright © 1974 *by D. Reidel Publishing Company, Dordrecht-Holland.*

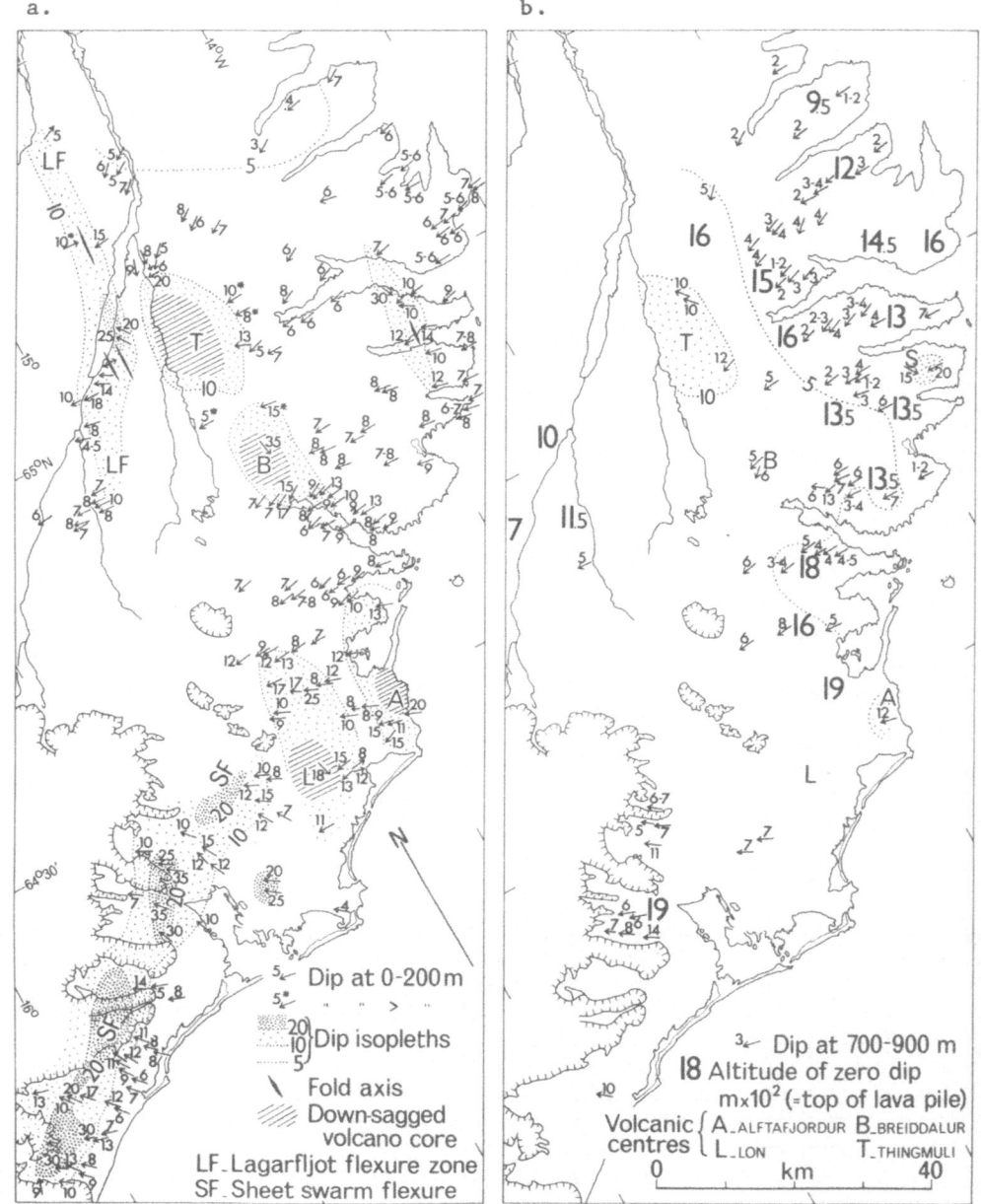

Fig. 1. The dip of the volcanic rocks in eastern Iceland, measured
at two levels and contoured; b also gives the extrapolated
altitude of zero dip.

a.

b.

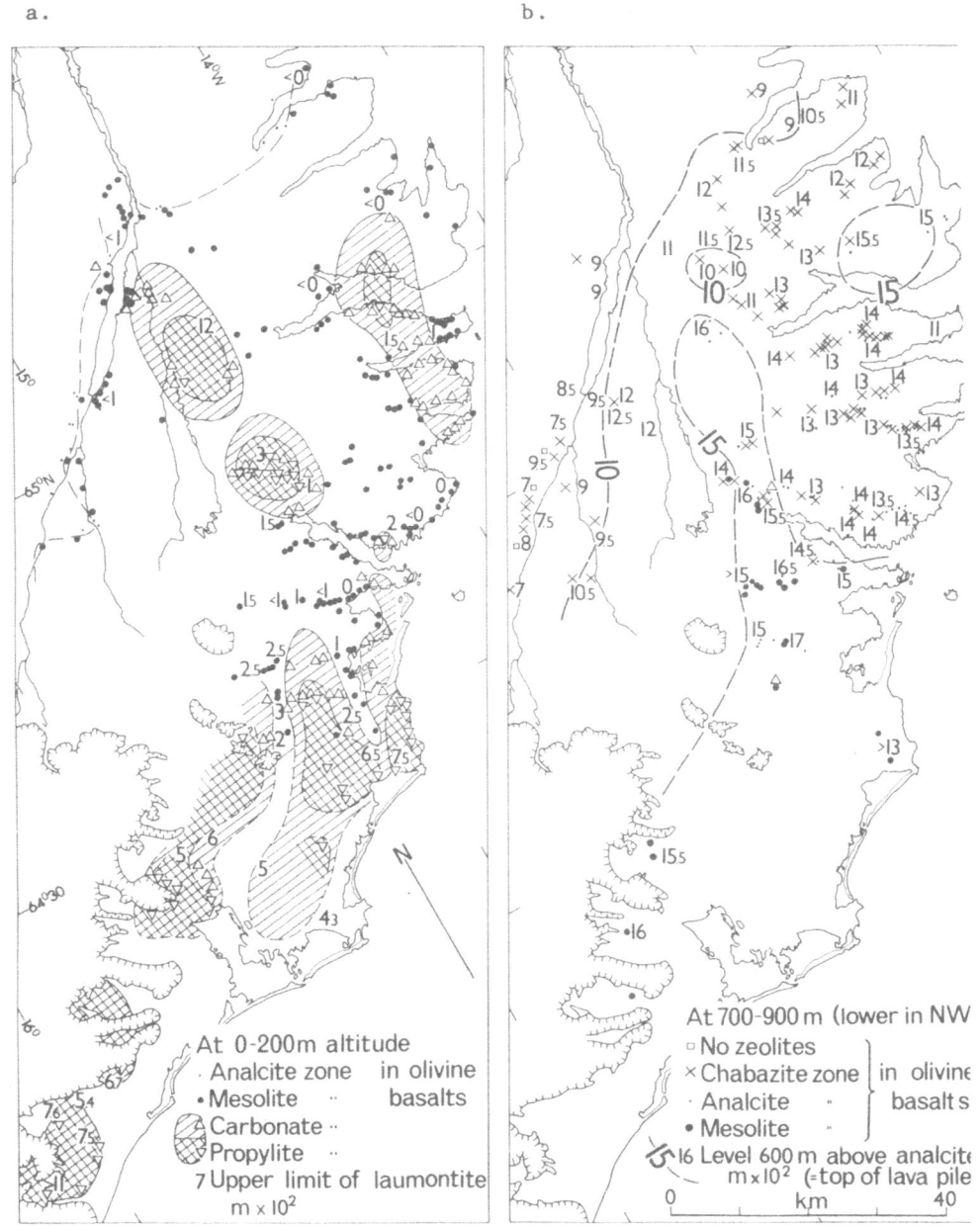

At 0-200m altitude
· Analcite zone in olivine
· Mesolite " basalts
Carbonate ··
Propylite ··
7 Upper limit of laumontite
m × 10²

At 700-900 m (lower in NW
□ No zeolites
× Chabazite zone in olivine
· Analcite " basalts
· Mesolite "
15 16 Level 600 m above analcite
m × 10² (=top of lava pile
0 km 40

Fig. 2. The distribution of amygdale minerals at two levels in
the basalts of eastern Iceland.

a.   b.

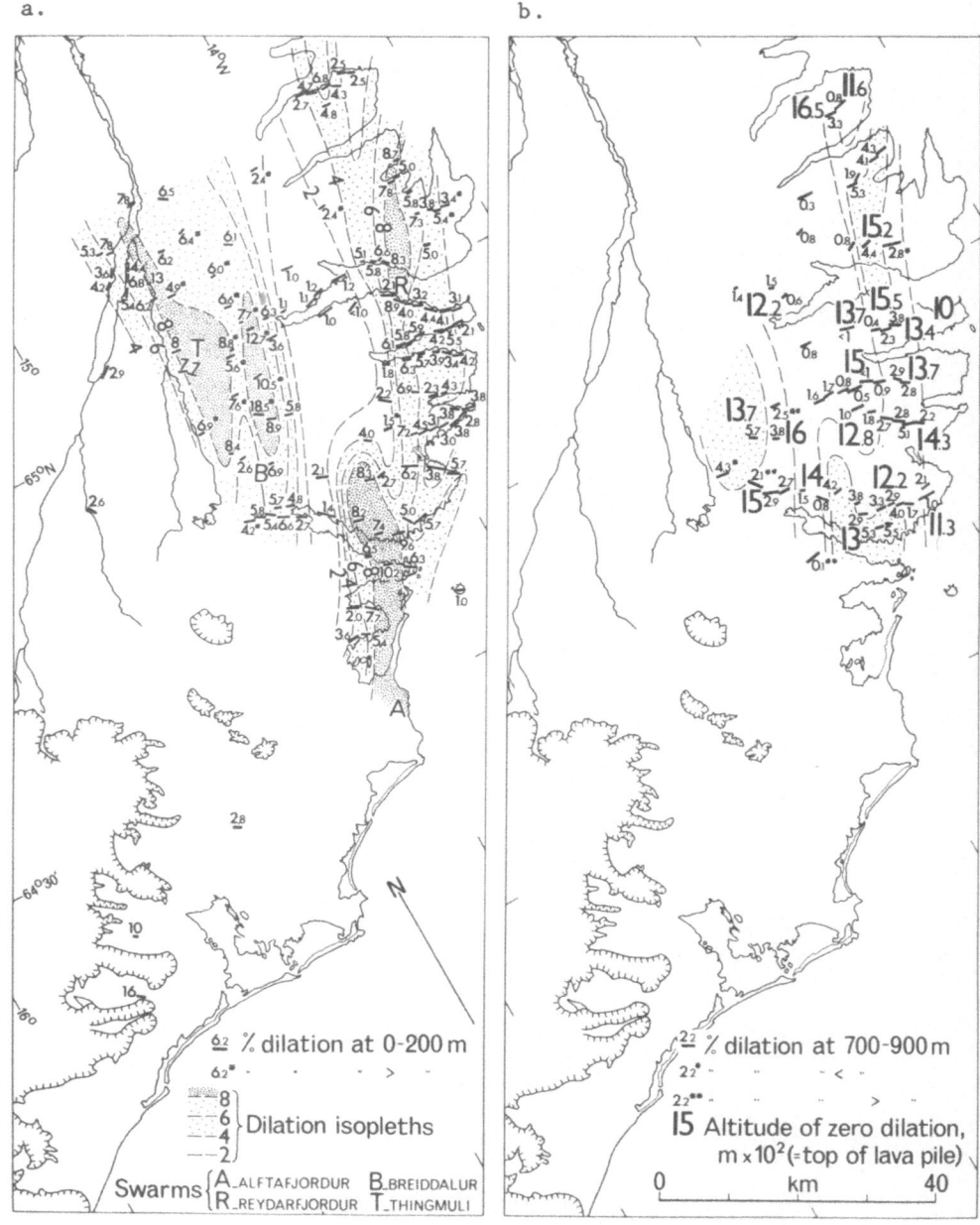

Fig. 3. The dyke swarm intensity in eastern Iceland, measured at
two levels and contoured; b also gives the extrapolated altitude
of zero intensity.

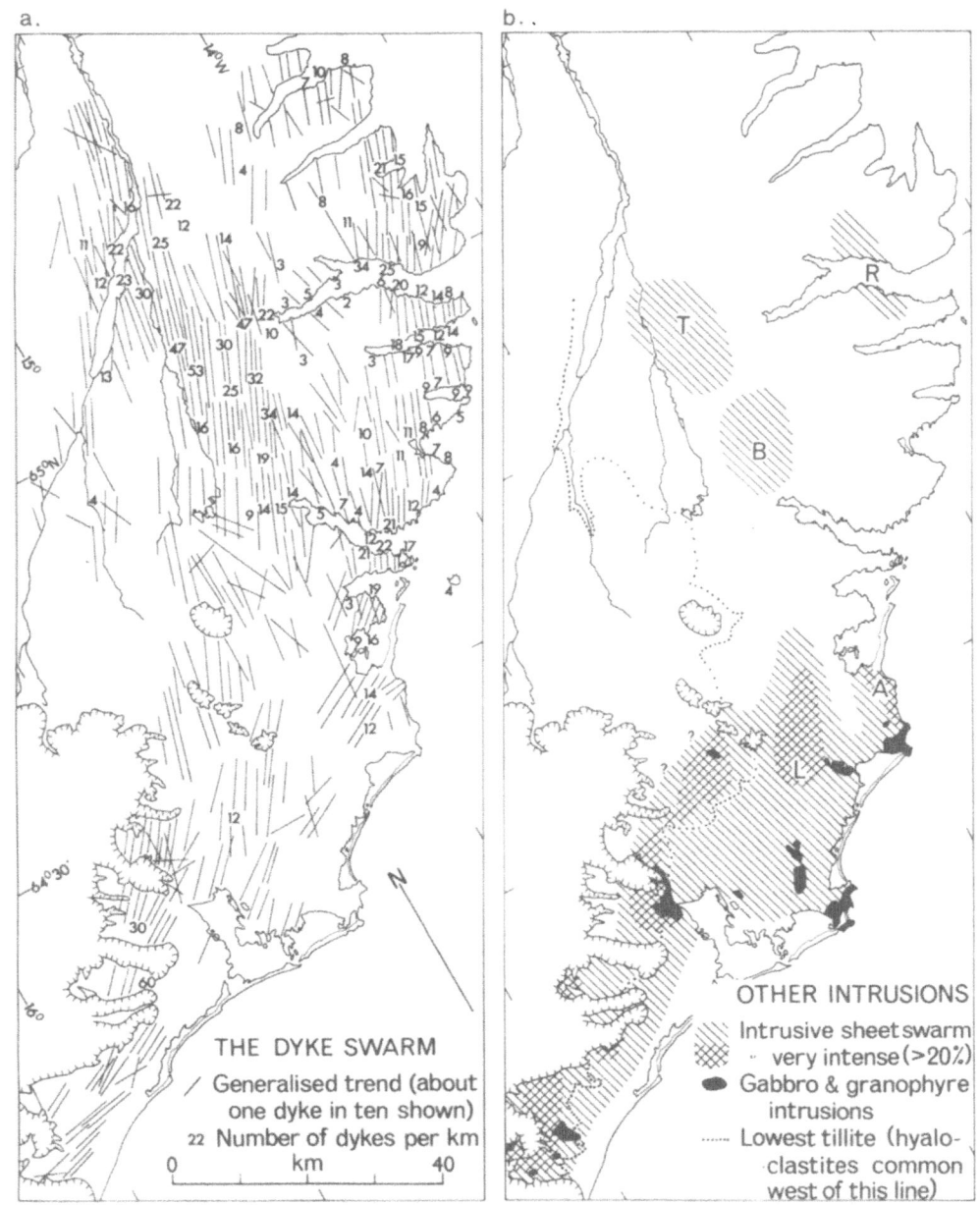

Fig. 4. The distribution of dykes and intrusive sheets in eastern
Iceland; b also shows the known intrusions of gabbro and granophyre.

volcanoes--"silicic volcanic centre" would be more appropriate for
some--in contrast with the flood basalts within which they occur
[7,8]. Several volcanic centres have been described, namely
Reydarfjordur [4,9], Breiddalur [10], Thingmuli [11] and Alftafjordur [12,13]. The silicic and associated basaltic rocks of each
centre occupy a volume of the order of 100 to 500 km^3.

The flood basalts contain three lithologic types which were
successfully distinguished in the field for stratigraphic mapping
[4,9,10], namely olivine basalts, tholeiites, and feldspar-
porphyritic basalts containing abundant bytownite phenocrysts. The
first are now known to be olivine tholeiites, but the lithologic
name is retained here. The volcanic centres contain tholeiite, and
the silicic rocks include extremely fine-grained lavas intermediate
between tholeiite and rhyolite called icelandites [11] or icelandes-
ites [14], as well as rhyolites and acid pitchstones. Olivine
basalts and porphyritic basalts are scarce in the volcanic centres.

The Quaternary succession differs from the Tertiary in
containing great volumes of basaltic hyaloclastites; also pillow
lavas, tillites, and grey diamictites regarded as solifluxion
sheets resting on a frost-shattered and frost-upheaved basalt floor.
In the southern part of the area the Tertiary-Quaternary succession
appears to be continuous, but in the northern part there is a zone
of structural complexity--the Lagarfljot flexure zone--which
interrupts the continuity.

It is of great importance in understanding the structure and
erosional evolution of the area to know the original position of
the top of the crust, and three independent methods have been used
to estimate its position, based on measurements of dip, the
intensity of the dyke swarm, and the position of the zeolite zones.
These are considered below and are shown to give consistent values.

2. DIP OF THE VOLCANIC ROCKS

In the flood basalts the dip is remarkably uniform in direction
and amount, and decreases regularly upwards from 6 to 9° or near
sea level (Fig. 1a) to 2 to 6° at an altitude of 700 to 900 m
(Fig. 1b). These uniform dips are believed to be due to tilting
accompanying volcanism, although a small component of dip is
locally no doubt an original depositional slope. Departures from
uniformity are due to:

a. Down-sagging of the floor of a volcanic centre, seen for
    example along the E. side of Thingmuli and Breiddalur.

b. Flexuring in the intrusive sheet complex of S.E.Iceland and
    the sheet swarm north of the Lon centre, believed to be

caused by the weight of the sheet swarms there.

c. Probable tectonic flexuring in the Lagarfljot flexure zone.

d. Localised uplift or pushing aside of the country rock by major intrusions, for example the Sandfell laccolith [15].

The regular decrease in dip with altitude implies that there is an up-dip wedging-out of stratigraphic units. This wedging-out is well established by stratigraphic mapping, but in general is less pronounced for olivine basalt lavas than for tholeiites, which is believed to reflect the lower viscosity of the former. The proportion of olivine basalts is in consequence higher at or near the mountain summits than at sea level.

Many dip measurements have been made at various altitudes, and when a group of readings taken over a limited area are plotted against altitude, a scatter of points is obtained to which a straight line can generally be fitted. The extrapolated altitude of zero dip (Fig. 1b) is believed to approximate to the original top of the crust.

The dip varies greatly in direction and amount in the volcanic centres. Original depositional dips, thought to be as large as $17°$ at Breiddalur [10] and $22°$ at Alftafjordur [12] account for some variation, but each centre also has a core region in which steep and variable dips appear to be largely due to subsidence. This subsidence is a central down-sagging which appears only occasionally [13] to be contained within a caldera ring fracture.

3. DISTRIBUTION OF AMYGDALE MINERALS

The assemblages of amygdale minerals in the flood basalts define mappable zones which are nearly horizontal and are parallel with one another [16]. The olivine basalts and feldspar porphyritic basalts have assemblages different from those in the tholeiites, but zones in one lava type are parallel with those in the other. Table 1 summarises the sequence and average thicknesses of the zones. The uppermost zone of no zeolites in the olivine basalts is found in the uppermost Quaternary rocks and also high on some mountain summits well down in the Tertiary succession, for example Skagafell ($14°18'$W, $65°09'$N).

The zeolite zones are thought to be developed at different temperatures and to represent fossil geoisotherms parallel with the original top of the crust. This surface is believed to lie some 600 m above the top of the analcite zone, and its estimated position is given in Fig. 2b. The zeolites are in part derived from mud washed into the amygdales, and the stratified mud is seen in

all stages of zeolitisation, commonly with a small dip showing
that tilting of the lavas had commenced before the mud was lithified.

TABLE 1. Summary of sequence and average thicknesses of zones
of amygdale minerals in the flood basalts of eastern Iceland

Zones in olivine basalt lavas	Average thickness	Zones in tholeiite lavas
zone of no zeolites	c. 150 m	
...................................zone of no minerals		
chabazite-thomsonite	450 m	
....		
analcite	150 m	
...............................mordenite-chalcedony		
	300 m	
mesolite-scolecite.......................................		
	c. 600 m	zone of abundant zeolites (additional to mordenite)
....		
	laumontite	zone

Each volcanic centre has a propylitised core in which albite
and chlorite and, in an inner zone, epidote are abundant. The rocks
have a distinctive pale green colour but from a distance they often
resemble rhyolites. Similar alteration is associated with the sheet
complex in S.E.Iceland. Pyrite is widely disseminated, laumontite
and carbonates are common, and a lime garnet also locally occurs.
The propylite zone is enclosed by a carbonate zone (Fig. 2a) rich in
calcite often platy on 0001, aragonite often in large masses
paramorphed by calcite, and dolomite. The famous Helgustadir
Iceland Spar mine (13°51'W, 65°02'N) is situated in this zone.

These propylite and carbonate zones delineate fossil high
temperature geothermal fields. Each has a volume of the order of
100 km^3, extending generally to less than 700 m above sea level.
These zones locally overlie flood basalts and have either a mush-
room shape or are "perched" like a perched water table. Larger
intrusions also have a propylite aureole [12,17].

The deepest (laumontite) zone in the flood basalts (Fig. 2a)
is below sea level over most of the northern half of the area but
rises to 1200 m at Thingmuli and in the extreme S.W. The top of
this zone, which is on average about 1700 m deep, could be used
to estimate the position of the original top of the crust, but
laumontite clearly fluctuates in height and occurs also in the
propylite zones where the thermal gradient may have been much
higher than usual.

4. THE DYKE SWARMS

Dykes are abundant, and the dyke intensity has been measured in
more than 200 well-exposed strips of country totalling more than
250 km long; these strips contain 2700 dykes totalling 11 km wide.
The intensity is given in Fig. 3 as the percentage of rock made up
by basic dykes, and Fig. 4a as the number of dykes per km. There
are several swarms, one for each volcanic centre, in which the
maximum sea level intensity exceeds 8%.

The dyke intensity everywhere falls with altitude, and when
intensity values measured over a limited area are plotted against
altitude, a scatter of points is obtained to which a straight line
can generally be fitted. The extrapolated altitude of zero
intensity (Fig. 3b) is believed to approximate to the original top
of the crust.

5. INTRUSIVE SHEET SWARMS

Intrusive sheets, mostly basic and averaging less than 1 m thick,
occur as swarms containing thousands of members in several areas.
They and their origin are discussed elsewhere([19] and the following paper of this volume). The intrusive sheet complex in the
deeply eroded Quaternary hyaloclastite belt along the edge of the
Vatnajokull (Fig. 4b) is particularly intense and in places (e.g. at
$16°11'W$, $64°09\frac{1}{2}'N$) sheets constitute about 100% of the rock. The
sheets dip in the same direction as, although more steeply than,
the rocks they cut. Each volcanic centre also has a swarm of sheets,
preferably not called cone sheets [19], emplaced mainly in the
acid volcanic rocks.

The stratification shown by chalcedony ("onyx") filling
amygdales in the flood tholeiites is normally horizontal. However
in the flexured basalts associated with the intrusive sheet swarms
it is sometimes not horizontal (Fig. 5). Tilting of these basalts is
attributed to subsidence caused partly by the weight of the lavas
and partly by the weight of intrusive sheets. It is thought that
tilting by the former takes place while the lavas involved are
within 10 or 20 km of the spreading zone and ends when no more lavas
are added above. The formation of zeolites and chalcedony proceeds
rapidly under this full cover. A further tilting may then take place
if sheets continue to be emplaced.

6. BROAD RELATIONSHIPS

Two kinds of relationships are discussed here. The first concerns
the position of the original top of the crust, and the second
concerns the non-parallelism of strike lines and dyke trends. The

three independent methods used to estimate the position of the first
give closely comparable results, normally not differing by more
than about 200 m as is evident from a comparison of Figs. 1b, 2b
and 3b, which tends to confirm the broad validity of the methods.

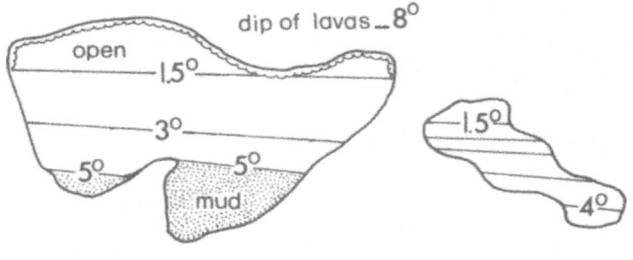

Fig.5. Variation in
dip during deposition
recorded by bedded
chalcedony in
amygdales seen at
15°36'W, 64°23'N.

scale
10 cm

A belt of high summits runs the length of eastern Iceland,
ranging in height from 1150 m in the extreme N.E. to 1400 m at the
edge of the Vatnajokull in the S.W. A thickness of rocks ranging
from 0 m in the extreme N.E. to about 500 m in the S.W. has been
eroded from above them. Both original top and summit level fall
northwestwards in the vicinity of the Lagarfljot flexure zone--
there is evidence that the zeolite zones are downflexed here--and
there is a less clear fall eastwards from the belt of maxima.

The second relationship is a more significant one and stems
from the fact that strike lines (isochrons) for the lavas at sea
level from Fig. 1a are not parallel with the predominant trend of the
dykes which cut them (dyke "isochrons"): the latter consistently
make an angle of about 30° (measured clockwise) with the former,
(Fig. 6). Only a small part of this discrepancy can be accounted for
by variations in the depth of erosion.

The generally nearly constant width of each submarine magnetic
strip anomaly indicates that the spreading rate is nearly constant
along the length of the spreading axis. The non-parallelism of
isochrons for lavas and dykes in eastern Iceland however appears to
be most readily explained as resulting from magmatism being concentrated along a short length of axis, this "locus" migrating south-
wards at the rate of about 10 km per m.y.

Fig. 7 illustrates the position of lava and dyke isochrons at
a given erosion level for a number of situations in all of which
the spreading axis is fixed in position. In b and c the locus is
fixed in position, but in c the spreading rate accelerates with
time. In d and e the locus moves and in e the spreading rate
accelerates with time. Other possibilities also exist.

The situation in eastern Iceland is similar to that of either

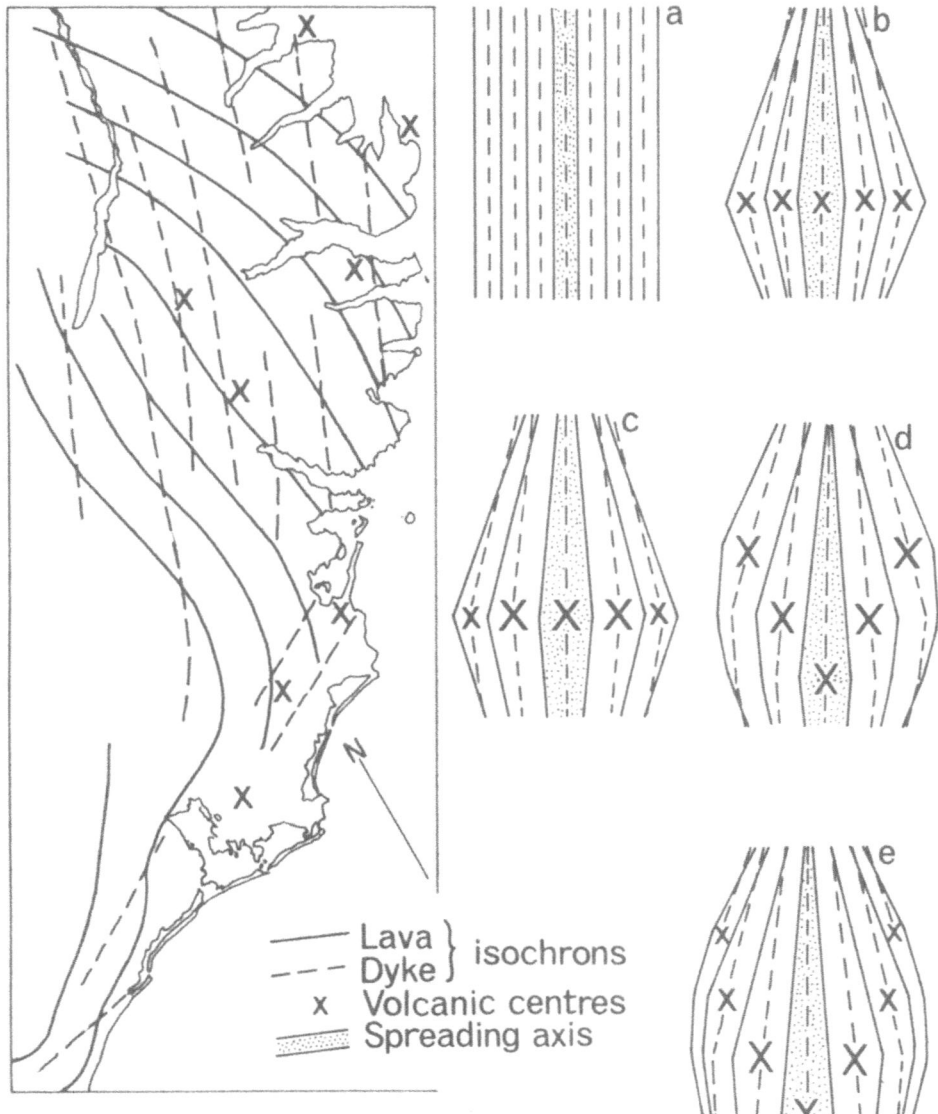

Fig. 6 (left). Isochrons for lavas and dykes in eastern Iceland showing their consistent non-parallelism. x's mark volcanic centres.

Fig. 7 (right). Distribution of lava and dyke isochrons for five different spreading axis situations.

d or e; the author tends to favour the second alternative, but a choice can only be made by closer dating of the rocks involved. The progressive south-westwards migration of acidic volcanic centres [8] supports the idea of a moving locus; if, as is suggested in the following paper, acidic centres develop at loci of most intense basaltic magmatism, then a younging line of acidic centres corresponds with the line of x's on Fig. 7d and e. The active Torfajokull acidic centre lies on a present spreading axis at the south-westward continuation of the Tertiary line.

Iceland is already known to be in several ways an anomalous section of the mid-Atlantic spreading axis, and this postulated moving locus on the spreading axis adds yet another anomaly to the list.

REFERENCES

1. N. H. Gale, S. Moorbath, J. Simons and G. P. L. Walker, Earth Planet. Sci. Lett., 1, 284, 1966.
2. S. Moorbath, H. Sigurdsson and R. Goodwin, Earth Planet. Sci. Lett., 4, 197, 1968.
3. P. Dagley and others, Nature, 216, 25, 1967.
4. I. L. Gibson, D. J. J. Kinsman and G. P. L. Walker, Greinar 4, (2), Soc. Sci. Islandica, Reykjavik, 1966.
5. T. Tryggvason and D. E. White, Am. J. Sci., 253, 26, 1955.
6. G. P. L. Walker, Quart. J. Geol. Soc. Lond., 118, 275, 1962.
7. G. P. L. Walker, Bull. Volcanol., 27, 1, 1964.
8. G. P. L. Walker, Bull. Volcanol., 29, 375, 1966.
9. G. P. L. Walker, Quart. J. Geol. Soc. Lond., 114, 367, 1959.
10. G. P. L. Walker, Quart. J. Geol. Soc. Lond., 119, 29, 1963.
11. I. S. E. Carmichael, J. Petrol, 5, 435, 1964.
12. D. H. Blake, Sci. in Iceland, 2, 43, 1970.
13. D. H. Blake, Geol. Mag., 106, 531, 1969.
14. K. Grönvold, Structural and petrochemical studies in the Kerlingafjöll region, Central Iceland. D. Phil. thesis, Oxford University, 1972.
15. L. Hawkes and H. K. Hawkes, Quart. J. Geol. Soc. Lond., 89, 379, 1933.
16. G. P. L. Walker, J. Geol., 68, 515, 1960.
17. R. W. Johnson, Sci. in Iceland. Anniv. vol, 55, 1968.
18. G. P. L. Walker, J. Geol. Soc. Lond., 130, (in press), 1974.

ERUPTIVE MECHANISMS IN ICELAND

George P.L. Walker

Department of Geology,
The Imperial College of Science and Technology,
London SW7 2BP, England.

ABSTRACT. The idea is developed that the fate of an uprising batch of magma and whether or not it reaches the surface is heavily dependent on various physical constraints. It is shown that many features of the geology of Iceland can be explained by the operation of these constraints. In particular a mechanism is proposed by which many batches of magma are diverted to form the swarm of intrusive sheets (such as is exposed in S.E.Iceland) which is postulated to constitute crustal layer 3 in Iceland.

1. INTRODUCTION

This paper develops the idea that whether or not a batch of uprising magma reaches the surface is heavily dependent on various physical constraints, and that many of the features of the volcanism and structure of Iceland and adjacent areas can be explained by their operation. Some constraints depend on properties of the environment, and others on properties of the magma.

In what follows the physical constraints are first considered, and then some of the relationships which are thought to depend on them are discussed. The ideas are developed for eastern Iceland but are also applied to various aspects of the geology of Iceland as a whole, following an earlier account elsewhere [1]. The paper finishes with speculative ideas on the role of gabbro intrusions vis a vis the low intensity of volcanic activity in the Reykjanes/ Thingvellir zone, and the significance of the very restricted composition range of ocean floor basalts. The ideas expressed here have many corollaries, some of which are listed, which provide a means of testing them.

Kristjansson (ed.), Geodynamics of Iceland and the North Atlantic Area. 189-201. *All Rights Reserved.*
Copyright © 1974 by D. Reidel Publishing Company, Dordrecht-Holland.

TABLE 1. Crustal layers in Iceland.

Layer	Average thickness, km	p-velocity, km s^{-1}	Density g cm^{-3}
0	0.5	2.8	2.1-2.5
1	1.0	4.2	2.6
2	2.15	5.1	2.65
3	6	6.5	2.9

2. THE PHYSICAL CONSTRAINTS

The following are the physical constraints which are judged to have
a powerful control on the fate of uprising batches of magma. Rock
names are as in the preceding paper.

2.1 Density of the crust

The seismic refraction study by Palmason [2] reveals the layered
structure of the crust in Iceland, and the seismic velocities yield
approximate values for the densities of the constituent layers.
Table 1 gives average values for the thickness and density of each
layer. Considerable variations exist; these average values are
however a convenient basis for discussion and are taken as the
"average Icelandic crust", the average density of which is 2.78
g cm^{-3}. For the significance of crustal density see section 2.5.

2.2 The surface barrier

The young lava flows and associated rocks in the uppermost few
hundred meters of the crust in the active zones of Iceland are
extremely permeable, but below this level the permeability of the
rocks drops because of the infilling of voids by zeolites and other
secondary minerals and the conversion of pumice to clays. In areas
of low relief the permeable surface zone is waterlogged to very near
the surface, and the groundwater can freely move laterally and also
to a lesser extent vertically through the zone. This waterlogged
zone presents a formidable barrier to the passage of magmas and is
referred to in the following discussion as the "surface barrier".

The abundant gaping fissures (gjas) in Iceland are believed to
represent the sites of "abortive eruptions" where uprising magma
failed to reach the surface, and if so they may be a testimony to
the effectiveness of the surface barrier. The 1961 eruption in
Askja was not abortive, but the appearance of lava at the surface
was preceded by two weeks of vigorous solfataric activity [3], and

it is supposed that the magma took this time to penetrate the
surface barrier. Many examples are known in eastern Iceland of
normal dolerite dykes which pass upwards into multiple sideromalane
injections or palagonite breccia, due to the chilling of the dykes
where they entered the surface barrier of their day.

Impressive demonstrations of the ability of groundwater to
influence the styles of eruptions are seen in the area of rather
low relief lying between Hekla and the Vatnajokull. The several
postglacial basaltic eruptions which have taken place in this area
were highly explosive (phreatomagmatic, of surtseyan type) due to
the entry of copious quantities of water into their fissures. Good
examples are Eldgja, Valagja, Ljotipollur, the Veidivötn and, in
part, Lakagigar 1783. At Eldgja and Veidivötn as on Surtsey in 1964
the explosive phase changed to a quietly effusive one after water
was denied access to the fissure. There is evidence that the
eruptions were characterised by a high volumetric rate of effusion
--the Lakagigar eruption averaged 5000 m^3 s^{-1} over the first 50
days [4]--and one can speculate that but for this, the magma might
not have succeeded in breaking through to the surface at all.

2.3 Situation on land or under the sea

The surface barrier is more effective under the sea than on land
because the highly permeable pillow lavas and hyaloclastites are
certain to be waterlogged, and water under high pressure will enter
any fissure which opens. The exsolution of gases which, in
eruptions on land, aids the magma through the surface barrier will
not help it below deep water since the gases will not be exsolved
under the high confining pressure there [5].

2.4 Width of the conduit

It is yet uncertain what controls the width of fissure or conduit
through which a magma rises; the significance of it is that the
narrower the conduit, the greater is the cooling surface area and
the greater the frictional energy loss per unit volume of magma
rising through it at a given rate. Both of these handicap the magma
in its uprise.

2.5 Density of magma

The more mafic a magma is, the higher is its density. The density
of a rhyolite melt at 900°C is 2.1 to 2.3 g cm^{-3}, and of a tholeiite
one at 1100° about 2.65 [6,7,8,9]. An olivine basalt melt is likely
to have a higher density than a tholeiite one, probably nearer 2.70.
These figures apply to pure melts. A melt containing crystals will

TABLE 2. Probable viscosity ranges of Icelandic lavas

Viscosity in poises	
10^8	rhyolite
10^5	icelandite
10^4	tholeiite
	olivine basalt

in general have a higher density than one without, by an amount
which depends on the nature and content of the crystals. Thus
a tholeiite melt containing 40% of bytownite/anorthite (density
2.69 at magmatic temperatures) should have a density about 0.02
higher than its non-porphyritic equivalent, and one containing
40% of olivine (density 2.8-2.9 at magmatic temperatures) should
have a density 0.06 to 0.10 higher than a non-porphyritic one.

The significance of density is twofold. Firstly, magmas
(including porphyritic melts) which have a density higher than the
2.78 of average Icelandic crust--all ultramafic and some picrite
basalt melts come into this category--should be generally incapable
of reaching the surface. Secondly, magmas which have a density
lower than 2.78 have a hydrostatic pressure sufficient to carry them
to the surface, and this pressure increases the further their
density drops below 2.78. The simple situation is assumed here that
magmas originate from immediately below the crust and that the
hydrostatic pressure is solely due to the weight of the crust. If a
magma originates within the crust it will be under a lower hydrostatic pressure appropriate to the lower average density of the
overlying crust.

2.6 Viscosity and rheology of magma

The more mafic a magma is, the lower is its viscosity and the
greater the ease with which it can rise to the surface along a
fissure. There is good evidence that the most fluid Icelandic lavas
are the olivine basalts as is shown for instance by the thin-ness
of their flow units, their relatively coarse grain, and their strong
tendency to form pahoehoe rather than aa. The tholeiites are more
viscous, the icelandites more viscous still, and the rhyolites most
viscous of all. Probable viscosities at the liquidus temperature
are given in Table 2.

Some rhyolite magma at Domadalshraun (section 5) formed
spatter, indicating an unusually low viscosity of not more than
about 10^5 poises; there is good evidence that this particular acid

magma was heated in contact with basic magma, and it must have been superheated.

It is supposed that most Icelandic lavas are erupted as Newtonian fluids, but many soon become non-Newtonian and their rheology approximates to that of a Bingham fluid as cooling proceeds. Any magma which is a Bingham fluid before eruption stands a much lower chance of attaining the surface, particularly through a fissure, than if it is Newtonian. The fact that icelandite and rhyolite lavas have a generally low or zero content of phenocrysts is taken to reflect this fact, since only above or very near their liquidus temperature are they likely to be Newtonian.

2.7 Volumetric rate of uprise

The volumetric rate of uprise is probably not an independent variable: it may be controlled by the hydrostatic pressure and be a function of magma and crustal densities, and it is presumably also affected by the viscosity. Its significance is that, other things equal, the higher it is the better able a magma will be to penetrate to the surface, keep its conduit open, and erupt there.

2.8 Gas content of magma

Dissolved gases begin to exsolve from a rising magma at a moderate depth (commonly less than 1 km for basalt) below the surface and their significance is that, because gas bubbles effectively increase the volume and so decrease the density of a magma, they greatly aid its rise to the surface particularly through the surface barrier. The greater the content of dissolved gases, the deeper is the level at which exsolution begins and the greater is the volume increase, and hence the more strongly do the gases contribute to carrying the magma through to the surface.

3. THE EFFECT OF CONSTRAINTS IN THE FLOOD BASALTS

Consider magmas which have a density less than the 2.78 g cm^{-3} crustal average. They have a hydrostatic pressure sufficient to carry them to the surface. If such a magma has a density which is lower than that of all the crustal layers, it rises to the surface or congeals on the way up. If on the other hand the density/depth curve for a magma intersects that for the crust (Fig. 1), the magma will rise to the level where the two curves cross and will then either continue to the surface or form an intrusive sheet at this level. The magma is particularly likely to form a sheet if the volumetric rate of uprise is so low that the magma fails to penetrate or keep open a passage through the surface barrier and

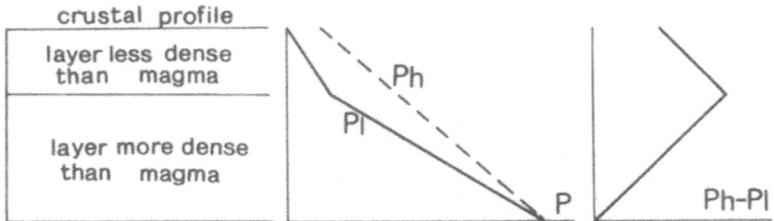

Fig. 1. Curve Pl gives the lithostatic pressure with depth, and Ph
the hydrostatic pressure in a magma. The right hand diagram shows
how Ph - Pl reaches a maximum at the level where magma and crust
have the same density.

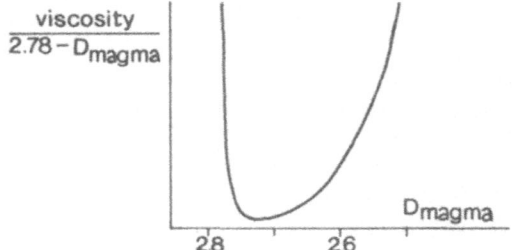

Fig. 2. Since increase in viscosity and decrease in hydrostatic
pressure reduces the likelihood of a magma reaching the surface,
a small value of (visc/2.78 - density of magma) provides a crude
index of this likelihood; a small value favours the magma.

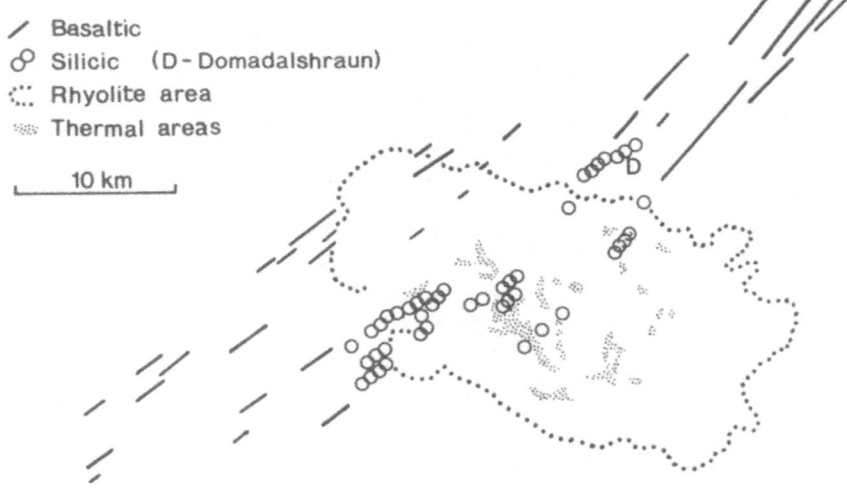

Fig. 3. Outline geological map of the Torfajokull silicic volcanic
centre showing basaltic and silicic postglacial eruptive fissures.

clogs the fissure there.

It has often been stated that intrusive sheets or sills in general form at the level where the magmatic pressure equals the superincumbent load [10]; however it seems unlikely, except for the most mafic magmas, if this situation is ever realised in Iceland where at almost all levels the hydrostatic pressure should exceed the lithostatic. The level at which the magma has the same density as the crust is that at which the difference between its hydrostatic pressure and the lithostatic pressure is at a maximum, and it is believed that this excess hydrostatic pressure is the driving force for sheet intrusion.

The following relationships in the Tertiary/Quaternary volcanic area of eastern Iceland can be explained by this mechanism:

a. The occurrence of the swarm of intrusive sheets in S.E.Iceland [1]. This swarm is the result of many uprising batches of magmas forming sheets instead of surface extrusions.

b. The origin of crustal layer 3. This layer is believed to consist of a swarm of basic intrusive sheets, each sheet representing a magma batch which failed to penetrate to the surface. The sheet complex in S.E.Iceland is regarded as an exposed part of layer 3.

c. The cut-off in composition of lavas at the mafic and salic ends. Ultramafic lavas and basalts rich in ferromagnesian phenocrysts have a density too high to reach the surface. No ultramafic lavas are known in Iceland. Only one picrite basalt lava--it contains 26 vol.% of olivine [11]--is known in eastern Iceland, and although such rocks occur elsewhere they are not common. At the other extreme, icelandites and rhyolites occur hardly at all in the flood basalts, and this is attributed to their inability to reach the surface because of their relatively high viscosities (Fig. 2).

d. The high proportion of porphyritic sheets in the swarm. The proportion is higher than that of porphyritic lavas, and is attributed to the higher density of porphyritic lavas causing their preferential injection as sheets. Much of the associated gabbro is bytownite-rich, and was intruded as prophyritic magma very similar to that of the sheets.

The following relationships in the present active zones of Iceland can be explained by the same mechanism:

e. There are many non-eruptive fissures (gjas). These are regarded as the surface manifestation of magmatic events whereby intrusive sheets form instead of surface effusions.

f. The surface deformation and seismicity. Some of the surface
   deformation including the subsidence in the active zones [12,13]
   could result from the intrusion and cooling down of sheets.
   Some of the seismic activity could be due to the same cause.

g. The excess heat flow. The excess heat flow is most easily
   explained as coming from basic minor intrusions, the volume of
   basic magma involved being several times that which erupts [14].

4. THE EFFECT OF CONSTRAINTS IN THE SILICIC VOLCANIC CENTRES

In Iceland the icelandites and rhyolites are concentrated in well
defined areas: silicic volcanic centres or "central volcanoes". The
conventional explanation for the restriction of salic rocks to
these centres is that magmas of these compositions are only
available there, but an alternative view now presented is that they
may be generally available, but only in the volcanic centres is
basic magmatism sufficiently intense to permit or cause the salic
magmas to rise to the surface. The role of basaltic magma can be
either passive or active. In the passive role it supplies the
thermal energy required to mobilise the salic magma. In the active
role it provides the pathway to the surface as well, as is manifest
in composite minor intrusions and the resulting composite lava flows
[15,16,17]. This point will be returned to in section 5. The
ultimate origin of the salic magmas is another problem and is not
relevant to present purposes.

The acid rocks at these centres have a relatively low density:
2.4 to 2.5 for lavas, and well below 2.4 for pyroclastic rocks.
Uprising basic magmas having a high density will tend to be
"trapped" by the acid rocks and will form intrusive sheets; it can
be surmised that only those magma batches which have the ideal
combination of fairly low viscosity and fairly high hydrostatic
pressure to yield the maximum volumetric rate of uprise are likely
to penetrate to the surface. The following relationships can be
explained by this mechanism:

a. Each centre has an intrusive sheet swarm. The sheet swarm in the
   centre occurs at a level at which sheets do not occur in the
   adjacent flood basalts and projects upwards like a cupola. At
   some centres the sheets have been interpreted as cone sheets;
   this question is discussed elsewhere [1].

b. Crustal layer 3 is shallower at the volcanic centres [2]. As
   layer 3 is regarded as consisting of a swarm of intrusive
   sheets, the shallower depth to layer 3 at a volcanic centre is
   equated with the shallower sheet swarm found there.

c. The sheets at some centres (e.g. Breiddalur and Reydarfjordur)

are intermediate in composition. The acid rocks have been able
to trap magmas of density lower than basalt, and one can
speculate that basalt has been trapped lower down, below the
present erosion level.

d. Olivine basalts and feldspar-porphyritic basalts are scarce.
   They are present in much lower proportions relative to
   tholeiite than in the flood basalts, which could be a direct
   consequence of the higher densities of their magmas and the
   increased likelihood of their being trapped.

e. Each centre has a down-sagged core region [e.g. 18]. This down-
   sagging can be attributed to the weight of basic sheets there.

f. Each centre has a propylite core. The geothermal activity is
   attributed to the basic sheet swarm in the core.

The above relationships come from eastern Iceland. They are
complemented by the remarkable relationships shown by the modern
Torfajokull volcanic centre, now described.

5.   THE EFFECT OF CONSTRAINTS AT THE TORFAJOKULL CENTRE

Torfajokull is a modern acid volcanic centre situated east of Hekla
[19]. Acid rocks occur over an area of 450 km^2, the largest
concentration of such rocks in the country. Many were erupted into
ice, but only the postglacial activity is considered here. The case
is made that Torfajokull is an impressive modern demonstration of
the effect of the trapping of uprising basic magmas by low-density
acid rocks. Hekla shows similar relationships, although less clearly.

The postglacial eruptive fissures in the active zone N.E. and
S.W. of the Torfajokull centre are basaltic, but the fissures which
occur where the same zone transects the rhyolite area have all
erupted salic magmas (Fig. 3). This absence of basaltic activity
within the rhyolite area is attributed to trapping, and basaltic
magmatism is believed to have caused the uprise of the salic magmas
within the same area. Saemundsson [19] records an example where
basalt, dacite, rhyolite and andesite lavas have erupted simultaneously from different parts of the same fissure, and the author can
contribute further data by citing the postglacial composite lava of
Domadalshraun, erupted just outside the rhyolite area, where there
has been a simultaneous effusion on rhyolite and basalt from the
same part of the same fissure. Explosive ejection from the vents has
produced mix spatter and mix pumice.

The existence of dense rocks below the Torfajokull centre is
indicated by the positive gravity anomaly, and the periodic
intrusion of basic sheets could account for the excess heat flow

in the widely spread Torfajokull geothermal field.

6. THE SHEET SWARM AND THE DEPTH OF CRUSTAL LAYER 3

The intrusive sheet complex in S.E.Iceland is regarded as an exposed
part of crustal layer 3. So far as is known, the part of Iceland
in which it occurs is the most deeply eroded in the whole country,
which partly accounts for the exposure of the sheets there. Nevertheless the swarm is exposed evidently at less than 2 km below the
original top of the crust, contrasting with the average 4 km depth
to layer 3 elsewhere in eastern Iceland.

The discrepancy can be accounted for by the great thickness of
basaltic hyaloclastites in S.E.Iceland, constituting the principal
country rock into which the sheets were emplaced. These hyaloclastites have a lower density than basalt lavas and the high level
reached by the sheet complex can be directly attributed to this.

The depth to layer 3 is a direct function of the proportion of
uprising magma batches which fails to reach the surface and is
diverted to form intrusive sheets instead. If the proportion is
large then layer 3 is shallow. The existence in the upper part of
the crust of concentrations of relatively low density rocks,
whether they be hyaloclastites as in S.E.Iceland or acid volcanics
as in a volcanic centre, causes the proportion to be greatly
increased and reduces the depth of layer 3 in consequence. The sheet
complex is thought to project up into the hyaloclastites as a broad
and overhanging "ridge", the linear equivalent of the upwardly-
projecting mushroom-shaped sheet swarm "cupola" of a volcanic centre.

7. THE MECHANISM OF GABBRO INTRUSIONS

It has been postulated elsewhere [1] that gabbro intrusions may
form within an intrusive sheet swarm when the frequency of uprise of
basic magma batches is sufficiently high. An incoming batch of magma
then arrives before the preceding sheet has fully solidified or
cooled down, and the new magma is likely to be preferentially
intruded into this sheet, less energy being required to follow this
pathway than to create a new one. When this is repeated several
times in quick succession, the resulting thick multiple sheet
starts to take on the characters of a single thick intrusion.
Injection repeated many times produces a gabbro intrusion.

The gabbro intrusions which occur in the intense part of the
sheet complex in S.E.Iceland are broadly sheet-like in form and
show features indicative of multiple injection [20,21] apparently
consistent with this confluent sheet growth mechanism.

A corollary of this mechanism is that during a period of particularly vigorous magmatism (vigorous in terms of a high frequency of uprise of magma batches) when gabbro intrusions grow, surface volcanism may in consequence drop to a low intensity. One anomaly in Iceland which might conceivably be accounted for in this way is the low intensity of volcanic activity in the Reykjanes/ Thingvellir zone since 1000 A.D. Many eruptions took place in the preceding millennia, and the author's impression is that the past millennium has seen a real fall in eruption rate. If so, it is not matched by a corresponding cessation of surface deformation--the postglacial lava fields are cut by a large number of gaping and very youthful fissures--or of seismic or geothermal activity.

One can speculate that the past 1000 years has seen an increase in the frequency of uprise of magma batches in this part of Iceland, gabbro intrusions are now forming, and volcanism has been reduced to nearly zero in consequence. Whether this is so or not, the point to make is that the quantity of lava erupted in any section of an active zone is no valid measure of the intensity of magmatism, a better guide to which may be the rate of surface deformation or heat flow.

8.   EFFECT OF CONSTRAINTS ON SUBMERGED PARTS OF A SPREADING AXIS

Iceland is in many ways an anomalous section of the mid-Atlantic Ridge, and the operation of various constraints can account for some of the differences. The important environmental difference between Iceland and the submerged parts of the ridge is that the surface barrier is more effective under water, and the exsolution of gases which on land greatly aids magmas in their passage through the surface barrier is unlikely to take place at all below 1 km or more of water. Conditions under which magmas can reach the surface are therefore much more stringent under water. Furthermore, lava erupted under water cools faster since the surface area of pillow lava is several tens of times greater than that of the same volume of subaerial lava, and water is also a more effective coolant than air.

The following can be accounted for by the operation of these constraints:

a. The composition range of ocean floor volcanic rocks is smaller. The range of viscosity/density conditions under which magmas are capable of reaching the surface is much more restricted than on land, and normally only the olivine basalt magmas satisfy them.

b. Layer 3 constitutes a larger proportion of the thickness of the oceanic crust. A larger proportion of batches of uprising

magmas form intrusive sheets than on land, and such lava
flows as do form are smaller because of their more rapid
cooling.

c. The oceanic crust is thinner than the Icelandic crust. Density
relationships are such that intrusive sheets form at the upper
surface of the sheet swarm; the rocks there are more likely to
be waterlogged under the sea and in consequence the sheets
cool more rapidly, extend less far laterally, and have a
smaller volume.

An important corollary is that the lavas which appear on the
surface, on land and even more so under water, are in no way
representative of the magmas available for eruption: the crust acts
as a "magma filter". Full account must be taken of the filtering
action before drawing petrogenetic conclusions from the study of
volcanic rock suites.

ACKNOWLEDGEMENTS
The stimulus for the ideas expressed here came from a visit to
Heimaey during the eruption there in Feb. 1973. The financial
support of the British Natural Environment Research Council, and
the help given by Steingrimur Hermannsson and Icelandic geoscientists is gratefully acknowledged.

REFERENCES

1. G. P. L. Walker, J. Geol. Soc. Lond., 130, (in press), 1974.
2. G. Palmason, Crustal structure of Iceland from explosion
    seismology, Soc. Sci. Islandica, Reykjavik, 1971.
3. S. Thorarinsson and G. E. Sigvaldason, Am. J. Sci., 260,
    651, 1962.
4. S. Thorarinsson, Bull. Volcanol., 33, 910, 1970.
5. A. R. McBirney, Bull. Volcanol., 26, 455, 1963.
6. C. E. Tilley, Mineralog. Mag., 19, 275, 1922.
7. W. O. George, J. Geol., 32, 353, 1924.
8. E. B. Dane, Am. J. Sci., 239, 809, 1941.
9. Y. Bottinga and D. P. Weill, Am. J. Sci., 269, 169, 1970.
10. J. Bradley, Trans. R. Soc. New Zealand, 3, 27.
11. I. S. E. Carmichael, J. Petrol., 5, 435, 1964.
12. E. Tryggvason, J. Geophys. Res., 73, 7039, 1968.
13. E. Tryggvason, J. Geophys. Res., 75, 4407, 1970.
14. G. Bodvarsson and G. P. L. Walker, Geophys. J. R. Astr. Soc.,
    8, 285, 1964.
15. I. L. Gibson and G. P. L. Walker, Proc. Geol. Assoc. Lond.,
    74, 301, 1964.
16. D. H. Blake, R. W. D. Elwell, I. L. Gibson, R. R. Skelhorn and
    G. P. L. Walker, Quart. J. Geol. Soc. Lond., 121, 31, 1965.

17. G. P. L. Walker and R. R. Skelhorn, Earth Sci. Rev., 2, 93, 1966.
18. G. P. L. Walker, Quart. J. Geol. Soc. Lond., 119, 29, 1963.
19. K. Saemundsson, Natturufraedingurinn, 42, 81, 1972.
20. A. E. Annels, The geology of the Hornafjordur region, S.E. Iceland. Ph.D. thesis, University of London, 1967.
21. T. C. Newman, The geology of some igneous intrusions in the Hornafjordur region of S.E. Iceland. Ph.D. thesis, University of Manchester, 1967.

ON THE TOPOGRAPHY OF THE VOLCANIC ZONES IN ICELAND

S. Thorarinsson

Science Institute, University of Iceland
Reykjavik, Iceland.

ABSTRACT. It has previously been pointed out by the present writer
that the Pleistocene volcanic zones of Iceland are not only a
supramarine exposure of a part of the World Rift System, and as
such much more accessible for geological, geophysical, and topographical research than the submarine parts of the system. The
Icelandic zones are also the only supramarine part of the system
where a relatively recent volcanic activity has taken place under
water and thus under conditions similar to those on the submarine
part. This water was meltwater from the overlying ice cover during
those periods of the Pleistocene when the country was blanketed by
ice. The basaltic magmas extruded in Iceland are similar enough to
those of the submarine rift zones to make one expect similar shape
and inner structures of the volcanic edifices built up. In both
cases the formation of pillow lavas is a characteristic feature.
The main difference in shape one may expect is that the subglacial
edifices are likely to be somewhat more steepsided, being, at least
in some cases, built up against walls of ice.

The basalt volcanoes that were built up in Iceland during the
Pleistocene glacials are mainly of two types: serrated ridges built
up on long fissures, and tablemountains built up on short fissures,
changing during the initial or early phase of the eruptions into
circular vents. The basalt volcanoes on the ocean floors are of
similar types, viz. ridges and seamounts, whereas the corresponding
types built up in Iceland in ice-free areas during the interglacials
and in postglacial time are crater rows and shieldvolcanoes.

A striking image of a part of the eastern rift belt in South
Iceland, between Mýrdalsjökull and Vatnajökull, was obtained from
the satellite ERTS-1 on January 31, 1973, when the sun was only

Kristjansson (ed.), Geodynamics of Iceland and the North Atlantic Area. 203-205. *All Rights Reserved.*
Copyright © 1974 by D. Reidel Publishing Company, Dordrecht-Holland.

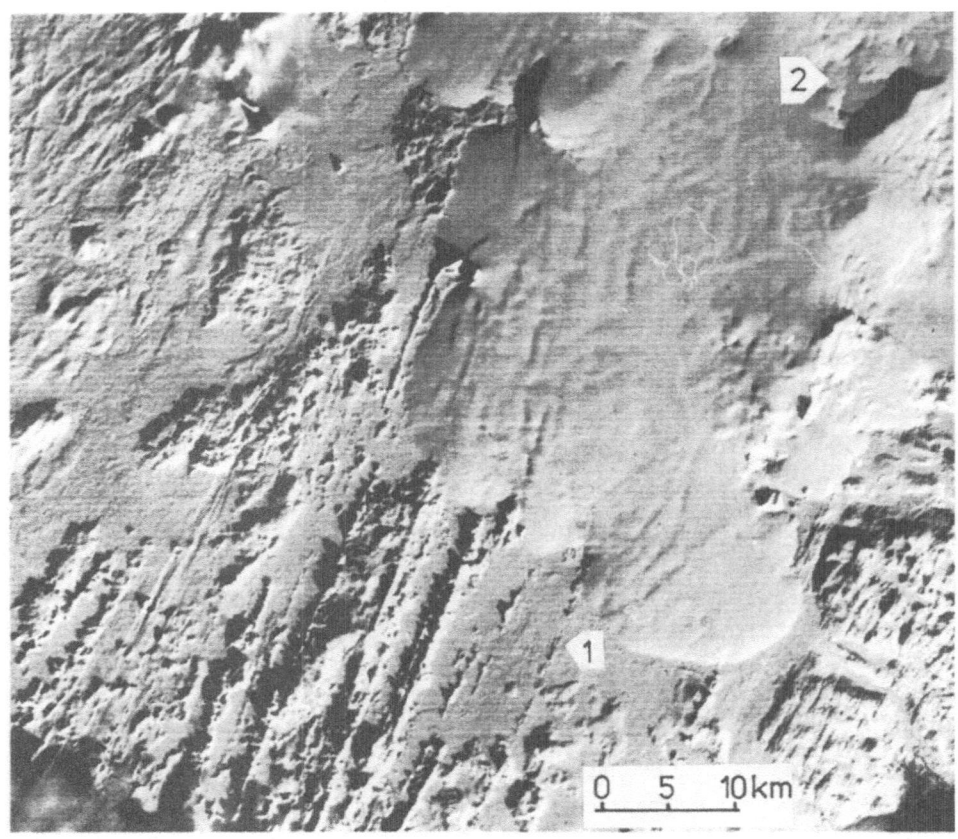

Fig. 1. NASA Earth Resources Technology Satellite (ERTS-1) image
of the volcanic ridge-rift belt SW of Vatnajökull. The image was
obtained on Jan. 31, 1973. The angle of the sun above the horizon
was 7° and the area was snow covered. 1: The Lakagígar crater row
of 1783. 2: The Grímsvötn caldera.

7 degrees above the horizon, hence the long shadows exaggerating
the relief. In view of what has been said above this picture ought
to give a fairly good idea of what it is like on, at least, some
parts of the Mid-Atlantic ridge-rift zone. One should keep in mind
the fact, previously stressed by the late G. Kjartansson and the
present writer, that although many faultlines run through the
Icelandic rift belts the straightlined valleys and ridges charac-
terizing the subglacially built up parts of the belts are predominantly constructional features. Most of the valleys are thus

not grabens, but formed between parallel ridges that were piled
up by subglacial volcanic activity. This holds probably true for
parts of the submarine Atlantic ridges as well.

In the alkali–olivine part of the volcanic belts of Iceland
the basalt volcanoes found are either ridges or crater rows.
Tablemountains and shieldvolcanoes are absent. In the tholeiitic
median zone tablemountains–shieldvolcanoes and ridges–crater rows
are found side by side. This raises the question why the same
type of magma sometimes produces shieldvolcanoes-tablemountains
and sometimes ridges–crater rows. On closer examination the formation of tablemountains–shieldvolcanoes seems to be, at least mainly,
restricted to limited periods of time. The last period of the formation of this type of volcanoes coincided with the last part of the
Würm glaciation and the first millenia of the postglacial time.
This means that these volcanoes were formed during a period of
glacial–isostatic uplift of the country. There seems thus to be
some causal relation between the isostatic uplift of the country
and the formation of this type of volcanoes. S. Steinthorsson has
recently suggested that this could possibly be explained by partial
melting processes caused by the pressure release (see paper by
Sigvaldason and Steinthorsson, this volume).

FOCAL MECHANISM SOLUTIONS OF INTRAPLATE EARTHQUAKES AND STRESSES
IN THE LITHOSPHERE*

Lynn R. Sykes° and Marc L. Sbar

Lamont-Doherty Geological Observatory of Columbia
University, Palisades, New York 10964, U.S.A.

ABSTRACT. Focal mechanism solutions of 33 intraplate earthquakes
are presented. These solutions, 47 additional mechanism solutions
from the literature, and measurements of stress in situ indicate
that the interiors of many lithospheric plates are characterized
by large horizontal compressive stress. The predominance of compressive stress is particularly notable for oceanic areas where
no examples of normal faulting (only thrust and strike-slip faulting) were found for crust older than about 10 to 20 m.y. The distribution of stress in lithospheric plates may be related to the
driving mechanism of plate tectonics or it may result mainly from
thermo-elastic stresses related to the cooling of the lithosphere.

1. INTRODUCTION

Focal mechanisms of earthquakes are one of the main sources of
information about stresses in the earth and about contemporary
crustal movements. The main purposes of this paper are to present
focal mechanism solutions of 33 intraplate earthquakes and to infer
the state of stress in lithosphere plates from them, from other
solutions available in the literature, and from other methods of
determining stress. Sykes and Sbar [22] summarized most of the
results presented here but did not publish the details of the focal
mechanism solutions.

---

*Lamont-Doherty Geological Observatory Contribution No. 2131.
°Also Department of Geological Sciences, Columbia University.

Kristjansson (ed.), Geodynamics of Iceland and the North Atlantic Area. 207-224. *All Rights Reserved.*

Most studies of focal mechanisms to date have concentrated on earthquakes along plate boundaries, i.e., <u>interplate</u> events. Very little work, however, has been done on <u>intraplate</u> earthquakes, and very little is known about the tectonic setting or the ultimate cause of intraplate events.

In this study a three-fold classification of earthquakes is used: 1) plate-boundary events, i.e., interplate earthquakes, 2) plate-boundary related events, and 3) intraplate shocks. Plate-boundary events appear to reflect relative plate motion. The slip vector as derived from focal mechanism solutions appears to be the most consistent parameter among plate-boundary events. Plate-boundary related events occur near but not on plate boundaries and appear to result from the stresses set up near plate boundaries by relative plate motions. Intraplate events, however, do not appear to be related to plate interactions, but instead the directions of principal stresses inferred from focal mechanism solutions appear to be the most consistent parameters among nearby shocks. In this paper we attempt to confine our attention to intraplate events <u>per se</u>. Nevertheless, some of the earthquakes studied that are close to plate boundaries (such as those in southern Africa and in the Basin and Range Province of the western United States) may be characterized better as plate-boundary related events.

Although intraplate earthquakes are not as frequent as shocks along plate margins, several large and damaging earthquakes, which appear to be best characterized as intraplate events, have occurred in eastern and central North America and in Australia, Africa and peninsular India. An understanding of events of this type, such as the Charleston, South Carolina earthquake of 1886, is vital to realistic assessments of earthquake risk in populous areas such as the eastern and central United States.

Sbar and Sykes [25] compiled data on the state of stress in eastern and central North America using focal mechanism solutions of earthquakes, post-glacial geological features and in-situ measurements of stress by the overcoring and hydrofracutring techniques. They concluded that a large part of that region was characterized by horizontal compressive stresses which were much larger than those expected merely from lithostatic (gravitational) loading.

The presence or absence of high stress is of great importance in earthquake prediction, seismic risk evaluation, the design of underground excavations, hydraulic mining, disposal of liquid waste underground, and the design of large dams and other major structures. The state of stress in the earth is also an important geophysical parameter which has received very little study to date. Knowledge of the stress field may furnish important clues to the driving mechanism of plate tectonics. A variety of models of the driving mechanism have been proposed. Measurements of stress may be a means of ascertaining which of these models is correct or whether some other factor, such as thermo-elastic stresses resulting from the cooling of the lithosphere, is responsible for the observed distribution of stress.

2. FOCAL MECHANISM DATA

Table 1 lists relevant parameters of 80 intraplate earthquakes for which focal mechanism solutions are available. Figure 1 a, b, c illustrates the focal mechanisms for the 33 new solutions obtained by us. Our analysis was confined to a study of the first motions of P and PKP waves. For some of the larger events the first motions (or polarizations) of the S wave were also used to restrain the inferred solutions.

For the better recorded events (usually those of larger magnitude) two orthogonal nodal planes can be used to divide the first motions of P into four quadrants. For some of the smaller events, however, it is only possible to state that the dislocation involved in the earthquake was predominantly normal, strike slip or thrust faulting. For example, event 42 (Fig. 1b) is a case involving a predominance of thrust faulting. From the data we used for event 42, however, it was not possible to infer either the strikes of the two nodal planes or the azimuth of the nearly horizontal principal stress (in this case the axis of greatest compressional stress, the P axis). As in other mechanism studies, the P and T axes of the focal mechanism solution are equated with the axes of maximum and minimum compressive stress.

Forsyth [16] and others have used the amplitudes of seismic surface waves to obtain focal mechanism solutions. Surface wave studies of the intraplate events reported here should provide more precise estimates of the strikes of nodal planes and of the azimuth of the nearly horizontal principal stress (P axis for thrust faulting and T, tensional axis, for normal faulting). Nevertheless, the data already available from first motions give a number of important insights about the stresses in the interiors of lithospheric plates.

3. STATE OF STRESS IN LITHOSPHERIC PLATES

Fig. 2 summarizes the focal mechanism solutions of 80 intraplate earthquakes. The solutions are classified as predominantly normal, strike slip or thrust faulting. The first-motion data are sufficient to support this classification with confidence for nearly all of the events reported. For some of the dip-slip events the strikes of the nodal planes and the azimuth of the nearly horizontal principal stress could not be determined. Hence, for these events only the type of faulting is indicated.

Table 1.   Parameters of Focal Mechanisms and Other Data for Intraplate Earthquakes

No.	Date	Latitude	Longitude	Depth (km)	Magnitude	Pole of 2 Nodal Planes az/pl		P axis az/pl	T axis az/pl	Type Faulting	Ref.
1	26 Nov 71	79.4N	17.7W	19	5.2	333/38	100/38	33/60	126/1	N	
2	04 Sept 63	71.4N	73.3W	33	5.9	3/20	196/70	358/65	186/25	N	1
3	02 Dec 70	68.4N	67.3W	27	4.9					(N)	
4	02 Oct 71	64.3N	86.4W	16	5.0					T	
5	05 Oct 65	65.4N	134.0W	8	5.2	181/40	314/38	336/01	250/63	T	
6	16 Apr 65	64.7N	160.1W	5	5.8	215/24	47/65	202/69	40/21	N	8
7	24 Nov 69	60.5N	58.7W	33	5.0	320/20	127/70	137/25	328/65	T	
8	07 Dec 71	55.0N	54.3W	33	5.4					T	
9	Oct–Nov 69	41.0N	74.6W	2	1.2	282/30	148/50	233/64	122/11	N–C	2
10	June–Sept 71	43.5N	74.3W	2	3.6	258/65	78/25	258/20	78/70	T–C	3
11	09 Nov 68	38.0N	88.5W	25	5.5	105/45	269/43	97/1	192/82	T	4
12	17 Sept 64	44.5N	31.3W	24	5.6	263/59	132/22	296/21	166/20	T	
13	06 Aug 62	32.3N	41.0W	0		319/33	196/40		(165)/3	N	10
14	11 Mar 67	19.1N	95.8W	33	5.5	63/0	153/52	212/34	94/34	SS–T	24
15	09 Sept 65	6.5N	84.4W	27	5.5	176/34	83/6	223/18	123/28	SS	24
16	21 Aug 63	14.3N	72.5W	33	5.4			45/0		T–SS	24
17	12 Mar 68	13.0N	72.6W	11	5.3	27/50	207/40	27/5	207/85	T	
18	03 Sept 68	20.6N	62.2W	33	5.5	299/55	119/35	299/10	119/80	T	
19	23 Oct 64	19.8N	56.0W	31	6.4	194/38	314/32	164/3	258/54	T–SS	24
20	20 Feb 68	12.4N	46.9W	13	5.6					(N)	
21	25 July 69	12.4N	40.7W	9	4.9					T	
22	09 Oct 66	12.6N	30.8E	11	5.1	358/36	258/15	301/36	42/15	N–SS	
23	30 Sept 71	0.5S	4.8W	20	6.0	342/30	124/53	147/12	24/66	T	9
24,25	23–25 Sept 63	16.7S	28.7E	0	5.8	270/58	115/30	138/75	283/10	N	10
26	15 May 68	15.8S	25.9E	33	6.1	272/50	136/32	181/65	298/10	N	12
27	02 Dec 68	13.9S	23.8E	7	6.0	112/26	292/62	116/72	296/18	N	12
28	29 Sept 69	32.9S	19.7E	33	5.9	134/8	223/9	90/10	181/6	SS	12
29	23 Mar 70	21.7N	73.0E	3	5.4	70/50	182/18	29/19	141/49	SS–T	6
30	10 Dec 67	17.4N	73.8E	33	6.0	21/0	111/15	157/10	65/10	SS	10
31	13 Apr 69	17.9N	80.6E	33	5.3	55/10	147/10	191/10	100/14	SS	5
32	15 Apr 64	21.7N	88.0E	36	5.5					(T)	
33	23 Apr 67	1.6N	80.2E	33	5.1					T	
34	10 Oct 70	3.6S	86.2E	30	5.9	36/7	128/9	172/3	82/11	SS	14
35	25 May 64	9.1S	88.9E	33	5.5	86/0	356/8	311/7	41/7	SS	10
36	31 Mar 70	3.8S	69.7E	33	5.5	170/33	54/33	111/52	203/01	N	
37	12 Sept 65	6.4S	70.8E	33	6.1	270/36	234/40	123/57	29/6	N	11
38	11 Nov 67	6.0S	71.4E	33	5.6	10/64	190/26	190/71	10/19	N	
39	11 Nov 67	6.0S	71.4E	33	5.7	36/56	216/34	216/79	36/11	N	
40	11 Nov 67	6.0S	71.4E	33	5.7	29/66	209/24	209/69	29/21	N	
41	02 Mar 68	6.1S	71.4E	33	5.6	19/52	199/38	199/83	19/7	N	
42	14 Sept 68	24.5S	80.4E	33	5.5					T	
43	08 Oct 68	39.8S	87.7E	33	6.0	124/60	304/30	304/75	124/15	N	
44	21 Nov 69	2.0N	94.6E	20	6.4	293/4	202/6	158/1	248/8	SS	14
45	26 Jun 71	5.2S	96.9E	25	5.8	280/50	166/20	320/18	206/51	T	
46	14 Oct 68	31.7S	117.0E	1	6.0	62/22	308/45	271/13	20/50	T	14
47	10 Mar 70	31.0S	116.5E	5	5.7	67/14	326/40	282/16	23/39	T–SS	14
48	24 Mar 70	21.9S	126.6E	33	6.2	34/40	262/38	235/0	144/61	T	14
49	27 May 71	53.8S	150.5E	33	5.6					T	
50	12 Nov 67	17.2S	172.0W	34	5.6	94/45	274/45	94/0	0/90	T	23
51	29 Jan 69	17.2S	171.5W	33	6.0	92/18	272/72	272/27	92/63	T	23
52	09 May 71	39.8S	104.8W	33	6.2	286/30	106/60	106/15	286/75	T	16
53	08 Feb 71	63.4S	61.2W	33	6.3					N	
54	25 Nov 65	17.1S	100.2W	29	5.8	81/42	292/44	276/0	10/74	T	15
55	13 Apr 67	7.0S	151.0W	33	5.2					T	
56	29 July 68	7.4S	148.2W	33	4.9					T	
57	24 Sept 66	12.0N	130.8W	33	5.3	213/30	105/29	255/1	161/42	T	
58	03 May 69	8.3N	175.5W	33	5.2					(T)	
59	28 Apr 68	44.8N	174.5E	39	5.5	107/40	230/32	77/05	174/56	T	
60	16 Mar 63	46.5N	154.7E	0	7.5	106/35	286/55	106/10	286/80	T	7
61	07 May 62	45.3N	146.7E	25	7.0	159/60	303/28	323/15	106/71	T	7
62	29 Aug 70	51.1N	135.2E	33	5.4					(T)	
63	15 Aug 67	44.8N	132.4E	33	5.3	113/30	340/50	313/11	63/64	T	
64	12 Nov 65	30.5N	140.2E	40	6.6	82/41	280/46	51/80	270/3	N	18
65	26 Aug 67	12.2N	140.8E	33	6.1	150/40	339/50	115/83	334/5	SS	18
66–69	see ref. 21	China								SS	21
70–79	see ref. 13	W. Europe								SS–N	13
80	21 Jan 72	71.9N	74.7W	6	4.5	30/30	210/60	30/75	210/15	N	17

az = azimuth, pl = plunge, T = thrust, SS = strike–slip, N = normal, C = composite mechanism, ( ) less reliable number, blanks in Ref. column indicates our solution.

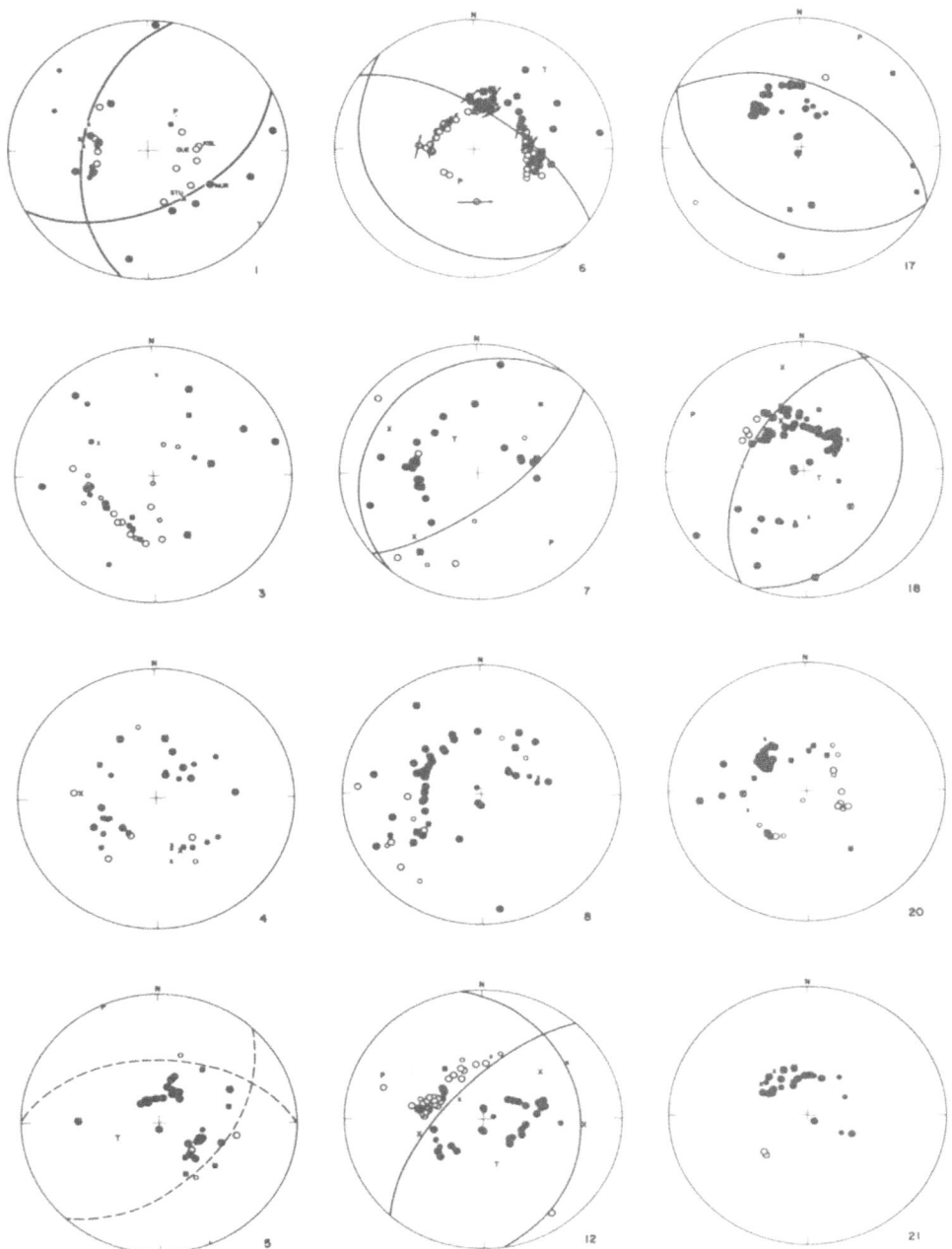

Fig. 1a. Focal mechanism solutions 1 to 21.

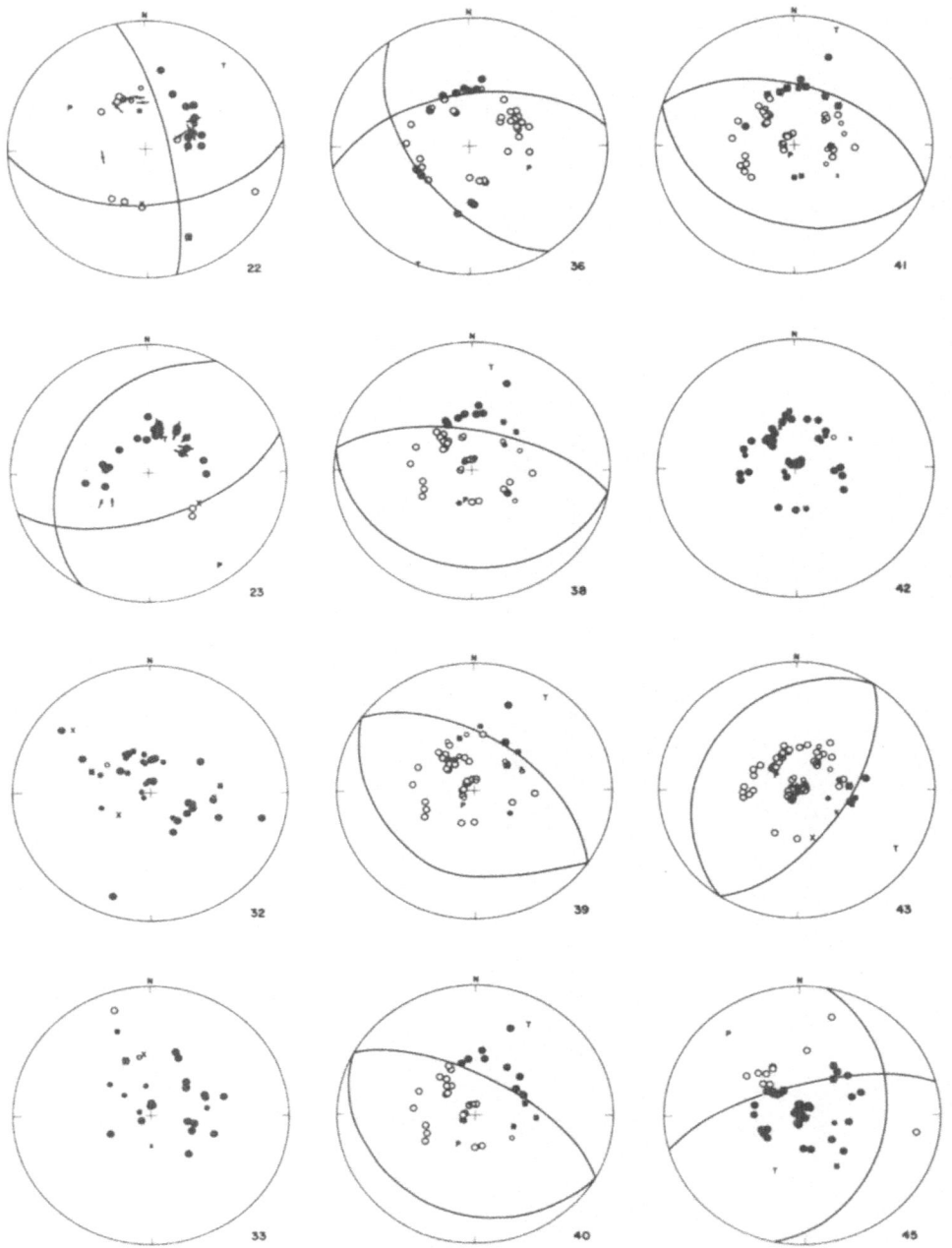

Fig. 1b. Focal mechanism solutions 22 to 45.

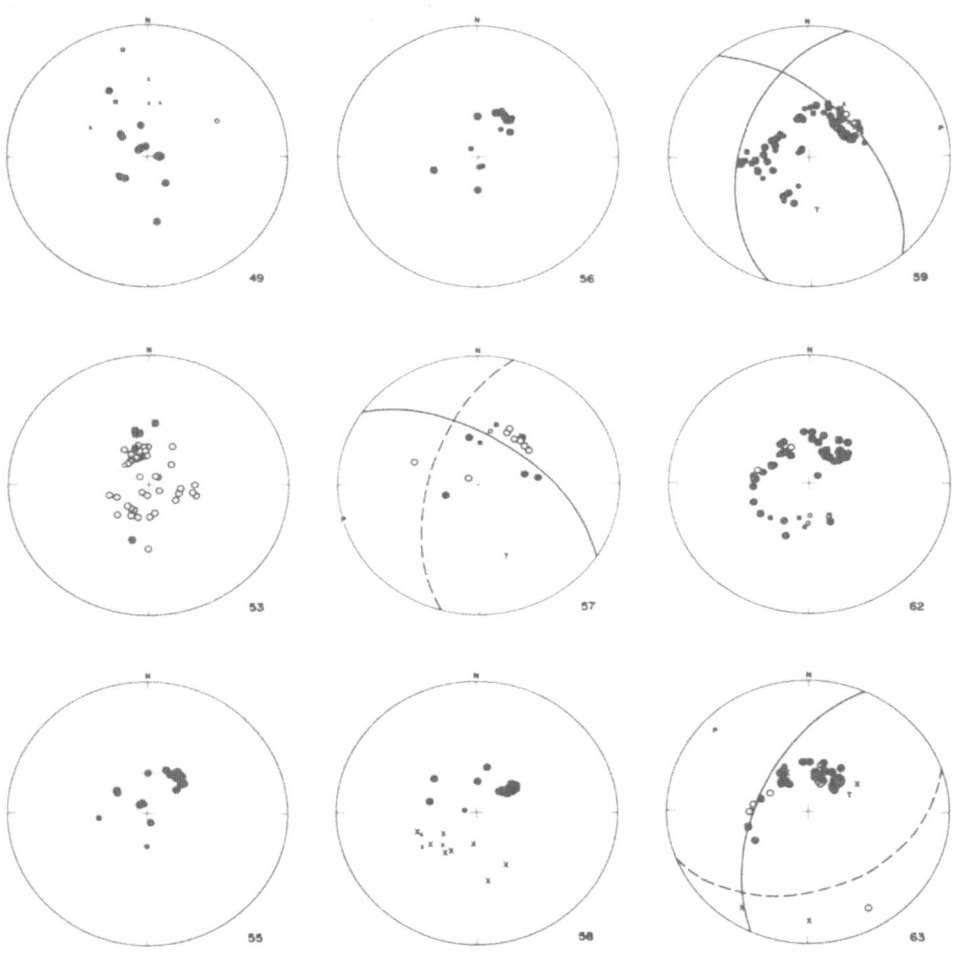

Fig. 1c. Focal mechanism solutions 49 to 63. Plot is an equal-area
projection of lower hemisphere of radiation field. Solid symbols
are compressional first motions; open symbols are dilatations;
X's are stations inferred to be near a nodal plane from their
signal character. Larger symbols denote more reliable readings.
Circles and X's are readings made by us; squares are data from
seismic bulletins. Arrows are first motions of S waves. P and T
are inferred axes of compression (maximum compressive stress)
and tension (least compressive stress) respectively. Numbers refer
to events in Table 1.

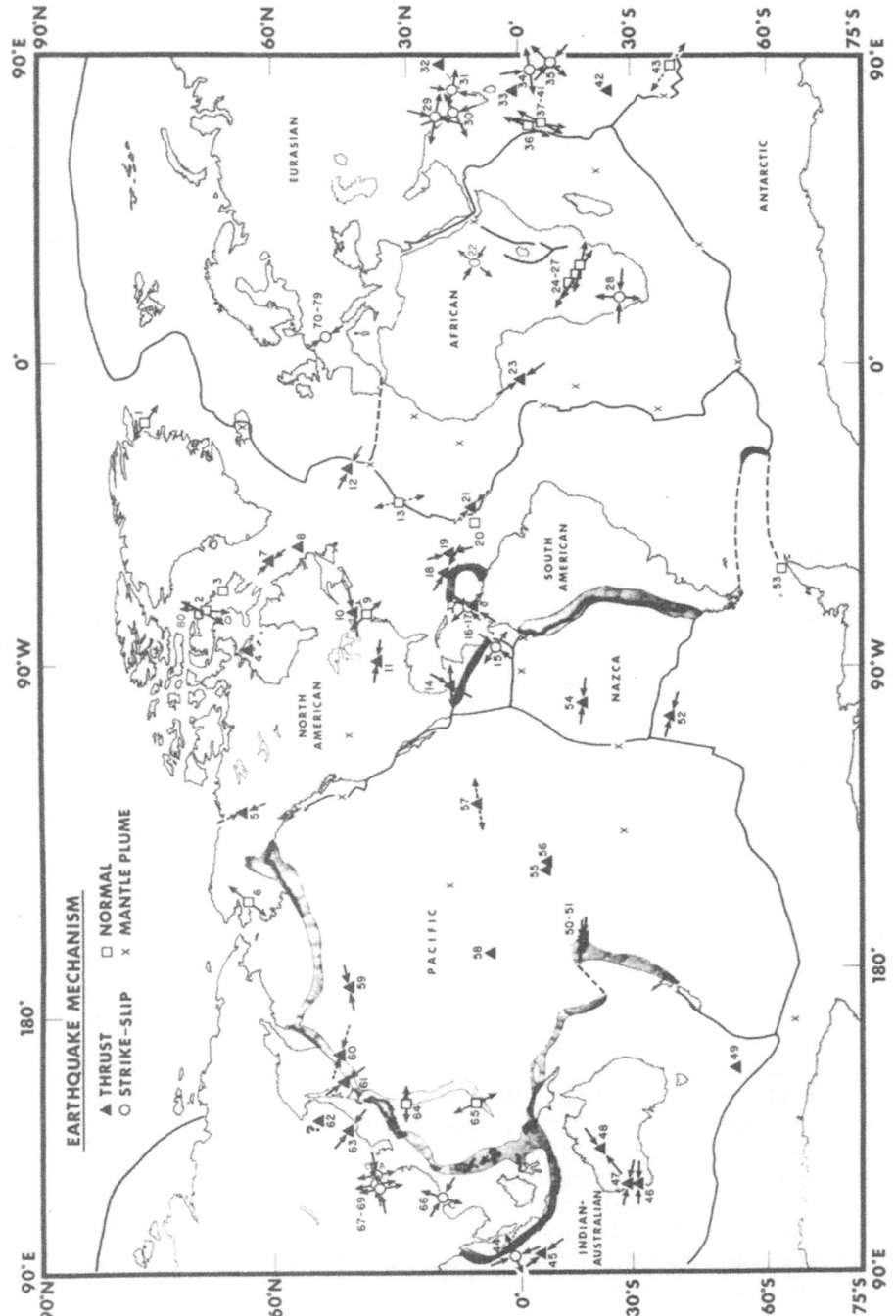

Fig. 2 (opposite). Worldwide summary of focal mechanisms of intraplate earthquakes. Numbers refer to events in Table 1. X's are
mantle plumes from [26]. Inwardly directed arrows indicate axis
of maximum compression (P axis) and outwardly directed arrows denote least compressive stress (T axis). Dashed arrows indicate
less reliable directions. Major plates are labelled. Hatching
indicates major subduction zones (Mediterranean-Himalayan zone
omitted), and solid line represents plate boundaries of extensional
and transform type. Solution in western Europe is representative
of about 10 mechanisms with similar orientation of P axes.

3.1 Oceanic Areas.

It should be remembered that while 80 solutions were sufficient
to provide at least a few observations for the interiors of most
of the major lithospheric plates of the world, the density of observations is still small. Nevertheless, a fairly simple pattern
of fault type and of stress distribution appears to be emerging
for most oceanic areas.

    In the Pacific and Nazca plates each of the solutions is
characterized by a predominance of thrust faulting. Thus, the
largest compressive stress is nearly horizontal. In the Atlantic
the solutions for all events located well away from the Mid-Atlantic
Ridge also involve a predominance of thrust faulting. Events occurring within oceanic lithosphere younger than 10 to 20 m.y. involve
normal faulting (Fig. 3). A similar pattern to the Atlantic appears
to be present in the northeastern Indian Ocean except that events
located within older oceanic lithosphere involve either strike-
slip or thrust faulting.

    Fig. 3 indicates that only thrust and strike-slip faulting
(and not normal faulting) were found for events occurring within
oceanic lithosphere older than 20 m.y. While the boundary between
normal and thrust faulting is not precisely delineated in Fig. 3,
the data do suggest that the transition takes place at an age of
about 10 to 20 m.y., which is approximately the thermal time constant for the formation or the destruction of the lithosphere
[27-28]. Obviously, additional mechanism solutions obtained either
from distant stations or from ocean-bottom seismographs would help
to delineate this transition in lithospheric stress. It might also
be possible to measure stress in situ in the oceanic crust by the
hydrofracturing technique in some of the holes to be drilled by
the JOIDES deep-sea drilling project. It is not possible to tell
at present if the stress field obtained from focal mechanisms
applies to the entire oceanic lithosphere since the depths of most
of the earthquakes are not known accurately and all of the earthquakes may be confined to the upper few kilometers of the oceanic

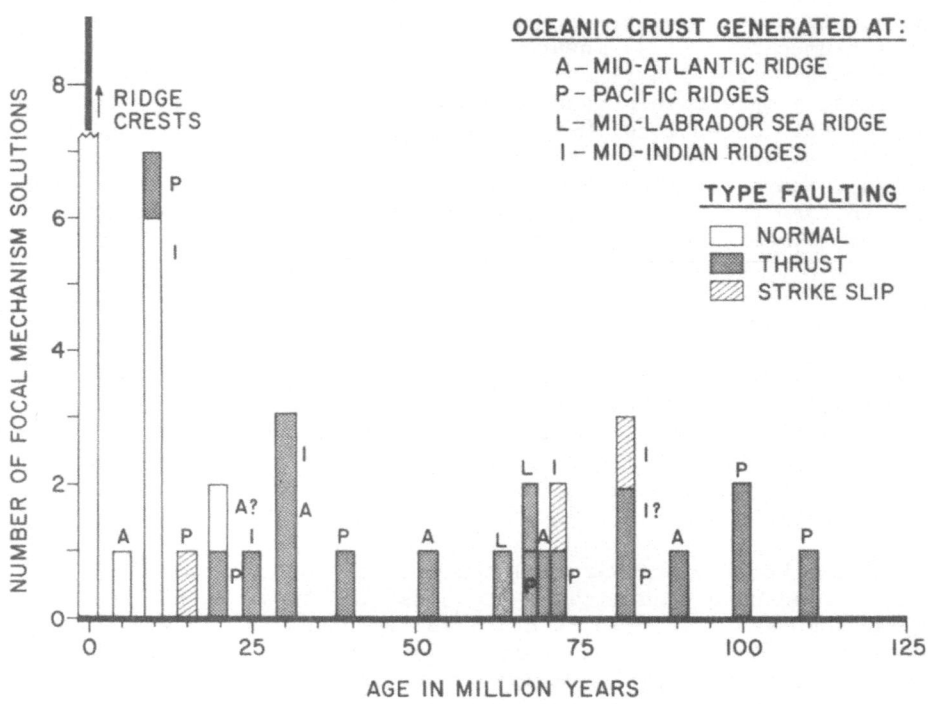

Fig. 3. Type of focal mechanism as a function of age of oceanic crust. Note predominance of normal faulting on or close to ridge crests and predominance of thrust faulting for ages greater than about 10 to 20 m.y.

crust. Hence, the focal mechanisms may only be indicative of the state of stress in the uppermost part of the oceanic lithosphere.

3.2.   Regions Behind Subduction Zones.

Each of the three main types of faulting occurs behind active subduction zones (Fig. 2). Nevertheless, a single stress pattern usually prevails behind a given subduction zone. Normal faulting is found behind the Bonin-Mariana (Fig. 2) and Tonga-Kermadec [29] arcs where inter-arc spreading either is occurring or has occurred very recently. Events located behind subduction zones were not included in the previous discussion of the state of stress in oceanic parts of plates since they probably are not typical of the stresses in the horizontal portions of the lithosphere. These events and possibly those in the Basin and Range province of the

U.S. may be described best as plate-boundary related events [49].

3.3 Continental Areas.

A greater variety of fault types and of directions of principal
stresses is found in continental areas than in oceanic areas
(Fig. 2). There are two quite different views emerging about the
uniformity of the stress field in continental areas.

From in-situ stress measurements by the overcoring technique
in Fennoscandia, Spitzbergen, Ireland, Portugal, France, Iceland,
British Columbia, Zambia and Liberia, Hast [30,31] concludes that
the earth's crust is generally characterized by horizontal compressive stresses that are much larger than those expected solely
from gravitational loading. Other overcoring measurements and
hydrofracturing experiments generally show similar results for a
large area in the eastern United States [25,33,34], the Colorado
Plateau [35,36], and parts of the U.S.S.R. [37,38]. Hast [32] concludes that the maximum compressive stress as measured by the
overcoring technique in mines does not change appreciably with
depth and that the direction of the stress field in areas like
Sweden, where it has been measured extensively, normally remains
nearly constant over large areas. Sbar and Sykes [25] showed that
the directions of maximum compressive stress inferred from overcoring, hydrofracturing, focal mechanisms and post-glacial faulting
generally agreed with one another for a large area in eastern North
America (Fig. 4). In this area they concluded that the maximum
compressive stress was oriented about ENE. Focal mechanism solutions, hydrofracturing and overcoring also give nearly identical
directions for the maximum compressive stress at Rangeley, Colorado
[35,36]. Focal mechanism solutions in the Basin and Range Province
of the United States [39] and hydrofracturing experiments at the
Nevada Test Site [40] indicate that the least compressive stress
in that province is nearly horizontal and trends WNW to NW. Focal
mechanisms [41] and evidence from the orientation of stylolites
[46] indicate that the maximum compressive stress is nearly horizontal and trends northwesterly in a large area of western Europe
north of the Alps.

Street et al. [42] concluded that there is a wide variation
in focal mechanism solutions for earthquakes in the central United
States. Their solutions for earthquakes in southern Illinois and
in central Missouri are in agreement with the ENE trending pattern
of compressive stress inferred by Sbar and Sykes [25]. Solutions
for earthquakes along the New Madrid fault zone in the Mississippi
embayment, however, indicate normal faulting. This sense of motion
agrees with that inferred from geological evidence for Cretaceous
and Cenozoic crustal movements in the Mississippi embayment [43].
This evidence does not contradict the inference of Sbar and Sykes

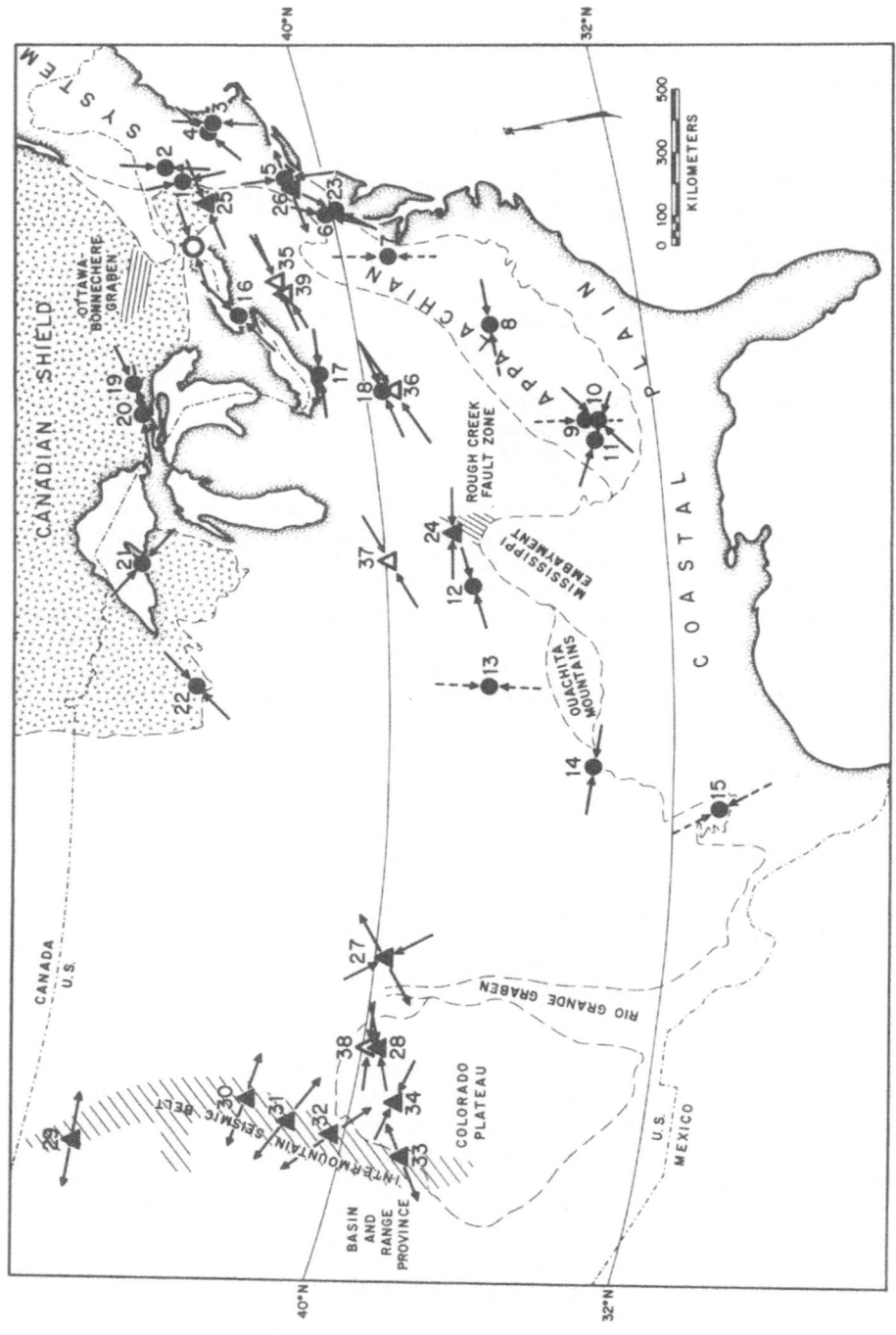

Fig. 4 (opposite). Map of part of North America with selected tectonic features showing fault plane solutions of earthquakes (solid triangles), strain relief in-situ stress measurements (solid circles), hydrofracture in-situ stress measurements (open triangles), and a "pop-up" near Chippewa Bay, New York (open circle). After Sbar and Sykes [25]. Strike of horizontal component of maximum or minimum compressive stress is shown at each locality. Arrows denoted by dotted lines are less reliable. Numbers refer to table in [25].

---

that the maximum compressive stress is nearly horizontal and is oriented about ENE in the area extending from west of the Appalachian Mountain system to the middle of the continent and from southern Illinois to southern Ontario. Nevertheless, this presumed uniform pattern obviously does not appear to include the Mississippi embayment, which Burke and Dewey [44] interpret as a failed rift (aulacogen) that extended north from a triple junction located near the mouth of the Mississippi River.

Very little published data on the state of stress are available for the region between the Mississippi River and either the Colorado Plateau or the intermountain seismic belt (Fig. 4). Hence, it is difficult to ascertain the state of stress in this area.

Sbar and Sykes [25] found that the observed directions of principal stress in the Appalachian region and near the adjacent continental margin were not the same as those farther west (Fig. 4). Thus, the region of ENE-directed compressive stress inferred by them appears to be delimited on the southwest by the Mississippi embayment and on the east by a different, but as yet poorly defined, stress regime. Similarly, a region of easterly directed compressive stress on the Colorado Plateau (Fig. 4) appears to be bounded on the east and west by an extensional stress regime in the Basin and Range Province and along the Rio Grande Graben [25]. Obviously, many new measurements of stress are needed to ascertain whether principal stresses are, in fact, nearly uniform in orientation over broad areas and if so, how the stress regime changes from region to region.

4. SOURCE OR SOURCES OF STRESS FIELD

Sources of the stress field in lithospheric plates can be divided into three broad classes: stresses related to the driving mechanism of plate tectonics, those related to the cooling of the lithosphere (Fig. 5, lower right), and stresses remaining in the lithosphere from geological processes that are not longer active.

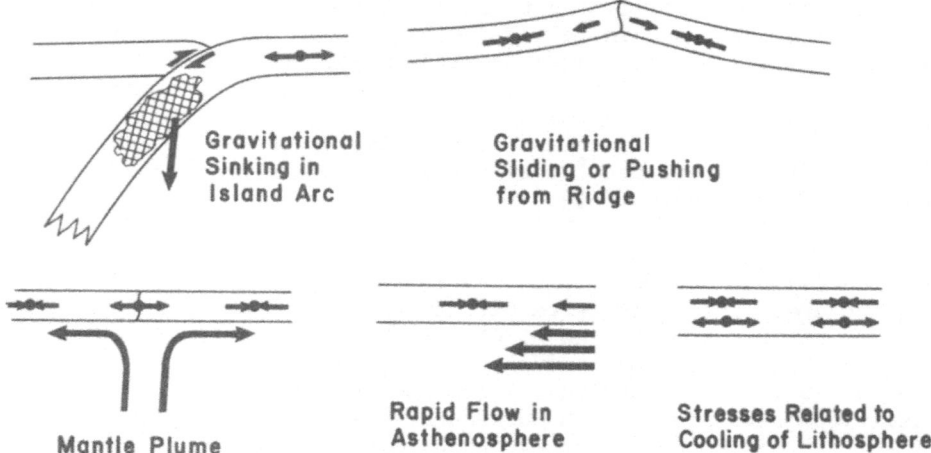

Fig. 5. Schematic diagrams of various mechanisms for generating
stresses in lithospheric plates. Single arrows denote relative
motions and double arrows denote either compressional (inward) or
tensional (outward) deviatoric stresses. Hatching in upper left
denotes mass excess caused by either cooler material or increased
elevation of phase boundaries in sinking slab.

4.1 Ancient Stress Fields Persisting to Present.

Some authors propose that present-day stresses in eastern
North America were generated by tectonic activity in the Pre-
Cambrian or Paleozoic. Dikes of Mesozoic age in western New York
[25], however, appear to have originated in response to an extensional stress field trending nearly the same as the present-day
directions of maximum compressive stress. Hence, the present
stress regime in that area appears to be post-Mesozoic in age and
not to be a remnant stress field from an older geologic era.

4.2 Thermo-Elastic Stresses.

Clearly, the state of stress within the oceanic lithosphere has
changed from horizontal extension to horizontal compression within the past 10 to 20 m.y., as it moved away from the ridge crest
where it was generated. It does not appear to be a coincidence
that this amount of time is also nearly identical to the thermal
time constant for the formation of the lithosphere at spreading
ridges and of its reabsorption by the mantle at subduction zones
[27,28]. The gradual cooling and thickening of the oceanic lithosphere as it moves away from a ridge crest may generate a thermo-

elastic or "frozen in" stress which may account for the observed types of faulting and the distribution of stress in Fig. 2. The upper most part of the oceanic lithosphere cools to an ambient temperature very soon after it is generated. New hot material added to the base of the lithosphere would cool with time and would place the uppermost part of the plate in compression as illustrated in the lower left part of Fig. 5.

According to this simple model of stress generation, the directions of maximum compressive stress should be approximately perpendicular to lines of constant age (isochrons) for the oceanic crust. The inferred directions of maximum compressive stress for solutions 12, 18, 52, 54, 57 and 60 in Fig. 2 are in general agreement with those predicted by this hypothesis. Solutions 50, 51, and 59 are also in approximate agreement, but the isochrons are either contorted or are not well determined for these areas [19]. The directions of compressive stress for solutions 34, 35, 44 and 45 in the northeastern Indian Ocean also are approximately perpendicular to isochrons inferred from magnetic anomalies and deep-sea drilling [20]. The directions of the maximum compressive stress for solutions 7, 19 and 23, however, appear to be more nearly parallel than perpendicular to isochrons although the isochron orientation near solutions 19 and 23 is not well defined. Also, it should be remembered that the azimuth of the horizontal principal stress (in this case axis of maximum compression) is one of the least well-determined parameters of these focal mechanism solutions. These azimuths need to be determined more precisely using the amplitudes of surface waves.

Overcoring measurements made by Hast [31] indicate that the maximum compressive stress in Iceland is nearly horizontal and is oriented nearly radially about the center of the island. It is difficult to reconcile these measurements with the idea that seafloor spreading is occurring in Iceland or with focal mechanism solutions of earthquakes which indicate that extensional tectonism oriented WNW is occurring either as normal or strike-slip faulting along the plate boundary that passes through Iceland [45]. It may be possible to interpret Hast's measurements of horizontal compression as a result of thermo-elastic stresses within lithospheric plates or as stresses related to a hot spot [26] rather than as compression along a plate boundary. Focal mechanisms of earthquakes located off the main plate boundary would be a means of checking these hypotheses.

This existence of tensional deviatoric stress oriented nearly perpendicular to several continent margins that have opened by seafloor spreading since the Mesozoic (Fig. 2, solutions 1, 2, 9, 29, 30, 31 and 80) may be related to the slow cooling and subsidence of the oceanic crust near the margin. This mechanism, however, does not account for horizontal compression oriented nearly perpendicular to other continental margins as in solutions 46 and 47 in western Australia.

4.3 Stresses Generated by Driving Mechanism of Plate Tectonics.

The observed stresses within plates may also be generated totally or in part in response to the driving mechanism of plate tectonics (Fig. 5). The radial orientation of maximum compressive stress in Iceland as determined from overcoring measurements [31] agrees with the stress distribution proposed by Morgan [26] for a mantle plume or hot spot beneath Iceland that pushes the North America and Eurasian plates apart. Nevertheless, other mechanisms, such as thermo-elastic stresses generated by the cooling of the thick pile of volcanic material in Iceland, could also produce the observed stress distribution. Also, the directions of principal stresses inferred for other earthquakes (e.g., 18, 19, 23) in Fig. 2 do not correspond to those anticipated if the plumes proposed by Morgan (X's in Fig. 2) are the sole driving forces of the plates.

Likewise, the orientation of principal stresses for other focal mechanism solutions (i.e., 18, 19, 23, 34, 35, 44-47) do not agree with those inferred for a mechanism that involves only gravitational sliding or pushing of the plates from the present mid-oceanic ridges. The presence of large horizontal compressive stresses seaward of several deep-sea trenches as inferred from focal mechanisms (Fig. 2, solutions 45, 50, 51, 59 and 60) and from gravity and topographic anomalies [47] indicates that gravitational sinking at subduction zones is not the principal mechanism generating stresses within lithospheric plates. The observed stresses in the lithosphere, however, do appear to be consistent with viscous drag of the plates from below as illustrated in Fig. 5 (lower center). They do not appear to be compatible with changes of plate curvature and membrane stresses on an elliptically-shaped earth [48] as the sole mechanism generating lithospheric stress.

Thus, it is quite possible that a number of mechanisms may generate stresses within lithospheric plates. Focal mechanism solutions and in situ stress measurements should provide a constraint for estimating the relative importance of the various sources. For example, the compressional stresses within most oceanic parts of the lithosphere, whether they are generated thermo-elastically or by drag from below, appear to dominate that component of tensional deviatoric stress transmitted through the plate by graviational sinking at subduction zones. Thus, it is important to obtain an estimate of the magnitude of the stress in the oceanic lithosphere and to obtain more accurate measurements of the directions of the principal stresses.

ACKNOWLEDGEMENTS

We thank Drs. P. Richards and C. Scholz for reviewing the manuscript and V. Cormier and T. Johnson for providing two of the
unpublished focal mechanism solutions used by us. This work was
supported by the National Science Foundation under grant GA-37093X
and the Advanced Research Projects Agency of the Department of
Defense through the Air Force Cambridge Research Laboratories
under contract F19628-71-C-0245.

REFERENCES

1. L. R. Sykes, Bull. Seism. Soc. Am., 60, 1749, 1970.
2. M. L. Sbar, J. M. W. Rynn, F. J. Gumper and J. C. Lahr, ibid.,
    60, 1231, 1970.
3. M. L. Sbar, J. Armbruster and Y. P. Aggarwal, ibid., 62,
    1303, 1972.
4. W. Stauder and O. W. Nuttli, ibid., 60, 973, 1970.
5. A. R. Banghar, ibid., 62, 603, 1972.
6. H. K. Gupta, I. Mohan and H. Narain, ibid., 62, 47, 1972.
7. W. Stauder and G. A. Bollinger, ibid., 54, 2199, 1964; idem.,
    56, 1363, 1966.
8. V. Cormier, personal communication.
9. T. Johnson, personal communication.
10. L. R. Sykes, J. Geophys. Res., 72, 2131, 1967; idem., 75,
    5041, 1970.
11. A. R. Banghar and L. R. Sykes, ibid., 74, 632, 1969.
12. N. Maasha and P. Molnar, ibid., 77, 5731, 1972.
13. L. Ahorner, Geol. Rundschau, 61, 915, 1972.
14. T. Fitch, M. H. Worthington and I. B. Everingham, Earth
    Planet. Sci. Lett., 18, 345, 1973.
15. J. A. Mendiguren, J. Geophys. Res., 76, 3861, 1971.
16. D. Forsyth, Nature, 243, 78, 1973.
17. M. Hashizume, J. Geophys. Res., 78, 6069, 1973.
18. M. Katsumata and L. R. Sykes, ibid., 74, 5923, 1969.
19. R. L. Larson and C. G. Chase, Bull. Geol. Soc. Am., 83,
    3627, 1973.
20. J. R. Heirtzler et al., Science, 180, 952, 1973.
21. P. Molnar, T. J. Fitch and F. T. Wu, Earth Planet. Sci. Lett.,
    19, 101, 1973.
22. L. R. Sykes and M. L. Sbar, Nature, 245, 298, 1973.
23. T. Johnson and P. Molnar, J. Geophys. Res., 77, 5000, 1972.
24. P. Molnar and L. R. Sykes, Bull. Geol. Soc. Am., 80, 1639,
    1969.
25. M. L. Sbar and L. R. Sykes, ibid., 84, 1961, 1973.
26. W. J. Morgan, Bull. Am. Assoc. Petrol. Geol., 56, 203, 1972;
    Geol. Soc. Am. Mem., 132, 7, 1973.
27. B. L. Isacks, J. Oliver and L. R. Sykes, J. Geophys. Res.,
    73, 5855, 1968.

28. D. P. McKenzie, Geophys. J. R. Astr. Soc., 18, 1, 1969.
29. D. E. Karig, J. Geophys. Res., 76, 2542, 1971.
30. N. Hast, Tectonophysics, 8. 169, 1969.
31. N. Hast, Phil. Trans. R. Soc. Lond. A., 274, 409, 1973.
32. N. Hast, Modern Geology, 4, 73, 1973.
33. V. E. Hooker and C. F. Johnson, in: Proc. 4th Rock Mechanics
     Symposium, Pub. Dept. Energy, Mines and Resources, Ottawa,
     Canada, 1967.
34. B. Voight, Am. Assoc. Petrol. Geol. Mem., 12, 955, 1969.
35. C. B. Raleigh, J. H. Healey and J. D. Bredehoeft, Geophys.
     Monog. Am. Geophys. Union, 16, 275, 1972.
36. B. C. Haimson, in: 14th Symposium on Rock Mechanics, Am.
     Soc. Civil Eng., p. 689, 1973.
37. P. N. Kropotkin, Phys. Earth Planet. Inter., 6, 214, 1972.
38. I. A. Turchaninov et al., ibid., 6, 229, 1972.
39. R. M. Hamilton, B. E. Smith, F. G. Fischer and P. J. Poponek,
     Bull. Seism. Soc. Am., 62, 1319, 1972.
40. B. C. Haimson, in: 3rd Int'l Congress on Rock Mechanics,
     Denver, Colorado, in press, 1974.
41. L. Ahorner, Geol. Rundschau, 61, 915, 1972.
42. R. N. Street, R. B. Herrmann and O. W. Nuttli, Science, 184,
     1285, 1974.
43. R. G. Stearns and C. W. Wilson Jr., Relationships of Earth-
     quakes and Geology in West Tennessee and Adjacent Areas,
     Tennessee Valley Authority, Chattanooga, Tenn., 1972.
44. K. Burke and J. F. Dewey, J. Geol., 81, 406, 1973.
45. F. W. Klein, P. Einarsson and M. Wyss, J. Geophys. Res.,
     78, 5084, 1973.
46. K. Schäfer, Trans. Am. Geophys. Union, 55, 443, 1974.
47. A. B. Watts and M. Talwani, Geophys. J. R. Astr. Soc., 36,
     57, 1974.
48. D. L. Turcotte, ibid., 36, 33, 1974.
49. C. Scholz, M. Barazangi and M. L. Sbar, Bull. Geol. Soc. Am.,
     82, 2979, 1971.

SEISMICITY OF ICELAND*

Sveinbjörn Björnsson   and   Páll Einarsson

Science Institute,   Lamont-Doherty Geological
University of Iceland,   Observatory and Department
Reykjavik, Iceland.   of Geology, Columbia
                                University, N.Y., U.S.A.

ABSTRACT. Earthquakes in Iceland located teleseismically are mainly confined to a zone off the northern coast (the Tjörnes Fracture Zone), a narrow E-W zone in S-Iceland (including the Reykjanes Peninsula), and the eastern volcanic zone. Destructive historic earthquakes of magnitude up to 7 or 8 are known to have occurred in these zones, except in the eastern volcanic zone and the western part of the Reykjanes Peninsula. Local instrumental observations show that significant seismic activity occurs in other parts of the country. Smaller earthquakes and earthquake swarms occur frequently on the Reykjanes Peninsula and in the western and the eastern volcanic zones.

Detailed investigations of the seismicity of the Reykjanes Peninsula have been carried out since 1971. The mid-Atlantic plate boundary can be traced as a narrow zone of epicenters extending from the southwestern tip of the peninsula 40 km eastwards along the volcanic zone. Most of the earthquakes occur at 2 to 5 km depth and to date no earthquake has been located in seismic layer 4 (V_p = 7.2 km/sec) lying at a depth of 8-9 km under the region. Focal mechanism solutions on the Reykjanes Peninsula consistently have a horizontal tensional axis with a NW trend. The compressional axis varies between the vertical and horizontal direction giving rise to both normal and strike-slip faulting.

Seismic activity in other parts of Iceland is not as well known in detail but strike-slip earthquakes seem to prevail in the zones of large earthquakes in S-Iceland and off the northern coast.

---

*Lamont-Doherty Geological Observatory contribution 2152.

Kristjansson (ed.), Geodynamics of Iceland and the North Atlantic Area. 225-239. *All Rights Reserved.*
Copyright © 1974 *by D. Reidel Publishing Company, Dordrecht-Holland.*

Volcanic eruptions are generally accompanied by earthquakes. During the Heimaey eruption in 1973 earthquakes were found at a depth of 20-30 km or well within the anomalous upper mantle.

1. INTRODUCTION

Studies of earthquakes have played an important role in defining plate boundaries in the oceans. If Iceland were under deep ocean water, the distribution of epicenters and focal mechanism solutions of teleseismically located earthquakes would simply suggest that Iceland was a spreading ridge segment bounded by two transform faults in the north and south (Fig. 1). To observers in Iceland, however, this would be a gross simplification, obscuring the real tectonic picture. There are two main zones of rifting and volcanism instead of the expected one and the arrangement of fissures and faults not only suggests tensional dip-slip motion but also strike-slip motion along some parts of the volcanic zones. On the Reykjanes Peninsula, most of the volcanic fissures and surface faults have north-easterly strike and are arranged en echelon in a zone with a trend of N 75° E. Earthquake studies by Klein et al. [1] indicate that this zone is an obliquely diverging plate boundary with both ridge and transform fault characteristics. In S-Iceland where earthquake epicenters and focal mechanism studies suggest an E-W transform fault connecting the western and eastern volcanic zones [2] surface fractures associated with larger earthquakes in the past indicate right-lateral strike-slip motion on northerly trending faults. The seismic zone north of Iceland has been interpreted as a transform fault zone joining the Kolbeinsey Ridge and the northern volcanic zone [2,3]. According to Saemundsson [4] this zone is composed of several N-S trending subsiding troughs and volcanic chains arranged in a 75 km wide zone. On the south side this zone is bounded by a fault swarm of WNW trending right-lateral faults with some dip-slip component.

These examples show that the tectonics of Iceland is not as simple as suggested for the Mid-Atlantic Ridge system in general. Some complexity may be due to the superposition of a plume of uprising mantle material under Iceland but one has also to consider that Iceland is the only part of the system exposed to direct observations and comparable detailed investigations of the submarine portion of the ridge system may reveal similar complexities.

The position of Iceland across the Mid-Atlantic plate boundary provides a unique opportunity to study processes of crustal accretion and deformation, and to interpret these processes in terms of the tectonic stress regime near the crest of a mid-ocean ridge. In a detailed study of this kind, earthquakes will play a major role as in the investigation on global scale.

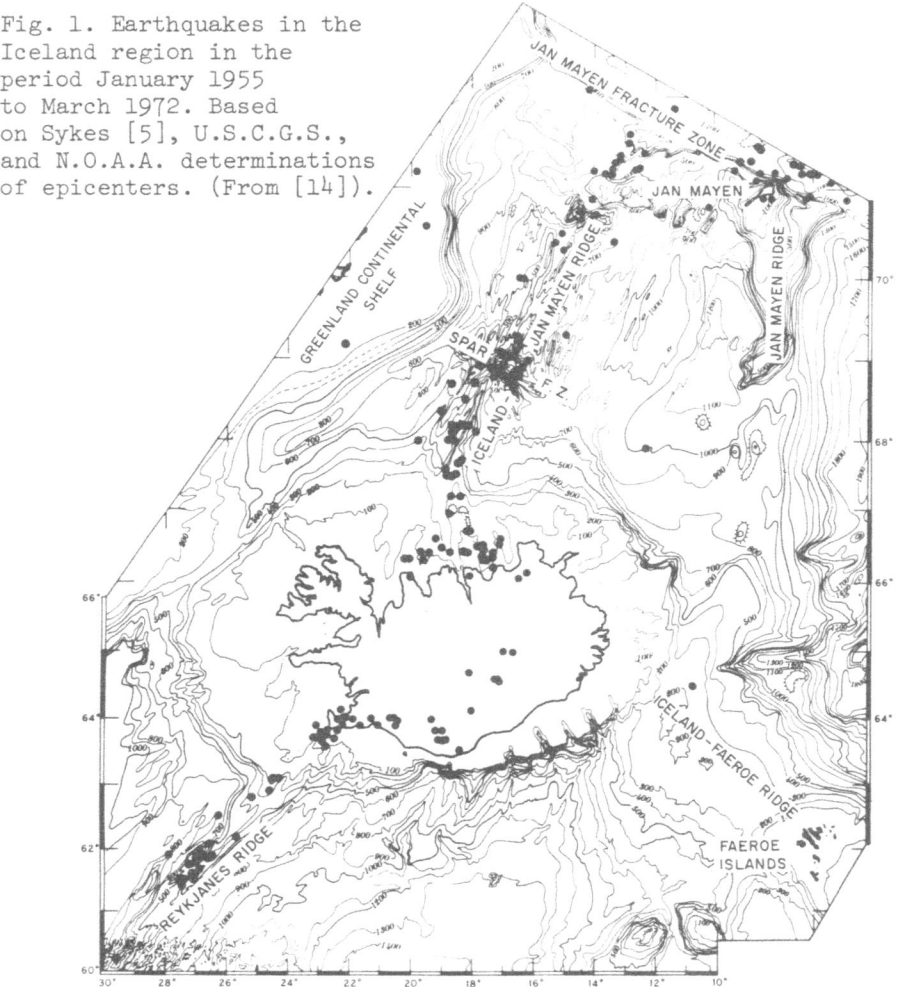

Fig. 1. Earthquakes in the
Iceland region in the
period January 1955
to March 1972. Based
on Sykes [5], U.S.C.G.S.,
and N.O.A.A. determinations
of epicenters. (From [14]).

The long term seismicity of Iceland is fairly well known
from historical data and instrumental records in this century.
Local stations have, however, been so few that epicentral locations may be up to 20 km in error and depth of earthquakes is
shallow but not known. Although useful on large scale, these
data are not sufficient for detailed studies. In order to obtain
better data a dense network of short period stations is presently
under construction in Iceland. Work with this network and a
number of portable stations has already demonstrated its validity
for the investigation of tectonic processes on the Reykjanes
Peninsula [1] and it is expected that details of the plate boundary
within Iceland can be delineated with the aid of this network
within the next decade.

The purpose of this paper is to review the available evidence
on the seismicity of Iceland, the long term record based on
historical accounts and instrumental observations in this century,
activity recorded by local permanent stations and microearthquake
activity mapped by portable stations. Focal mechanism solutions
and their tectonic implications are discussed but the data are still
fragmentary. Better data will be available in the near future.

2. DISTRIBUTION OF LARGER EARTHQUAKES

The epicenters of earthquakes which occurred in the Iceland region
during the period 1955-1972 are shown in Fig. 1. These locations
are taken from Sykes [5] and the Preliminary Determinations of
Epicenters by the U.S.C.G.S. and later N.O.A.A. and include most
earthquakes larger than about 4.3 in magnitude. Generally the
accuracy is regarded to be better than 30 km. In the ocean, seismic activity appears to be confined to the crest of the Reykjanes
Ridge in the southwest and the Kolbeinsey Ridge (or the Iceland-
Jan Mayen Ridge in Fig. 1) north of Iceland. Earthquakes in Iceland
located teleseismically are mainly confined to a zone off the
northern coast, a narrow E-W zone in southern and southwestern
Iceland and the eastern volcanic zone. No earthquakes have been
located in the northern volcanic zone, the western volcanic zone
or the Snaefellsnes zone (location, Fig. 2) during this period.

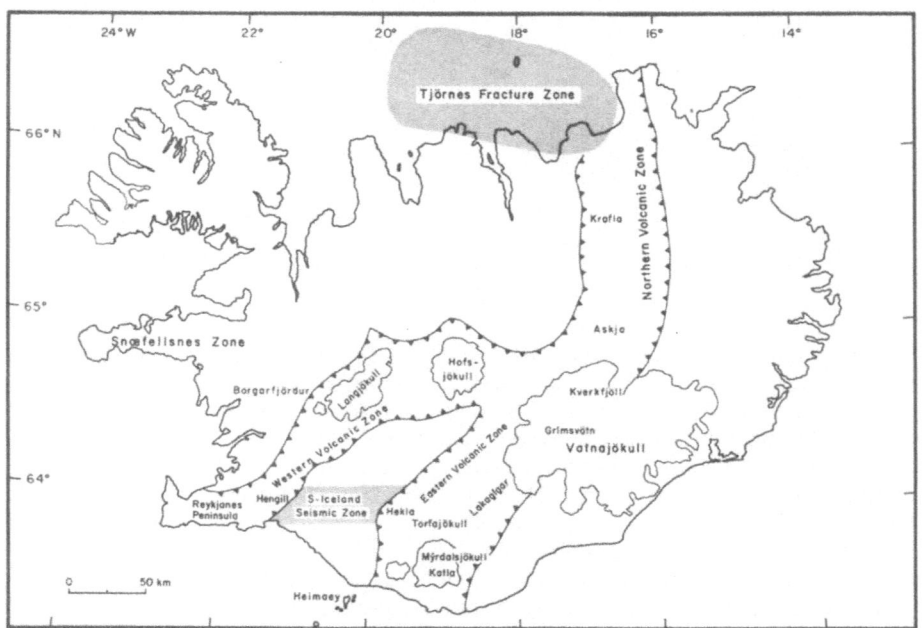

Fig. 2. Index map showing the relationship between the volcanic
and the seismic zones of Iceland.

Tryggvason [6] reviewed the seismicity of Iceland, mostly on the basis of local observations in the period 1929–1963 and historical accounts on destructive earthquakes. His epicentral locations for the years 1929–1963 are shown in Figs. 3 and 4. The locations are largely based on recordings of Icelandic seismograph stations. The accuracy of the epicenter locations is usually better than 20 km, although some epicenters off the north coast and on the Reykjanes Ridge may be in error by a larger amount. The magnitudes are based on a local magnitude scale, which was constructed to fit the surface wave magnitude scale of European seismological stations as closely as possible.

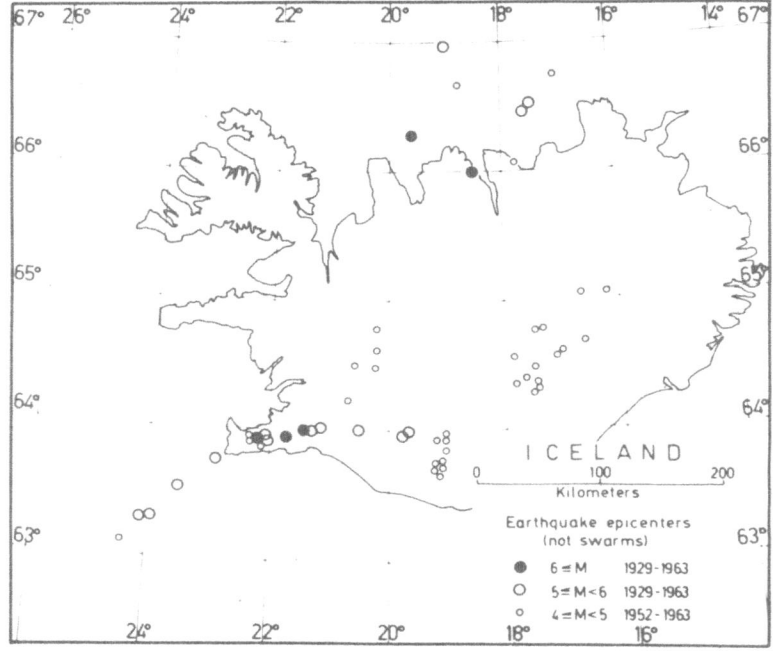

Fig. 3. Location of earthquakes, which do not belong to swarms, in the Iceland area from 1929 to 1963 [6].

The data on earthquakes which do not belong to swarms (Fig. 3) are complete for the magnitudes and time intervals shown, but a number of earthquake swarms (Fig. 4) of the lower magnitude intervals may not have been detected.

Tryggvason [6] has also estimated the magnitude and locations of major destructive earthquakes in Iceland since 1700 A.D. (Fig. 5). The amount of destruction is fairly well documented and the areal extent of destruction was used to obtain a crude esti-

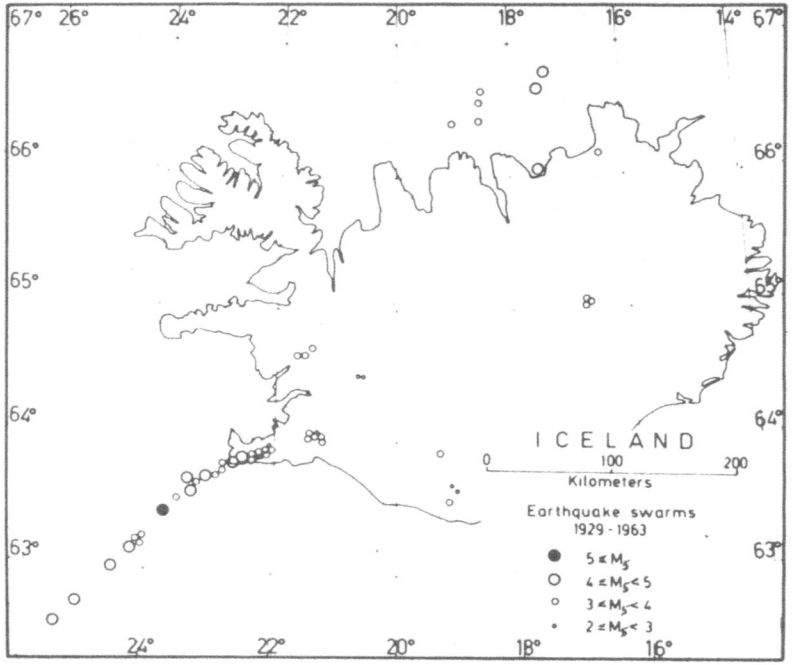

Fig. 4. Location of earthquake swarms in Iceland, 1929-1963. M_5 is the magnitude of the fifth largest earthquake of each swarm [6].

mate of the earthquake magnitude. Earthquakes occurring after 1900 were located with the aid of seismic stations in Iceland and abroad, but some of the epicenters were modified to fit macroseismic data. All earthquakes since 1900 exceeding magnitude 6 are believed to be included.

The data presented in Fig. 1 and Figs. 3 to 5 clearly indicate that relatively large earthquakes have occurred only in two well-defined areas in Iceland, one in southwest Iceland along a line from 64°N, 19.5°W to 63.9°N, 22.2°W, and another near the north coast of Iceland between 66°N and 67°N, 16°W to 20°W. The epicenters in south Iceland are concentrated in a very narrow zone, probably less than 20 km wide, while the epicenters off the north coast appear to be distributed over a much wider zone, possibly 100 km wide from north to south. Outside these two seismic zones, no earthquake of magnitude exceeding 5 was known to have occurred in Iceland until June 1974 when a swarm of earthquakes in Borgar-fjördur (Fig. 2), W-Iceland culminated in an earthquake of m_b =5.5.

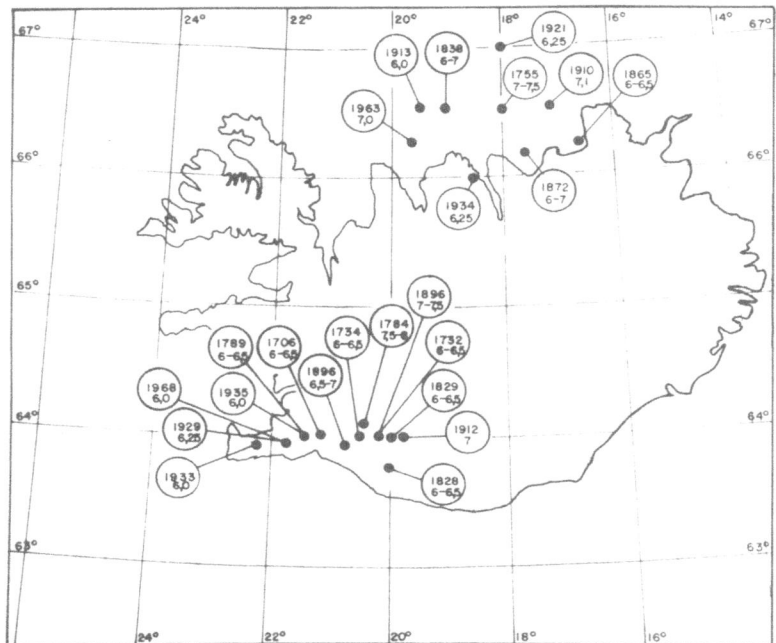

Fig. 5. Estimated magnitudes and locations of major destructive earthquakes in Iceland since the year 1700. Epicenters are shown as black dots. Encircled numbers indicate the year of occurrence and the estimated surface wave magnitude of the earthquake. Drawn after data given by Tryggvason [6].

The epicenter was near the eastern termination of the WNW trending fissure system of the Snaefellsnes volcanic zone (64.7°N, 20.3°W) (Fig. 2).

Local observations for the period 1929-1963 are in good agreement with the teleseismic locations for the period 1955-1972. Outside the seismic zones defined above earthquakes between 4 and 5 in magnitude are found in the western volcanic zone and the eastern volcanic zone. Activity in the eastern zone appears to correlate with regions of the central volcano complexes in the Torfajökull and Mýrdalsjökull region in the south, in the north-western part of the Vatnajökull ice sheet, including the volcanoes Grimsvötn and Kverkfjöll, and in the Dyngjufjöll region north of Vatnajökull with the Askja caldera (See Fig. 2 for locations). No activity is seen in the region of fissure eruptions between Mýrdalsjökull and Vatnajökull nor in the volcanic fissure zone north of Askja north to the seismic zone at 66°N.

3. EARTHQUAKE SWARMS

Most of the seismic events on the Reykjanes Ridge nearest to Iceland and on the Reykjanes Peninsula occur within earthquake swarms. The epicenter zone on the Peninsula is probably less than 10 km wide, coinciding with the zone of recent volcanic activity. Swarms are dominant in the west, giving away gradually to isolated larger earthquakes as the zone of major earthquakes is approached at 22°W. Swarms are frequent in the Hengill region (64.0°N, 21.3°W) at the triple junction of the Reykjanes volcanic zone, the western volcanic zone and the seismic zone of S-Iceland. Local observations in the period 1929-1963 also show swarm activity in the Torfajökull and Myrdalsjökull region, near Askja in the eastern volcanic zone, and at southwestern Langjökull in the western zone. A few swarms have occurred in the Borgarfjördur area between the Snaefellsnes zone and the western zone.

Earthquake swarms are frequent in the seismic zone off the north coast and some of the greater earthquakes there had numerous large earthquakes preceding the greatest event of the sequence [7].

In the seismic zone of S-Iceland swarms appear to be rare. The typical activity is rather that of large earthquakes followed by numerous aftershocks. In some of the major sequences the first earthquakes have originated near the eastern end of the seismic zone, followed by earthquakes further west in the next days or weeks until the whole zone from Hekla to Hengill was activated. The best recorded sequence of this kind occurred in the year 1896. Fig. 6 shows the areas of destruction and the time sequence of the major earthquakes. A detailed description of these earthquakes is given by Thoroddsen [7]. The magnitudes of the two largest events were estimated by Tryggvason [6] to be 7-7.5 and 6.5-7.

A similar migration of earthquake activity has been noted in conjunction with several eruptions of Hekla, e.g. the eruptions of 1597, 1766 and 1947, which were accompanied by fairly strong earthquakes and followed by destructive earthquakes in the western half of the seismic zone a few months later [8].

4. SURVEY OF MICROEARTHQUAKES

A reconnaissance survey of microearthquakes in Iceland was initiated by the Lamont-Doherty Geological Observatory of Columbia University, U.S.A. in 1967 and continued in cooperation with the National Energy Authority and the Science Institute of the University of Iceland in the summers of 1968, 1969 and 1970. Portable seismographs were operated at about 250 sites throughout most of Iceland, recording for two days at each site. The distribution of recording sites and the active zones of microearthquakes

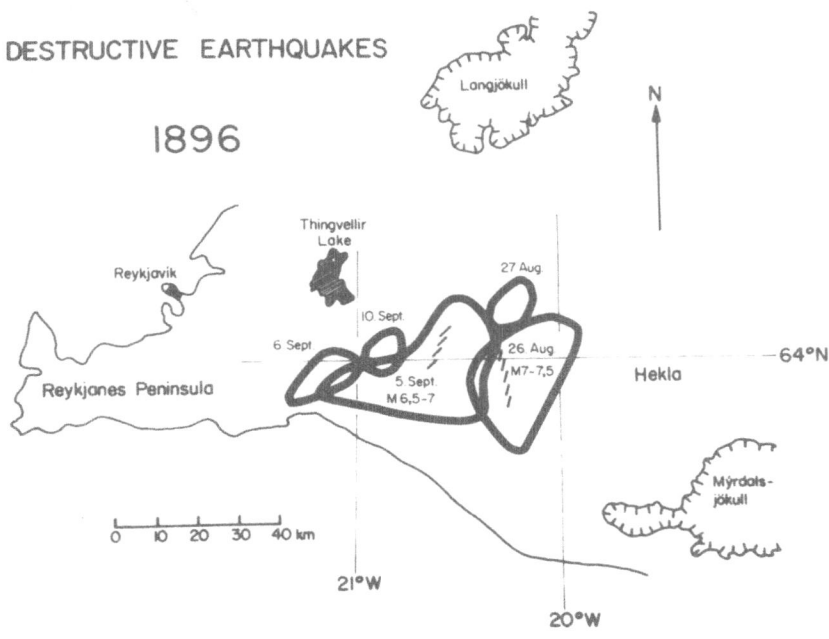

Fig. 6. A sequence of destructive earthquakes in S-Iceland in
the year 1896. Surface fractures of the two largest earthquakes
are indicated within the areas of destruction. Based on Thoroddsen [7]. Magnitudes of the two largest events were estimated by
Tryggvason [6].

of magnitude larger than -1 are shown in Fig. 7 [9,10]. The
observed microseismicity correlates well with the distribution of
earthquake swarms discussed above. The microearthquakes are mostly confined to the Reykjanes Peninsula, the western and the
eastern volcanic zone. There is a notable close correlation between
the active zones of microearthquakes and the major geothermal
areas. Nine of the zones found in the reconnaissance survey coincide
with high temperature geothermal activity. Later work by Conant [11]
and unpublished data of the authors (1970-1974), has confirmed that
most high temperature geothermal fields have fairly consistent
microearthquake activity of 3 to 30 events larger than magnitude
-1 recorded per day. In other parts of the volcanic zones activity
appears more sporadic with swarms of earthquakes lasting from a
few weeks up to a few months. The magnitude of these earthquakes
is generally less than 3 and often the swarms have therefore not
been observed clearly by the permanent network although sensitive

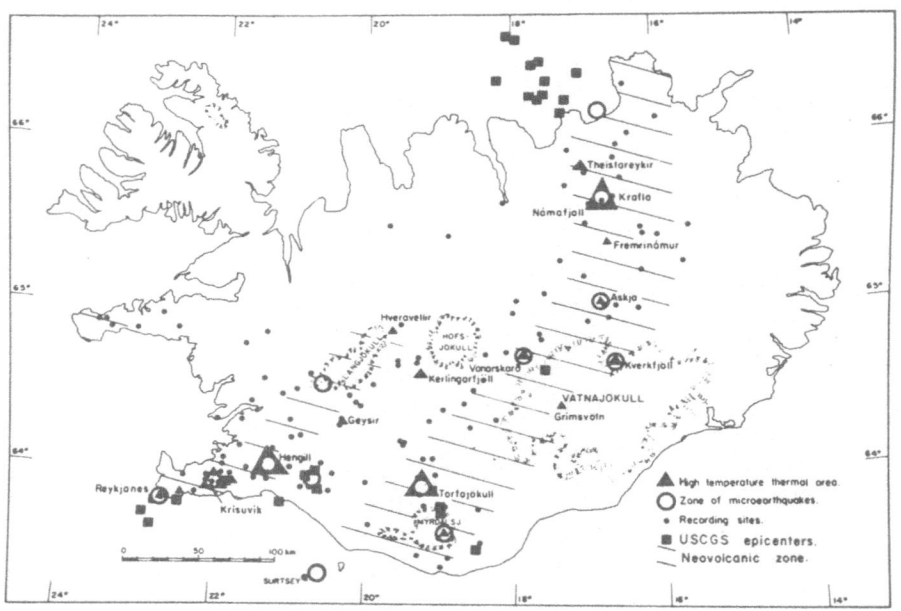

Fig. 7. Summary of regional microearthquake studies in the summers
1967 through 1969. The recording sites are shown as black dots,
epicentral zones of microearthquakes as open circles. Black triangles are high temperature thermal areas in the active volcanic
zones. Epicenters reported by U.S.C.G.S. during the period are
shown as black squares. (From [10]).

seismographs brought into the epicentral area have recorded up to
1000 events per day. A detailed study of microearthquakes was
carried out in the summer of 1968 with two tripartite arrays
operated at the geothermal areas Krisuvik on the Reykjanes Peninsula and Hveragerdi on the southern border of the Hengill region
[10]. Earthquakes located with a precision better than 1 km in depth
and distance showed close spatial correlation with regions of high
temperature and thermal alteration at the surface. Most of the
well-located earthquakes occurred at 2 to 6 km depth in the uppermost part of layer 3, a crustal layer with P-wave velocity of about
6.5 km/sec. These were the first well-determined depths of seismic
activity in Iceland and, for that matter, along most of the mid-
ocean ridge system.

      After the reconnaissance survey of microearthquake activity
in Iceland, the study was concentrated on the Reykjanes Peninsula,
which showed the highest seismic activity. A network of six permanent stations was established in 1971 and activity within the

network was investigated in greater detail by arrays of portable
stations in the summers of 1971 and 1972. The first results of
this work were published by Klein et al. [1]. The normal level
of microearthquake activity was found to be about 20 events per
day. The spatial pattern of microearthquake seismicity as observed in this study is essentially a linear zone following the trend
of the volcanic zone from the southwest tip of the Peninsula about
40 km eastward to Lake Kleifarvatn. The zone of brittle deformation,
as revealed by a sample of 300 hypocenters located with a typical

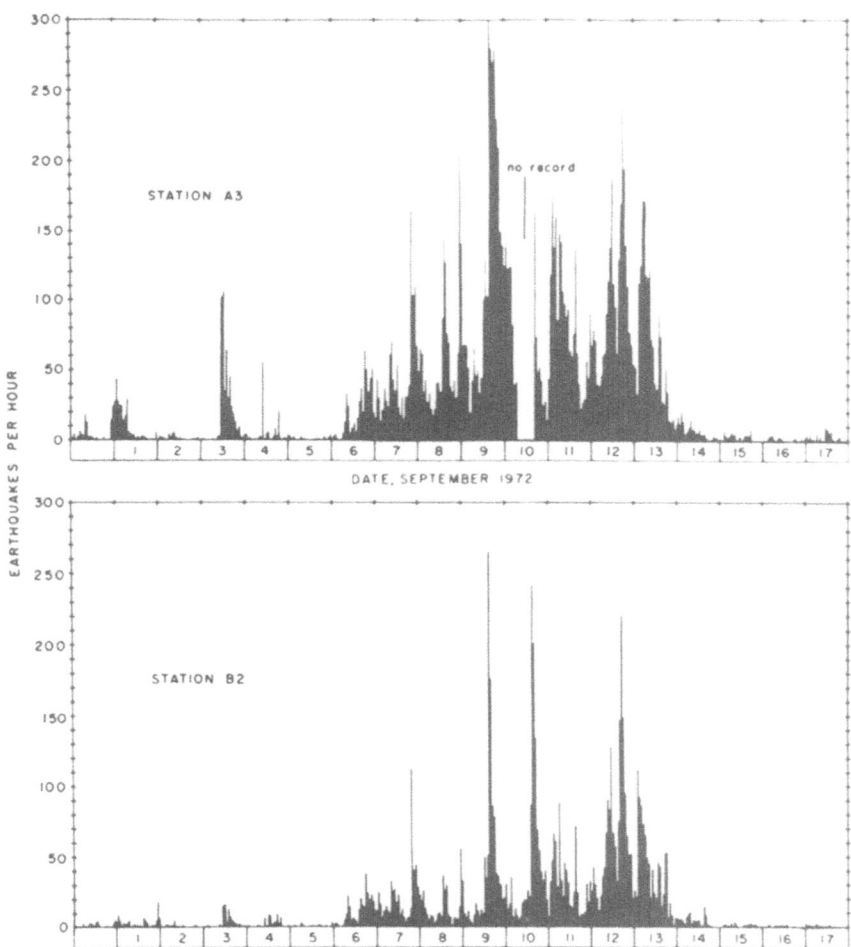

Fig. 8. Microearthquake rates showing swarm activity in the
Reykjanes area in September 1972. The recording stations A3 and
B2 were in the epicentral area, 9 km apart. Each line represents
counts for 1 hour. Earthquakes included in the count have peak
trace amplitudes exceeding 5 times the average noise level for
each station. (From Klein et al. [1]).

accuracy of 0.6 km, is about 2-5 km wide and 40 km long. Ninety
percent of the located earthquakes had depths between 1.5 and
5 km. Swarms account for most of the microearthquake activity
observed. An example of a major swarm is given in Fig. 8, which
shows earthquake rates as recorded at two stations in September,
1972 [1]. The swarm lasted about 8 days, but its structure in
time is complicated in detail as indicated in the figure. Most
prominent is the grouping into peaks of activity of subswarms
with duration of a few hours. The subswarms emanate from various
source regions, and so the relative counting rates at the two
stations vary. Some subswarms show the gradual rise and decay
characteristic of swarm sequences, whereas others appear like
main shock-aftershock sequences.

5. FOCAL MECHANISM OF EARTHQUAKES

Few earthquakes in Iceland have been recorded so widely that a
reliable focal mechanism solution could be obtained. Stefánsson
[12] and Sykes [3] independently examined a magnitude 7 earthquake in Skagafjördur north of Iceland. They found a strike-slip
solution with nearly vertical nodal planes. The axis of least
compression had a trend of N 62° E and that of maximum pressure
N 27°W. The corresponding fault motion could either have been a
sinistral fault striking slightly east of north or a dextral fault
striking WNW. The former has a strike similar to the Kolbeinsey
Ridge and the northern volcanic zone; the latter correlates well
with the trend of epicenters in the Tjörnes seismic zone and was
suggested by Sykes [3] to indicate a transform fault between the
northern volcanic zone and the Kolbeinsey Ridge. This interpretation has been supported by Saemundsson [4] who demonstrated
evidence for right-lateral displacements on the WNW-trending
Húsavík fault. According to Saemundsson [4] these faults are the
southern boundary of the 75 km wide Tjörnes Fracture Zone which
is composed of several north-trending subsiding troughs and
volcanic chains arranged in an en echelon pattern.

Ward [2] obtained focal mechanism solutions for two earthquakes in S-Iceland. A magnitude 5.5 to 6.0 earthquake east of
Lake Kleifarvatn on the Reykjanes Peninsula gave a relatively well-
determined mechanism solution with one vertical nodal plane striking about N 87° E indicating left lateral movement, the other plane
striking about N 4° W and dipping about 75° E indicating movement
in the right lateral sense. Another earthquake of magnitude 5 in
S-Iceland indicated a similar mechanism but the solution is not
so reliable. Both earthquakes indicate that the least compressive
stress on the Reykjanes Peninsula and in S-Iceland is nearly
horizontal striking NW and the maximum stress is striking NE.
Ward [2] interpreted his solutions and the distribution of earthquakes in this area as evidence for a west-north-west striking

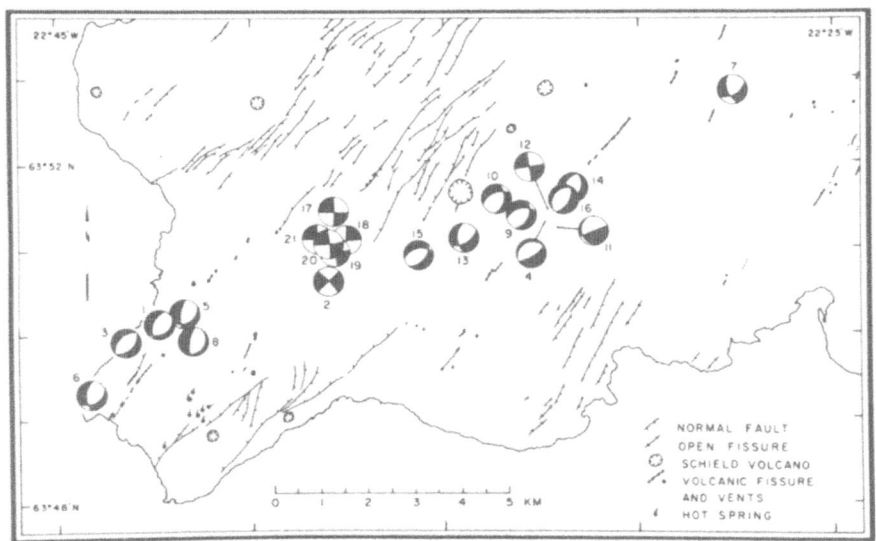

Fig. 9. Upper hemisphere focal solutions for 21 earthquakes in
the Reykjanes area on the Reykjanes Peninsula. The compressional
quadrants are shown in black, the dilatational quadrants in white.
Structural and geologic features are taken from Jónsson [15] and
unpublished maps by J. Jónsson. Younger lava flows mask older
faults. The density of fissures and faults seen at the surface
therefore depends on the age of the lava flow and is not indicative of the intensity of tectonic activity, (From Klein et al.
[1]).

transform fault zone. Surface evidence of such a fault is lacking.
A few of the larger earthquakes within this zone in the past, however, are known to have been associated with right-lateral strike
slip motion on northerly trending surface faults (Fig. 6) [7,6].

    Klein et al. [1] obtained focal mechanism solutions for
small earthquakes on the Reykjanes Peninsula (Fig. 9). The common
feature of the solutions is the horizontal NW trend of the least
compressive stress. The intermediate and maximum compressive stresses vary considerably over short distances, leading to faulting of
the normal and strike-slip types, often both types of solutions
occurring close to one another. The least compressive stress is
nearly perpendicular to the tensional surface faults and fissures
and to the trend of the Mid-Atlantic Ridge farther south. A stress
field of this type is likely to result in ground motion with a
component of separation as well as shear along the seismic zone.

Thus the tectonic picture of the Reykjanes Peninsula revealed by their study is one of an obliquely diverging plate boundary showing both ridge and transform fault characteristics.

6. EARTHQUAKES AND VOLCANIC ERUPTIONS

Most volcanic eruptions in Iceland are accompanied by swarms of earthquakes, felt for from several hours up to several days before the eruption. These earthquakes are generally small, but fairly strong earthquakes, exceeding intensity VI, have been observed at some volcanoes, notably at the beginning of eruptions of Katla and during eruptions of Hekla.

Intense earthquake swarms preceded the fissure eruption of Lakagígir in the eastern volcanic zone in 1783 and the eruptions of 1724-1729 and 1875 in the northern volcanic zone. Otherwise, swarms are rare in these regions as stated above.

A swarm of over 200 earthquakes less than 3.5 in magnitude was recorded 30 to 16 hours before the eruption broke out on the island Heimaey in 1973 [13]. These earthquakes were not felt on the island. Several earthquakes were recorded and felt in the last 6 hours before the outbreak, and only a few during the first week of the eruption. After that time the number of events was fairly consistently about 4 per day, falling gradually as the production of lava decreased. These earthquakes were found to originate at 20 to 30 km depth several km southeast of the crater. They are the deepest earthquakes located with confidence in Iceland, lying 8-18 km below the upper mantle boundary ($V_p = 7.2$ km/sec) where partially molten material is expected. These depth determinations may provide important constraints in discussions on the depth of magma generation under Iceland. They may also provide an important criterion to distinguish harmless swarms of tectonic nature from volcanic swarms, which are likely to be precursors of eruptions. The former are very frequent in the volcanic zones but have not been found to originate at depths greater than 5-8 km, the deepest ones in the seismic layer 3 ($V_p = 6.5$ km/sec) above the upper mantle boundary.

7. GENERAL CONCLUSIONS AND FUTURE PLANS

Earthquakes located teleseismically and large destructive earthquakes have occurred mainly within two zones, one in southern and southwestern Iceland and the other off the north coast. The transform motion between the Reykjanes Ridge and the eastern volcanic zone of Iceland and between the Kolbeinsey Ridge and the northern volcanic zone respectively is apparently taken up by these seismic zones. This motion is not simple and the details

are not well known. On the Reykjanes Peninsula, which is the best
known part, the plate boundary has both transform fault and
spreading ridge characteristics.

Significant seismic activity occurs outside of these two
zones, but the earthquakes are usually too small to be detected
teleseismically. For the purpose of studying these earthquakes a
dense network of over 20 seismic stations is presently under
construction in Iceland. Data from this network are expected to
yield information on the details of the tectonics, and can also
be used for studying problems such as the structure of the crust
and upper mantle under Iceland, and the forecasting of earthquakes
and volcanic eruptions.

ACKNOWLEDGEMENTS

This work was partly supported by the National Science Foundation
(Grant GA 43382), the Icelandic Science Foundation and the NATO
Research Grants Programme (Grant No. 715). Drs. P. Richards and
P. Pomeroy made many helpful suggestions for the improvement of
the manuscript.

REFERENCES

1. F. Klein, P. Einarsson and M. Wyss, J. Geophys. Res., 78,
   5084, 1973.
2. P.L. Ward, Bull. Geol. Soc. Am., 82, 2991, 1971.
3. L.R. Sykes, J. Geophys. Res., 72, 2131, 1967.
4. K. Saemundsson, Bull. Geol. Soc. Am., 85, 495, 1974.
5. L.R. Sykes, Bull. Seism. Soc. Am., 55, 501, 1965.
6. E. Tryggvason, Bull. Seism. Soc. Am., 63, 1327, 1973.
7. Th. Thoroddsen, in: Landskjálftar á Íslandi, Hid Íslenzka
   bókmenntafélag, 269 pp, 1899 and 1905.
8. S. Thórarinsson, in: The Eruptions of Hekla in Historical
   Times, The Eruption of Hekla 1947-1948, I, Soc. Sci. Isl.,
   p. 82 and 119, 1967.
9. P.L. Ward, G. Pálmason and C. Drake, J. Geophys. Res.,
   74, 665, 1969.
10. P.L. Ward and S. Björnsson, J. Geophys. Res., 76, 3953,
    1971.
11. D. Conant, Earthquake Notes, 43, 19, 1972.
12. R. Stefansson, Tectonophysics, 3, 209, 1966.
13. S. Björnsson and P. Einarsson, in press 1974.
14. G. Pálmason and K. Saemundsson, in: Annual Review of Earth
    and Planetary Sciences, 2, 25, 1974.
15. J. Jónsson, in: Iceland and Mid-Ocean Ridges, Soc. Sci. Isl.
    Rit, 38, 142, 1967.

VERTICAL CRUSTAL MOVEMENT IN ICELAND*

Eysteinn Tryggvason

Department of Earth Sciences, University of Tulsa,
Tulsa, Oklahoma, U.S.A.

ABSTRACT. The rift zones of Iceland are subsiding at relatively
constant rate. The absolute subsidence has not been determined
but the rate of tilt of the flanks of the rift zones is 0.2 to
0.7 microradians per year and the width of the zone that is being
tilted appears to be more than 40 kilometers. It is estimated
that the central part of the rift zone is subsiding 0.5 to 1.0
centimeter per year relative to surrounding areas. The ground
deformation associated with this subsidence is characterized by
gentle bending of the earth's crust with no measureable fault
displacement except during earthquakes and possibly during a
period of one or two years after earthquakes.

1. SEA LEVEL FLUCTUATIONS

Vertical movements of the earth's crust in Iceland have played
an important role in the formation and evolution of this volcanic
island. Evidences of large scale vertical crustal movements are
obvious all over the island although the amount of movement is
usually difficult to ascertain and the rate at which this movement took place is even more difficult to estimate.

---

* This work has been supported by grants GP-5365, GA-987,
  GA-4112 and GA-24152 from the National Science Foundation,
  Washington.

Kristjansson (ed.), Geodynamics of Iceland and the North Atlantic Area. 241-262. *All Rights Reserved.*
Copyright © 1974 by D. Reidel Publishing Company, Dordrecht-Holland.

A strong indication of a general subsidence of Iceland is
the near total lack of marine sediments in the Tertiary formations
although faulting and tilting of these formations indicate large
scale relative vertical movements. The only Tertiary marine sediments known in Iceland are the late Pliocene Tjörnes formation
consisting of shallow marine sediments interbedded by thin coal
layers. This formation covers about 20 square kilometers on the
western part of the Tjörnes Peninsula in North Iceland [1].
Pleistocene marine sediments are found at a few scattered locations, mostly near present sea level, although early Pleistocene
marine formations are found at 135 - 170 meters elevation on the
Snaefellsnes Peninsula in West Iceland, indicating that a small
area has been uplifted by that amount. Marine sediments of
recent age are found at low elevation along most of the coasts
of Iceland. Almost all of those were formed during the retreat
of the Pleistocene glaciers and the indicated uplift is due to
isostatic readjustment after the glacier load was removed [2].

A crude estimate of the subsidence of Iceland can be made
from bore hole information. A well, 1550 meters deep, on the
island Heimaey penetrated 180 meters of recent volcanic tuff and
below that about 640 meters of Pleistocene marine sediments.
Below these sediments the well penetrated layered Tertiary basalt
that had flowed on dry land. This terrestrial basalt continued to
the bottom of the well about 1530 meters below sea level [1].
Thus the subsidence in the area of Heimaey exceeds 1500 meters
since the formation of these lavas.

A well drilled for geothermal exploration on Reykjanes,
Southwest Iceland in 1969 reached a depth of about 1730 meters
below sea level. This well penetrated layers of basaltic lava,
sediments and palagonite tuffs. Several of the sediment layers
and all the lavas are dry-land formations and the palagonite tuff
layers are probably formed under glacial ice during the cold
periods of the Pleistocene. As the deepest tuff layer is found at
1500 meters depth, the indicated subsidence during the Pleistocene
period (about 3 million years) is no less than 1500 meters, but
can be greater. This gives an average subsidence rate of at least
0.5 millimeters per year [3].

The deepest well drilled in Iceland reached a depth of 2200
meters below sea level [4]. This well, located in Reykjavik,
penetrated numerous layers of flood basalt of dry land origin
all the way down to the bottom, indicating a subsidence of
at least 2200 meters since the formation of these lavas. The age
of these lava layers is not known, but they are probably of late
Tertiary age.

Vertical ground movements in Iceland in recent times have
been very significant as judged from raised shorelines. When the

glaciers retreated at the end of the ice age the sea level was as
much as 110 meters above present sea level in the South-Iceland
Lowland and 30 to 60 meters above present sea level in most other
areas in Iceland based on the highest post-glacial shorelines
which are about 11000 years old [1]. At this time (11000 years
B.P.) the actual sea level stood about 50 meters below present
sea level due to the large amount of water stored in the glaciers,
so the true uplift of nearshore areas in Iceland during the last
11000 years is somewhere between 80 and 160 meters.

The uplift of Iceland after the ice age was apparently completed about 9000 years B.P. as at that time peat had started to
form near or even below present sea level and the Thjórsa lava in
South Iceland flowed into the ocean below present sea level about
8000 years B.P. [1]. The rate of uplift from 11000 to 9000 years
B.P. has thus been of the order of 10 centimeters per year.
During the last 9000 years most of the coast line of Iceland has
been subsiding. Near Reykjavik the rate of subsidence has been
1.0 to 1.5 millimeters per year [1].

2. TILTING AND FAULTING

The Tertiary flood basalt which covers about one half of the surface area of Iceland, originally consisted mainly of nearly horizontal lava layers. The present tilt of the Tertiary lava layers,
which is widely observed in Iceland [10] indicates relative vertical crustal movements. This tilting is frequently 5 to 15 degrees
and occasionally 20 to 40 degrees [1]. Faults with vertical displacement are also very common in the Tertiary basalt in Iceland,
another clear indication of relative vertical displacement. However, the total amount of vertical crustal movement which is indicated by tilting and faulting has not been determined, although
indications are that it amounts to several kilometers in some areas
of Iceland.

Some of the active faults in Pleistocene and Recent times
show up in the topography and the total vertical displacement can
be estimated. The Bárdardalur fault on the west side of the North
Iceland neo-volcanic zone has a vertical displacement which exceeds
700 meters in its northern part, but the displacement decreases
southward. This displacement is younger than the "Grey Phase" west
of Bárdardalur, but its age has been estimated as early Pleistocene [11]. On the east side of the Tjörnes Peninsula in North
Iceland, a vertical displacement of about 1000 meters occurred
after the formation of the Tertiary basalt on Tjörnes [12].

Faults in Recent lavas (less than about 11000 years) are
very common along the volcanic zones of Iceland. Some of these
faults show vertical displacements which exceed 30 meters, such

as on Thingvellir in Southwest Iceland [7] and in Gjástykki in
North Iceland [13]. The recent faults are usually associated with
narrow grabens and tilting of blocks between faults. Because of
the tilting, the total vertical displacement is frequently much
greater than the displacement on individual faults.

Recent tilting of the ground surface can be observed in
several locations in Iceland. The present slope of the surface
of recent lavas is sometimes opposite to the flow direction.
Three cases from North Iceland are briefly discussed below.

The lake Mývatn lies in a depression which is the result of
recent subsidence. Most of the lake bottom is covered with lava
which flowed in a westerly direction from the crater row
Threngslaborgir, about 2000 years ago [14]. To the west of the
lake, the surface of the lava is on the average at about 280 meters
elevation, or three meters above the lake. The lake is shallow,
mostly 2.5 to 3 meters [15]. During the Threngslaborgir eruption,
about 2000 years ago, the slope of the lava surface must have been
in a westerly direction, but now the surface slope of the lava
below the western part of the lake is in easterly direction,
about one meter per kilometer. The ground tilting toward the east
during the last 2000 years must be greater than 1 meter per kilometer, and if the tilting is a continuous process, the rate of tilt
is of the order of one microradian per year.

The same lava flow that covers the bottom of lake Mývatn
flowed down the Laxárdalur valley and almost to the coast in
Adaldalur west of the river Laxá, a distance of some 60 kilometers.
Where the lava enters the relatively level bottom of the Adaldalur
valley, near Brúar (L1 on Fig . 2), its surface elevation is
about 25 meters above sea level. Ten kilometers farther to the
northwest the elevation is still about 25 meters and at the front
edge of the lava, about 18 kilometers northwest of Brúar the surface elevation of the lava is still close to 20 meters or less
than 10 meters lower than at Brúar. At the time of this lava flow
the surface must have sloped to north or northwest, in the direction of the flow. How steeply it sloped is not known, but the
present horizontal surface shows that some tilting towards south
or southeast must have taken place in Adaldalur during the last
2000 years.

A short distance east of Adaldalur, near 65°57' N, 17°10' W
is an old and dry river channel named Geldingadalur, apparently
formed during the last retreat of glaciers from this area. The
river that dug this channel flowed from east to west. Later, a
lava flow entered this channel from the east and flowed along the
whole length of the channel, about 5 kilometers. The age of this
lava flow is not known, but the glacial chronography of Iceland [1]
gives its age as less than 12000 years. Topographic maps show the

surface elevation of the lava east of the Geldingadalur channel
as less than 280 meters above sea level while the bottom of the
channel is at more than 300 meters elevation, probably close to
320 meters near the west end. Meltwater which occasionally flows
into the channel from south turns eastward when reaching the
bottom of the channel, regardless of inflow location, so the
present tilt of the surface along the whole channel is towards
the east. The present slope of the channel bottom as obtained from
topographic maps is approximately 4 meters per kilometer to the
east, while a tilt toward the west must have existed at the time
of the Geldingadalur lava flow, less than 12000 years ago. This
indicates a ground tilting towards east amounting to more than
0.4 microradians per year on the average.

These three examples, Mývatn, Adaldalur, and Geldingadalur,
all show recent tilting towards east or southeast. If it is
assumed that the tilting is going on at a constant rate, this rate
is at least 0.4 microradians per year, and possibly 1.0 microradian per year.

3. REPEATED PRECISION LEVELING IN ICELAND

A program of repeated precision leveling in Iceland under the
direction of the present author was started in 1966 and continued
until 1973. The field measurements have consisted of precision
optical leveling along profiles of closely spaced permanent bench
marks. A total of 20 locations has been selected for the repeated leveling and at least two levelings have been made at each
location. Four of these are 2 to 10 kilometer long profiles
crossing major active faults within the rift zones in Iceland.
Eight profiles, the longest about 2 kilometer long, are on or in
immediate vicinity of active volcanoes and another eight profiles,
all less than 2 kilometer long, are located within or in the
vicinity of the active rift zone (Figs. 1 and 2). The eight
profiles on or near active volcanoes will not be considered
further in the present paper.

Seven of the leveling profiles are more than one kilometer
long, consisting of 13 to 63 bench marks each. The other 13 profiles are 300 to 600 meter long consisting of 5 to 13 bench marks
each. The total number of bench marks in these 20 precision leveling profiles is about 390. Between 10 and 20 of these bench marks
are of no use in determining ground deformation because of unstable
foundation causing these bench marks to move irregularly, relative
to surrounding areas, probably due to freezing and thawing of
groundwater.

The leveling was conducted only during the months of June,
July, and August, so the adverse weather condition during most of

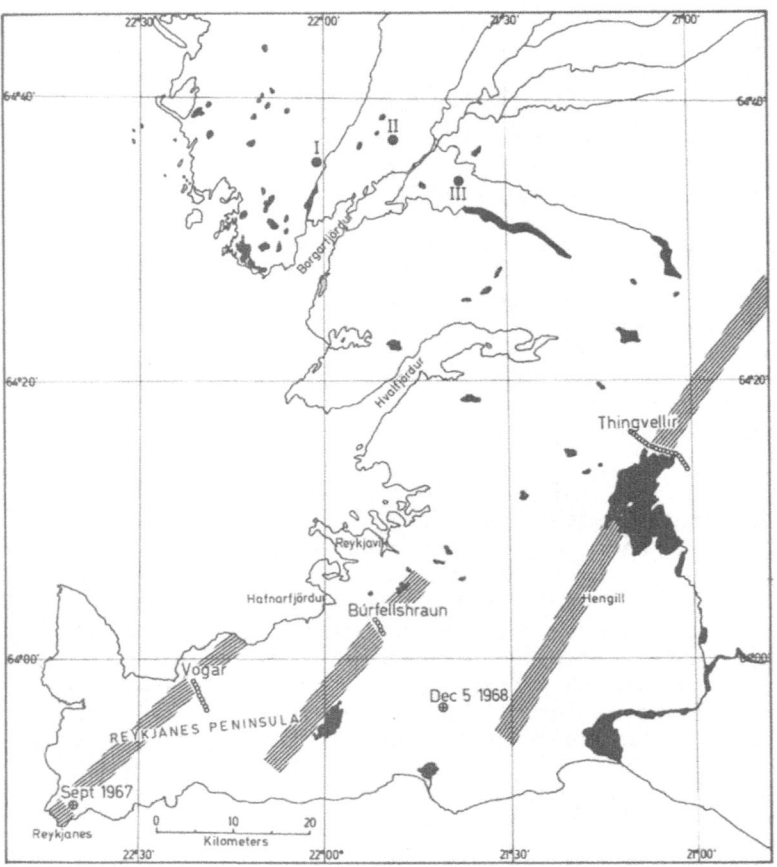

Fig. 1. Map of southwest Iceland showing the location of the
long leveling profiles at Vogar, Búrfellshraun, and Thingvellir
and the three short Borgarfjördur profiles marked I, II, and III.
Zones of most dense faults and fissures are shown by shading.
The approximate epicenters of the Reykjanes earthquake swarm of
September, 1967, and the Southwest Iceland earthquake of
December 5, 1968, are shown by small circles. Black spots are
lakes.

the remainder of the year was avoided. Furthermore, the leveling
was frequently delayed for hours or days if the weather condition
prevented accurate leveling. The leveling on each location was
normally repeated within a period of one or a few days, and if a
discrepancy exceeding 0.2 millimeters in the observed elevation
difference of two successive bench marks appeared, a third and
possibly a fourth leveling was made on the bench mark intervals
where discrepancy was observed.

The total error of leveling is estimated from the observed
difference between levelings made one or a few days apart,
assuming that no ground movement had taken place. The standard
error of leveling thus obtained is 0.15 $\sqrt{D}$ to 0.19 $\sqrt{D}$ millimeters where D is the distance along the profile of leveling in
kilometers.

The ground deformation is obtained from the comparison of
two levelings usually made one or more years apart. The error in
the observed ground deformation is caused by leveling errors of
the two levelings, and in addition by the erroneous movements of
the bench marks. This erroneous or irregular bench mark movement
represents a standard error of approximately 0.2 $\sqrt{D}$ millimeters
in the difference between two levelings. The total standard
error of the difference between two levelings, due to both observational error and irregular bench mark movement, is then estimated as roughly 0.4 $\sqrt{D}$ millimeters [5].

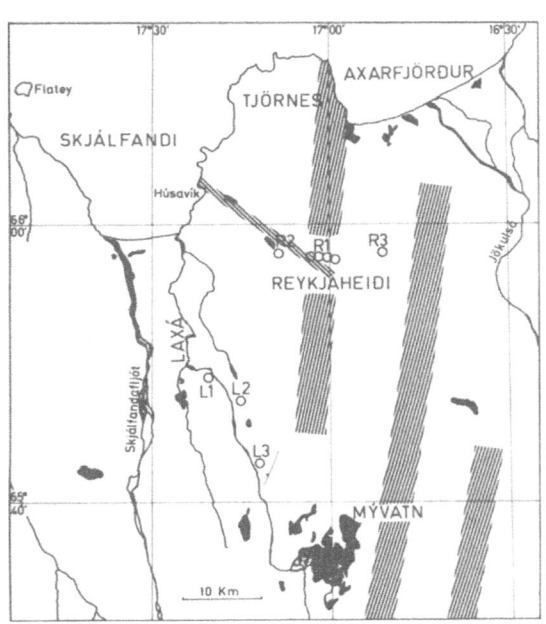

Fig. 2. Map of a part of
North Iceland showing the
location of the three
Reykjaheidi profiles (R1,
R2, and R3) and the three
Laxá profiles (L1, L2,
and L3). The Husavik
fault zone and fissure
swarms in the North
Iceland Rift Zone are
shown by shading.

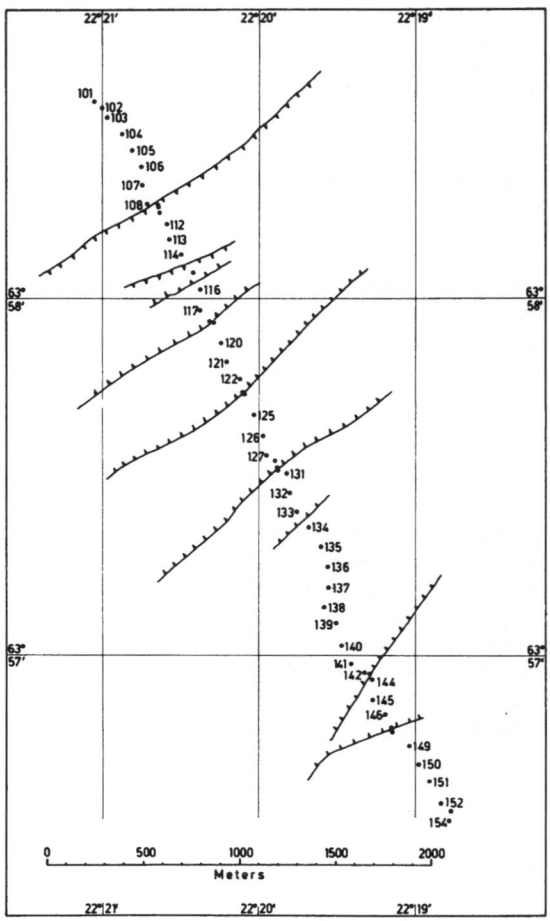

Fig. 3. Map of the
Vogar profile in South-
west Iceland, showing
bench mark locations
and fault scarps across
the profile.

4.   RESULTS OF REPEATED LEVELING IN SOUTHWEST ICELAND

4.1   The Vogar profile

The Vogar leveling profile is on the Reykjanes Peninsula and
extends from near the north shore about 4.2 kilometers inland in
azimuth of approximately 155° (Figs. 1 and 3). The profile lies
across a zone of intense faulting (fissure swarm) which extends
southwest to Reykjanes and northeast to the north shore of the
Reykjanes Peninsula (Fig. 1). The whole profile lies on post-
glacial lava from the shield volcano Strandarheidi. This volcano
has been active in early postglacial time, possibly 10000 years
B.P. [6].

The first leveling of the Vogar profile was made in June,

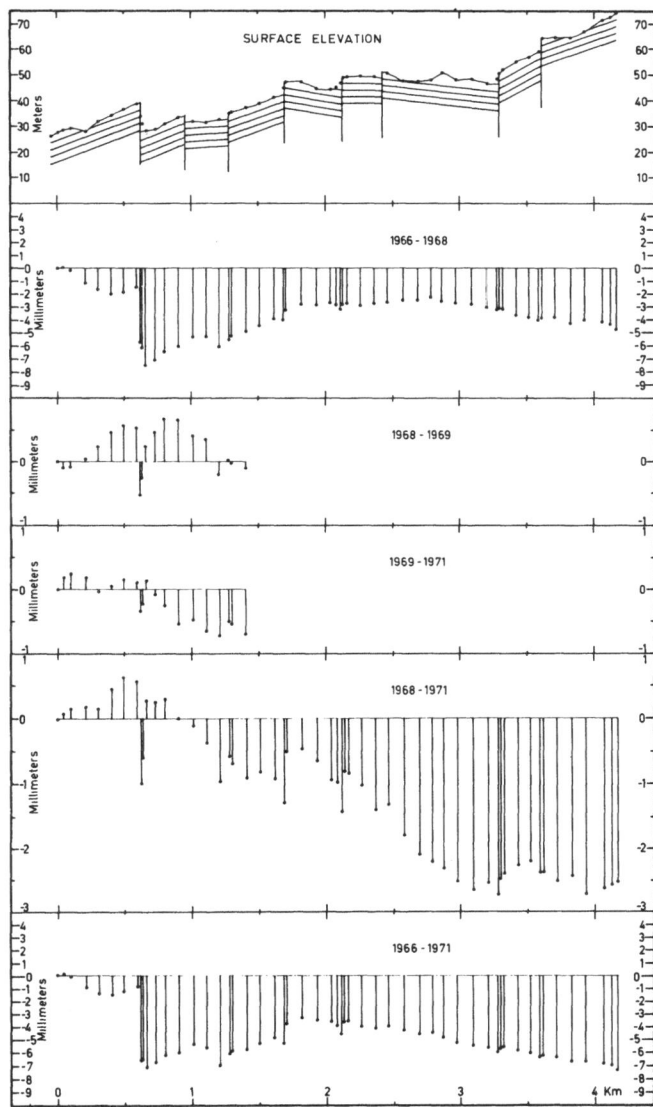

Fig. 4. Surface elevation and vertical displacement of the
bench marks of the Vogar profile. The northwest end of the profile is to the left. Vertical displacement of bench marks are
relative to bench mark 101, farthest to the left.

1966, and repeated in 1968 and 1971 and a fraction of the profile was leveled in 1969. The levelings show clearly at least eight faults with vertical displacement of from 2 to 11 meters (Fig. 4) and the blocks between the faults have been tilted.

Between the two first levelings, the Reykjanes earthquake swarm of September, 1967, occurred. The epicenter of this swarm is near the southwest end of the fissure swarm crossed by the profile. Therefore, it can be assumed that any movement between 1966 and 1968 on the Vogar profile may have been affected by the earthquake swarm. The deformation observed between 1966 and 1968 (Fig. 4) is rather complicated. There are apparently at least three processes involved in this deformation: (1) Fault displacements, primarily on one fault but minor displacements are indicated on two other faults; (2) subsidence of a zone around the fault where the displacement was observed; and (3) tilt of the whole area of the profile [6].

Between 1968 and 1969 the zone around the fault, where displacement had been observed the year before, was uplifted slightly, so the subsidence indicated the year before had been slightly reduced.

Between 1969 and 1971 and between 1968 and 1971, the principal deformation on the profile was a uniform tilting towards the southeast. There are, however, some indications of slight fault displacements on several faults, not including the fault of principal displacement between 1966 and 1968.

It is assumed that the fault displacements and the subsidence of the zone around the displaced fault is the result of the earthquake swarm of September, 1967 [5], but that the general ground tilting is due to some continuous process. The direction of this tilting is not accurately known, because the linear shape of the profile does not allow determination of tilt perpendicular to its direction. The tilt component in the direction of the profile, as determined from the part of the profile least affected by the earthquake swarm, is about two microradians in five years or a tilt rate of 0.4 microradians per year.

4.2   The Búrfellshraun profile

The Búrfellshraun profile lies about six kilometers southeast of Hafnarfjördur. It is about 2.5 kilometers long and has 32 permanent bench marks (Fig. 1). Leveling was conducted in 1966, 1967, 1968, 1969, and 1970, but no significant ground deformation was observed, except a fault displacement of about one millimeter, probably associated with the earthquake of December 5, 1968 [5,7].

4.3   The Thingvellir profile

The Thingvellir profile is almost 10 kilometers long in the
general direction of 120°, or perpendicular to the direction of
faults in the Thingvellir area (Fig. 5). It crosses a five
kilometer wide graben and the fault displacement on either side
of the graben is 30 to 35 meters at the location of the profile.
This displacement has taken place since the eruption of the
Thingvellir lava about 9000 years ago [8].

The profile was first leveled in 1966 and again in 1967,
1969, and 1971. The first leveling covered only the central part
of the profile, about 5.5 kilometers long and the last leveling did
not cover the westernmost 1200 meters of the profile. The observed
vertical displacement (Fig. 6) is a general subsidence of the
Thingvellir graben relative to the surrounding areas with no
observed displacements on any fault between times of leveling.
The greatest subsidence is observed east of the central part of
the graben and maximum tilt rate of approximately 0.25 microradians per year is observed in the vicinity of the major faults
Almannagjá and Hrafnagjá on either side of the graben.

The relative displacement during the last 9000 years can be
estimated from the ground elevation along the profile, noting that
the lava covering most of the profile flowed in westerly direction.
Such estimates show maximum subsidence in the eastern part of the
graben, in agreement with the present observations. The total
subsidence of the eastern part of the Thingvellir graben relative
to the west side of the fault Almannagjá is 50 to 70 meters in
9000 years, or an average subsidence rate of 5 to 8 millimeters
per year [7], while the subsidence of the eastern part of the
graben relative to the western side of Almannagjá during the six-
year period, 1966 to 1971, is only about 2.5 millimeters, less
than 1/10 of the average subsidence during the last 9000 years.
This may indicate that the subsidence of the Thingvellir graben
is now slower than it was a few thousand years ago.

There is, however, another possibility. It may be that the
total width of the subsiding zone is much greater than the width
of the graben. During fault displacement the zones on either side
of the graben are uplifted but the graben block subsides. The
total subsiding zone then continues to subside. In this case,
the present subsidence rate of the graben, relative to a point
far away from the graben, may be much more than 0.4 millimeters
per year.

Therefore, it cannot be concluded that the subsidence of the
Thingvellir graben is slowing down. It is even probable that the
present subsidence rate is almost equal to the average rate of
subsidence during the last 9000 years [7,9].

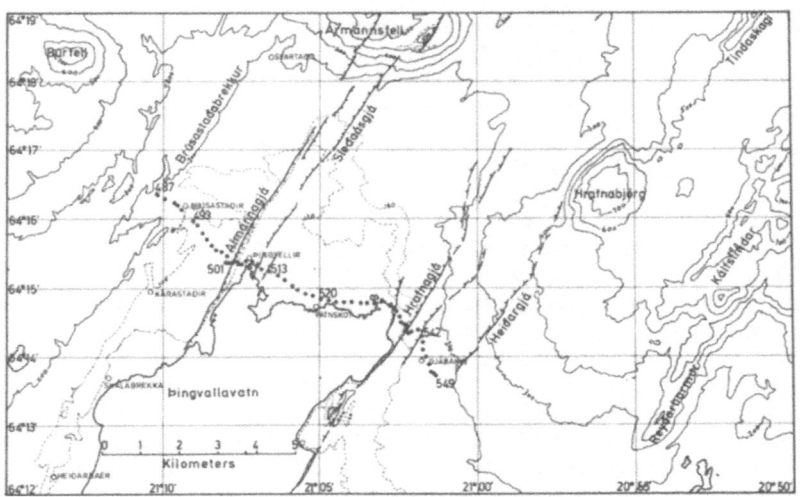

Fig. 5. Map of the Thingvellir area, showing the location of
bench marks and principal active faults.

4.4   The Borgarfjördur profile

Three short precision leveling profiles in the Borgarfjördur
district in West Iceland were first leveled in 1970 and releveled
in 1973. The location of these profiles (I, II, and III on Fig. 1)
was chosen to be an approximate continuation of the Thingvellir profile, but 40 to 55 kilometers outside the Thingvellir graben. The
observed tilt over a three-year period (Fig. 7) does not indicate
that the deformation associated with the Thingvellir graben extends
this far out, but a southerly tilt is indicated, expecially at
Langá and Gufá (I and II on Fig. 1) located north of the Borgarfjördur inlet. The tilt rate of 0.5 to 0.9 microradians per
year is rather high, and it may be suspected that some local
phenomenon, other than crustal deformation, is causing the observed
tilt. However, the similarity of the observed tilt at Langá and
Gufá, although at geomorphologically different setting, indicates
rather strongly that the area of the Borgarfjördur inlet is subsiding.

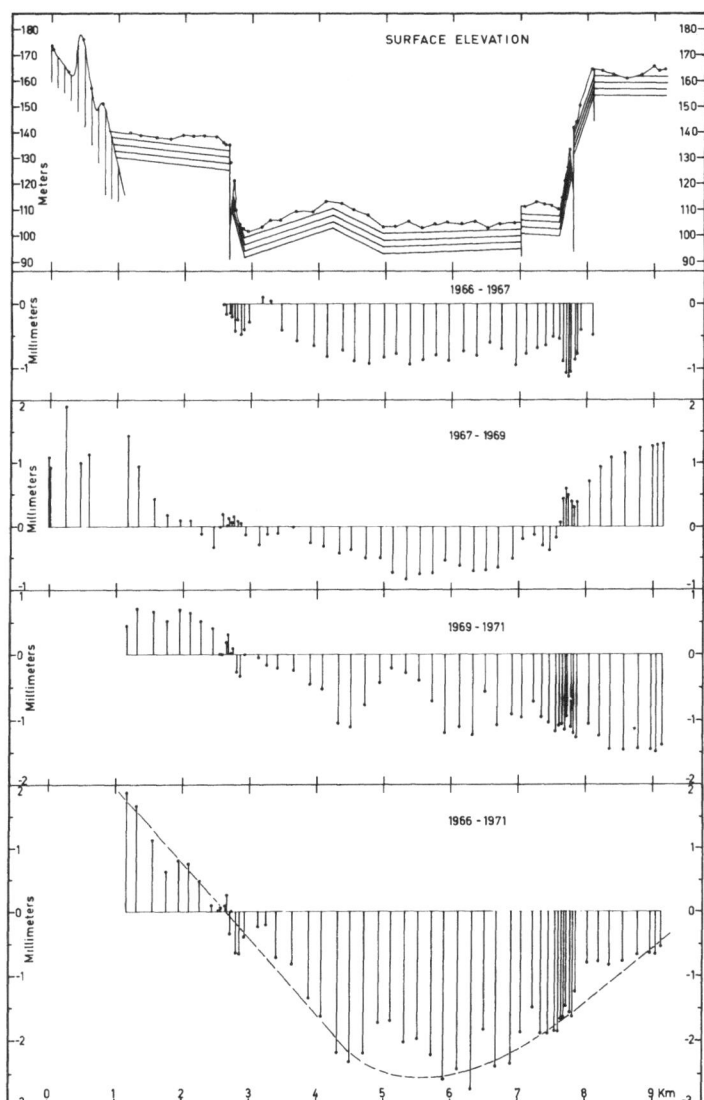

Fig. 6. Surface elevation and vertical displacements of bench marks on the Thingvellir profile. Northwest end of the profile is to the left. Bench mark movements are relative to bench mark 501, a short distance west of Almannagjá (see Fig. 5).

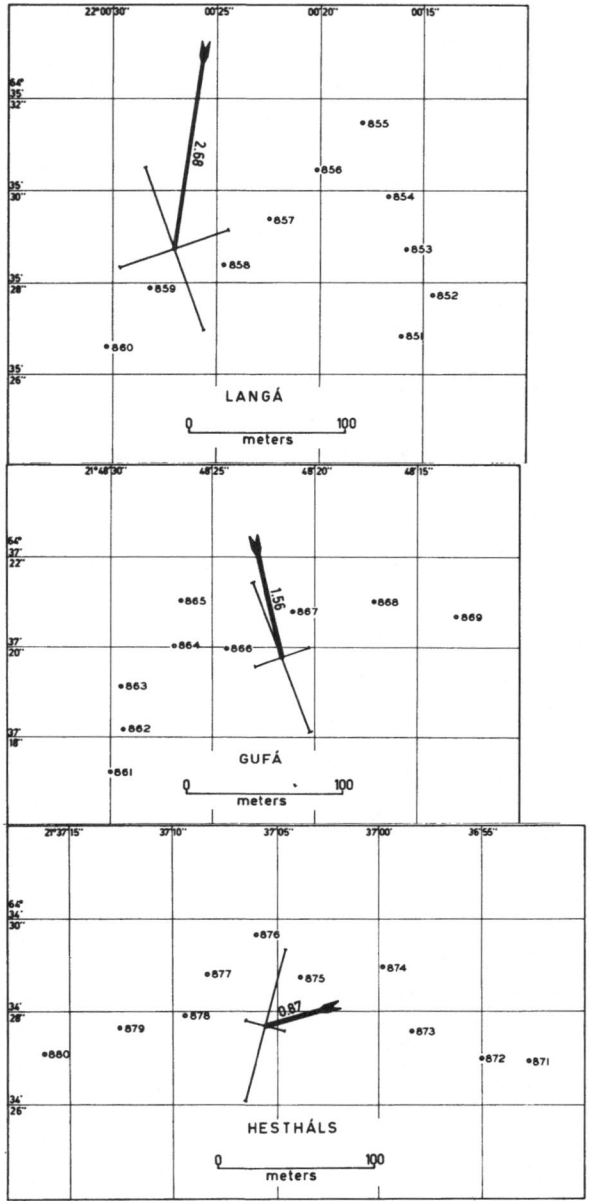

Fig. 7. The three short profiles in Borgarfjördur (I, II, and III on
Fig. 1), and the observed ground tilt between 1970 and 1973. The
amount of tilt is given in microradians and the maximum and minimum
axes of the error ellipse are shown at the 99 per cent confidence
level.

5.   RESULTS OF REPEATED LEVELING IN NORTH ICELAND

5.1. The Reykjaheidi I profile

A three-kilometer long profile, consisting of 30 permanent bench
marks, was first leveled in 1966 and again in 1968, 1970, and
1972. The location of the profile (RI on Fig. 2) is where the
Húsavik Fault Zone enters the north-south trending fissure swarm
of Theistareykir, the westernmost fissure swarm in the North
Iceland Rift Zone. The general direction of the profile is about
98° (Fig. 8). Most of the profile lies on early post-glacial lava
from the shield volcano Theistareykjabunga (Stóra Víti), but one
bench mark (301) is on Pleistocene palagonite tuff and the west
end of the profile (328, 329, and 330) lies on a lava whose age
is here estimated as 2000 years. This lava came from an irregular
crater row a few kilometers to the south of the profile.

      Relative vertical displacement of the Reykjaheidi I profile
in the past has not been estimated, although the ground has
obviously been tilted towards the east. The lava on which most of
the profile lies did flow from the east and the present ground tilt
is also generally towards the east. The age of this lava is not
known, but it flowed after the last retreat of glaciers from this
area, which may have occurred as early as 12000 years ago, but
probably not earlier than 11000 years ago [1].

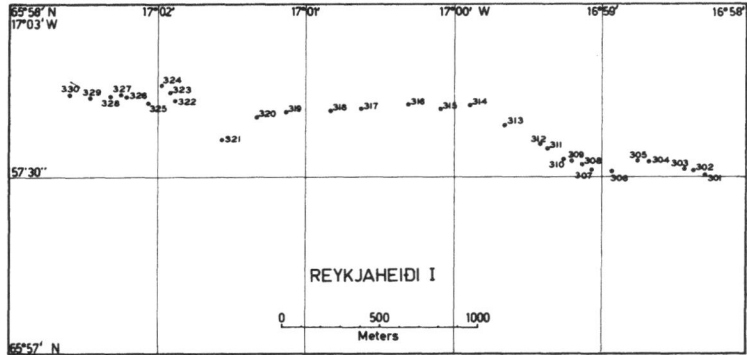

Fig. 8. Location of bench marks of the Reykjaheidi I profile.

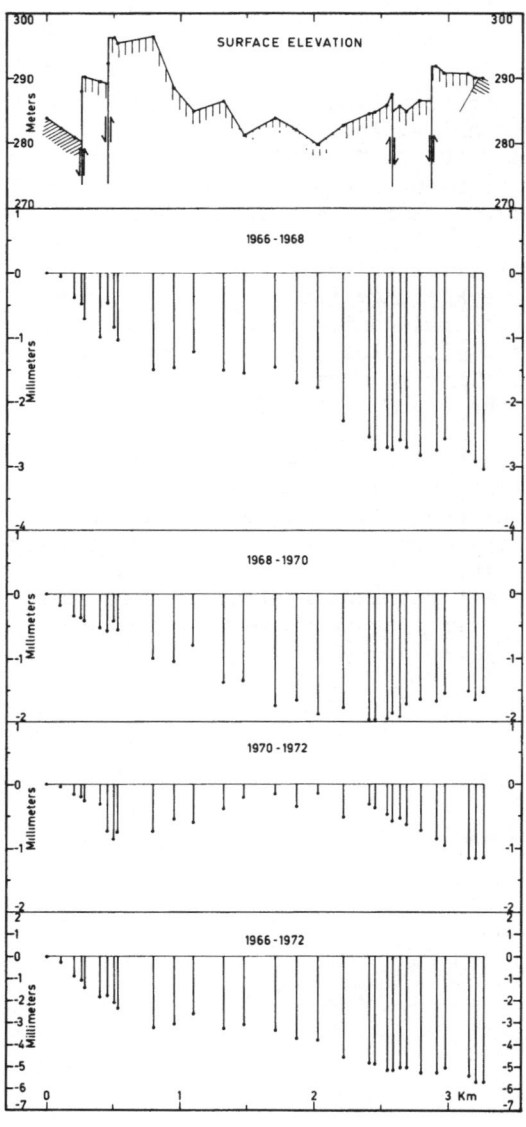

Fig. 9. Surface elevation and vertical displacement of bench marks of the Reykjaheidi I profile. Vertical displacements are relative to bench mark 330 on the west end of the profile and to the left on this illustration.

The ground tilting between 1966 and 1972 is generally towards the east, but it varies along the profile (Fig. 9). Tilt component perpendicular to the general direction of the profile is apparently small, although a southerly tilt is indicated on the central portion of the profile (Table 1).

The average annual tilt of the whole profile is about 0.3 microradians towards the east, but the east end of the profile has been tilted only about 0.16 microradians per year while the west end of the profile has been tilted 0.68 microradians per year towards the east (Table 1). If conditions on Reykjaheidi are similar to those on Thingvellir, this may indicate that the east end of the Reykjaheidi I profile is near the zone of maximum subsidence, but tilting probably occurs to a distance of several kilometers to the west of the profile.

TABLE 1

Least squares fit of observed tilt of the Reykjaheidi I profile between 1966 and 1972 in microradians. Components of computed tilt are along (east) and perpendicular to (north) the general direction of the profile.

Bench marks	Tilt towards east	Tilt towards north	Annual rate of tilt East	North
301-310	0.978 ± 0.247*	0.145 ± 3.083	0.163	0.024
311-320	1.313 ± 0.175	-3.575 ± 1.244	0.219	-0.596
321-330	4.110 ± 0.314	0.568 ± 1.169	0.685	0.095

* The errors are at the 99 per cent confidence level.

5.2 Short profiles in North Iceland

Five short precision leveling profiles were established in 1970 in North Iceland. Two of these profiles are on Reykjaheidi (R2 and R3 on Fig. 2), Reykjaheidi II about five kilometers to the west of Reykjaheidi I and Reykjaheidi III about four kilometers to the east of Reykjaheidi I. Three profiles are in the vicinity of the river Laxá at Brúar (L1 on Fig. 2), Geitafell (L2), and Hólar (L3). All of these profiles were leveled in 1970. The two Reykjaheidi profiles were releveled in 1972 and the Laxá profiles were releveled in 1973.

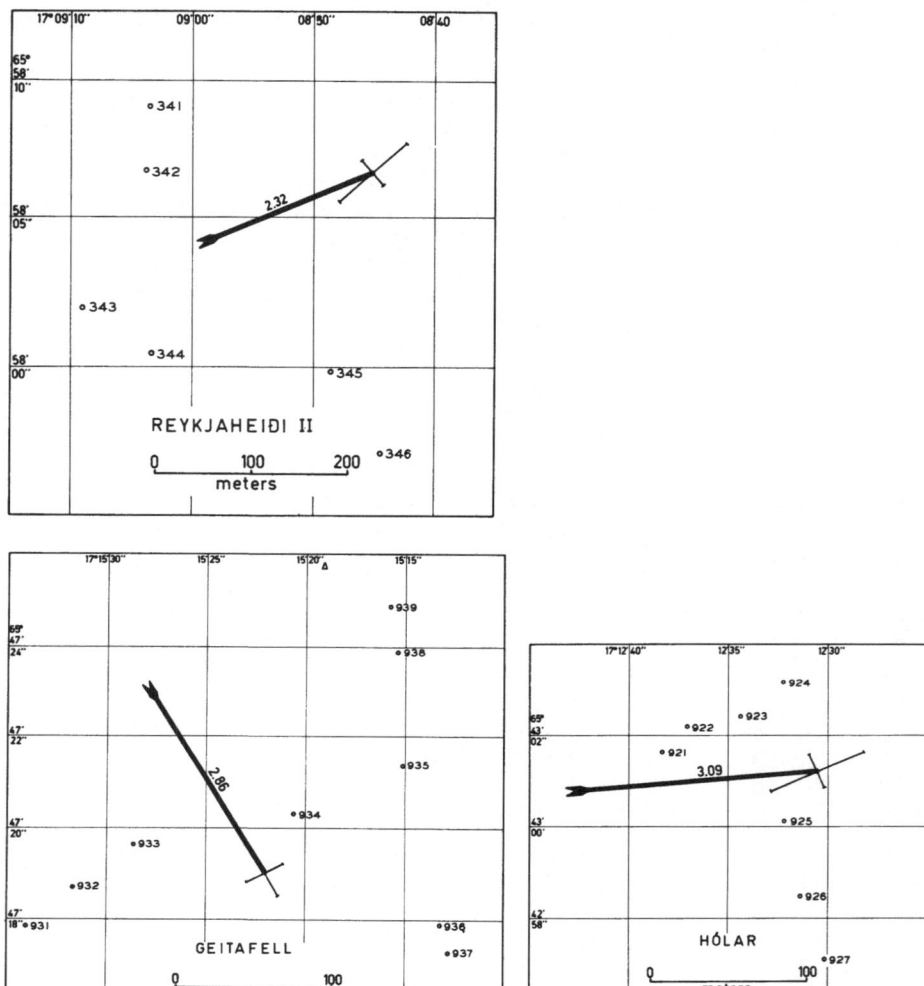

Fig. 10. The short profiles Reykjaheidi II (R2), Geitafell (L2), and Hólar (L3) in North Iceland. Arrows show the ground tilt between 1970 and 1973 in microradians computed by the least squares method. The observed tilt at Reykjaheidi II over a period of two years is multiplied by 3/2 to make it comparable to the observed three-year tilt at the other two profiles.

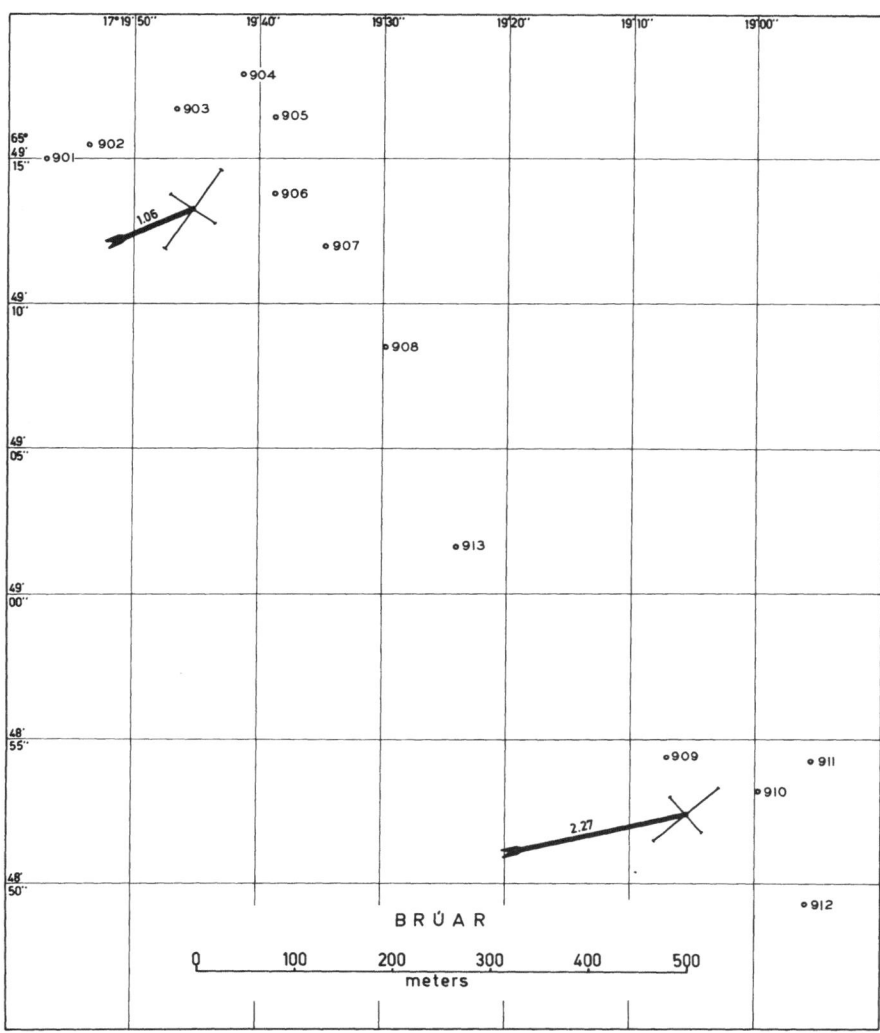

Fig. 11. The Brúar profile (L1) and the observed tilt for two parts of the profile between 1970 and 1973 in microradians.

The Reykjaheidi III (R3) profile consists of only five bench
marks along an east-west line, so north-south component of tilt
cannot be observed. The observed tilt component along the profile
over a two-year period is 0.8 ± 0.4 (99 per cent confidence)
microradians towards west, which indicates that a zone of maximum
subsidence lies between R1 and R3 (Fig. 2), probably closer to R1.

The Reykjaheidi II (R2) profile and the three Laxá profiles
all show a tilt towards east (Figs. 10 and 11). The north component of observed tilt is small, but positive on all the profiles
except Geitafell, where significant southerly tilt component is
observed. As there is no clear indication of systematic tilt
variation within the area of these leveling profiles, we may
accept the average tilt as the tilt of the west flank of the North
Iceland Rift Zone. The average tilt at the five locations illustrated on Figs. 10 and 11 (the Brúar profile divided into two)
over a three-year period is 1.979 microradians towards east and
0.141 microradians towards south. The south component is
insignificant, so the average annual tilt is about 0.66 microradians towards east or about the same as found near the west end
of the Reykjaheidi I profile.

6. CONCLUSION

The vertical ground deformation in the rift zone of Iceland is
characterized by a subsidence of the central part of the rift
zone. The center of this subsidence coincides with the graben
in the Thingvellir area in Southwest Iceland, but it does not
coincide with the swarms of fissures in the Vogar area in Southwest Iceland or the Reykjaheidi area in North Iceland. These
fissure swarms have an en echelon pattern (Figs. 1 and 2) but
the center of subsidence apparently follows the center of the rift
zone which has a different strike from that of the fissure swarms.

The rate of ground tilt on the flanks of the subsidence zone
is between 0.25 and 0.68 microradians per year at the locations
studied, lowest in the Thingvellir area, and highest in the
Reykjaheidi region of North Iceland. The width of the subsidence
zone is difficult to judge from the present observations. In
North Iceland the tilt towards the rift zone at the Laxá profiles
is similar to the maximum tilt on the flanks of the subsidence
zone, although these profiles are located roughly 20 kilometers
west of the central axis of the North Iceland Rift Zone. In
Borgarfjördur, tilt towards the Thingvellir zone is not detected,
at about 40 kilometers distance from the rift zone. This may
indicate that the subsidence zone extends to a distance of somewhere between 20 and 40 kilometers from the central axis of the
subsidence zone. Tilt towards south which is indicated on the
two Borgarfjördur profiles which lie north of the Borgarfjördur

inlet indicate that ground deformation is taking place outside
the subsidence zone associated with the principal rift zone in
Iceland.

The rate of subsidence in the central part of the subsidence
zone in North Iceland and in Southwest Iceland can be roughly
estimated from the tilt rate on the flanks of the subsiding zone
and the estimated width of this zone. Assuming that a 30 kilometer
wide zone on either side of the axis of the subsidence zone tilts
towards this zone, and the average tilt rate is one half of the
maximum observed tilt rate in each region. Then the rate of
subsidence of the central part of the North Iceland Rift Zone is
about 1.0 centimeters per year relative to areas outside the
zone of subsidence. On the Reykjanes Peninsula in Southwest
Iceland the maximum subsidence is apparently about 0.6 centimeters
per year and in the Thingvellir area in Southwest Iceland the
maximum subsidence rate is estimated as about 0.4 centimeters per
year using the same assumptions. It should be emphasized that
this estimated subsidence rate is relative to the land outside
the zone of subsidence. The result of the precision leveling
cannot be used to estimate vertical crustal movements relative
to sea level.

REFERENCES

1. Th. Einarsson, Jardfraedi (Geology), 335 pp, Mál og Menning,
    Reykjavik, 1971.
2. Tr. Einarsson, Jökull, 16, 157, 1966.
3. S. Björnsson (ed.), Reykjanes, heildarskýrsla um rannsókn
    jardhitasvaedisins, (Report on geothermal research in
    Reykjanes), mimeographed, 122 pp, Orkustofnun, Reykjavik,
    1971.
4. G. Pálmason, On heat flow in Iceland, in: Iceland and Mid-
    Ocean Ridges, ed. S. Björnsson, Soc. Sci. Islandica, 38, 1967.
5. E. Tryggvason, NATO Advanced Study Institute on Continuum
    Mechanics Aspects of Geodynamics and Rock Fracture Mechanics,
    D. Reidel Publ. Co., Dordrecht, (in press), 1974.
6. E. Tryggvason, J. Geophys. Res., 75, 4407, 1970.
7. E. Tryggvason, J. Geophys. Res., 73 7039, 1968.
8. G. Kjartansson, Natturufraedingurinn, 34, 101, 1964.
9. E. Tryggvason, Natturufraedingurinn, 43, 175, 1973.
10. Tr. Einarsson, Geophys. J. R. Astr. Soc., 10, 283, 1965.
11. J. Hospers, Geologie en Mijnbouw, 16, 491, 1954.
12. K. Saemundsson, Geol. Soc. Am. Bull., 85, 495, 1974.
13. O. Niemczyk and E. Emschermann, Sonderdreiecksmessung auf
    Island zur festellung feinster Erdkrustenbewegungen, in:
    Spalten auf Island, ed. O. Niemczyk, Verlag von Konrad
    Wittwer, Stuttgart, 1943.

14. S. Thorarinsson, <u>Geolog. Rundschau</u>, <u>57</u>, 1, 1967.
15. S. Rist, Mývatnsisar (Ice on lake Mývatn), in: <u>Hafísinn</u>,
     ed. M. A. Einarsson, Almenna Bókafélagid, Reykjavik, 1969.

CRUSTAL MOVEMENTS IN THE MYVATN- AND IN THE THINGVALLAVATN-AREA,
BOTH HORIZONTAL AND VERTICAL

Karl Gerke

Technical University,
Braunschweig, F.R.G. ·

1. INTRODUCTION

The geodynamic research work, which forms the subject of this
report, continues the work of Niemczyk and his team begun in 1938
in the north-east of the island, in the Myvatn area. Following in
his steps, geodetic and geophysical disciplines co-operated in
the project, for only from the analysis of their measurements,
such as triangulation or alternatively trilateration, levelling,
gravity survey, magnetic survey and seismic survey can reliable
statements about movements be made.

From 1964 to 1971, measurements of high accuracy were carried
out several times in order to be able to determine the sudden
secular and also periodical change of position of fix points –
both horizontally and vertically. As well as the repeat measurements in the improved large trigonometrical network of the Myvatn
area of 1938, the Gjastikki profile in the middle of the fracture
zone was increased to 2 quadrilaterals and 2 new quadrilaterals
in Klaustur were set up. In addition a 142-kilometer long
levelling of high accuracy over 160 bench marks was measured
straight across the survey area. Not only the large trigonometrical network, but also the precise levellings lead straight across
the young volcanic zone, from the stable basalt near Akureyri to
the palagonite mountains east of Grimsstadir. But also in the
Thingvellir area a new trigonometrical control network was marked
out and in 1967 and 1971 measured, which stretches across the
fracture zone and, i.a., is fixed in 2 old shieldvolcanoes in
the west and east. In the northern part of this fundamental network 2 quadrilaterals were marked and measured, whose fix points
are partially identical with those of Prof. Tryggvason, Tulsa and

Kristjansson (ed.), Geodynamics of Iceland and the North Atlantic Area. 263-275. *All Rights Reserved.*
Copyright © 1974 by D. Reidel Publishing Company, Dordrecht-Holland.

Prof. Mason, London. A precise levelling loop surrounds the
Thingvallavatn.

The analysis of the measurements in the trigonometrical network led to a theoretical work by Dr. Pelzer, in which criteria
for the proof of significant changes of position are developed.

At the same time as the afore-mentioned measurements addi-
tional gravity surveys were carried out by Profs. Schleusener and
Torge, TU Hanover, magnetic field and magnetic rock surveys by
Prof. Angenheister, Munich, and his colleagues Dr. Peters and
Dr. Schönharting as well as microseismic surveys by Dr. Steinwachs,
Hanover. A detailed report will be given of the results and
consequences of these projects.

It should be mentioned here that the hydrographic surveys
north of Iceland by the German research vessels "Komet" in 1971
and "Meteor" in 1972 are linked with the afore-mentioned projects.

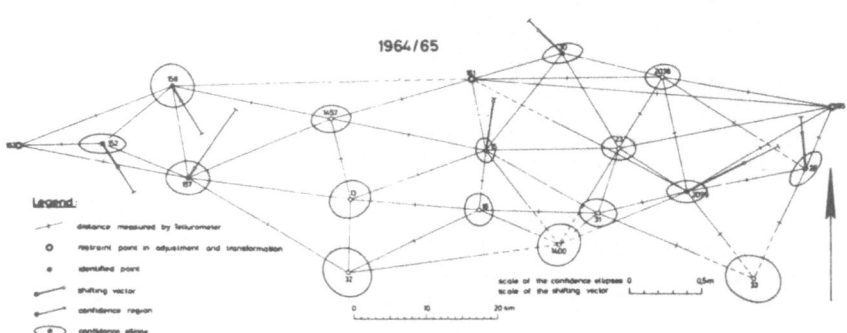

Fig. 1. Special triangulation network, North Iceland 1964/65.

2. MYVATN AREA

2.1 Geodetic control network

The original geodetic network of 1938 was measured with theodo-
lites and a simplified base-line measure. A geodetic network,
improved in its configuration, and enlarged, and which had a

number of identical fix-points from the old network, was then surveyed in 1965 by our group from the T.U. Braunschweig (Gerke, Pelzer) with the precision theodolite WILD T3 and microwave distance-measurer Tellurometer MRA 3. The research work was sponsored by the German Research Society (Deutsche Forschungs-gemeinschaft).

These survey elements were subjected to a combined adjustment of co-ordinates by fixing 3 stations regarded as geologically stable. The comparison of the co-ordinates from 1938 and 1965 showed differences, and therefore apparent point-dislocations which all lay within the confidence interval for a probability of 95%. Therefore no reliable conclusion could be made about the change of position of the trigonometrical stations [1,2,3]; see Fig. 1.

Out of these deliberations on the comparison of 2 independently surveyed geodetic networks for the purpose of significant statements about changes of position of fix-points ensued the publication of the team-member Dr. Pelzer [4]. In this work he gives criteria based on error theory and mathematical statistics.

In 1971 the whole geodetic network was again surveyed with the precision theodolite WILD T3 and microwave distance-measurer Tellurometer MRA 3 and Tellurometer MRA 4 as well as with the AGA-Laser-Geodimeter Model 8.

Although the adjustments to these surveys are completed, the error analyses are not yet fully completed. The results are to be written up as a doctoral thesis and published.

2.2 Precise levellings

In 1965 a level line 142 kilometers in length, marked out with 160 bench marks, which, leading from Akureyri to Grimsstadir, completely crossed the young volcanic zone, was measured in both directions as a precise levelling. For this the instruments ZEISS NI2 with plane parallel plate and JENOPTIK KONI 007 with invar rods were used. The standard deviation proved to be ±0.5 mm/km. Together with the gravity survey along this line it is possible to get the geoid profile [5].

This line will be surveyed again for the first time in 1975 or 1976, so that a report can be made on height changes that have occurred.

2.3 Deformation quadrilaterals

For the local determination of horizontal and vertical changes
of position quadrilaterals are more suitable than profiles.

2.3.1 Gjastikki

In the extended fracture zone of Gjastikki - north-east of
Myvatn - Niemczyk in 1938 marked out and surveyed a profile
established by 9 fix-points.

This profile was completed in 1967 with 2 quadrilaterals. Repeat
surveys of the directions, distances and heights were carried out
in the years 1965, 1967 and 1971. For this the following instruments
were brought into operation: Precision theodolite WILD T3 (1965),
ZEISS TH2 and JENOPTIK THEO 010 (1967/71), electro-optical distancer
Geodimeter Model 4 B (1965 and 1967) and Tellurometer MA 100 (1971).

The adjusted distances between profile points are set out in
the following table:

from-to	adjusted distances				differences of adjusted distances		
	1938	1965	1967	1971	65-38	67-65	71-67
300-301	276.749	.987	.986	.971	+0.238	-0.001	-0.015
301-302	213.759	.605	.609	.630	-0.154	+0.004	+0.021
302-303	436.281	.360	.342	.351	+0.079	-0.018	+0.009
303-100	226.560	.011	.030	.009	-0.549	+0.019	-0.021
100-304	105.999	.947	.948	.937	-0.052	+0.001	-0.011
304-307	1175.698	861.920	.915	.927	+0.061	-0.005	+0.012
307-305		313.839	.889	.892		+0.050	+0.003
305-306	463.424	.140	.130	.119	-0.284	-0.010	-0.011
306-200	69.888	70.015	.026	.036	+0.127	+0.011	+0.010
300-100	1153.350	1152.962	.966	.961	-0.388	+0.004	-0.005
100-200	1815.009	1814.860	.908	.911	-0.149	+0.048	+0.003
300-200	2968.358	2967.820	.872	.870	-0.538	+0.052	-0.002
300-200	1968:		2967.842		+	tension	
400-500	-	-	1005.116	.129	-	-	+0.013
500-600	-	-	1945.858	.878	-	-	+0.020
400-600	-	-	2938.293	.328	-	-	+0.035
100-500	-	-	986.230	.236	-	-	+0.006

Considering the fact that the surveys of 1938 have lesser
accuracy than the new surveys, the differences between 1965 and
1938 are to be judged with caution. The newer differences in the
distances from the combined adjustment of co-ordinates of the
quadrilaterals show partially a correspondence, partially changing
signs and different sizes.

From this can be deduced at present that movements not at all similar happen in the fracture zone, but that total tension and pressure zones with expansions and contractions are to be found. The total distance shows now after an apparent shortening by 44.5 cm in 27 years, a minimally significant extension of 5 cm in 6 years, i.e. approximately 1 cm/year. Moreover this is a provisional value and must be proved by further surveys in 1975 and thereafter.

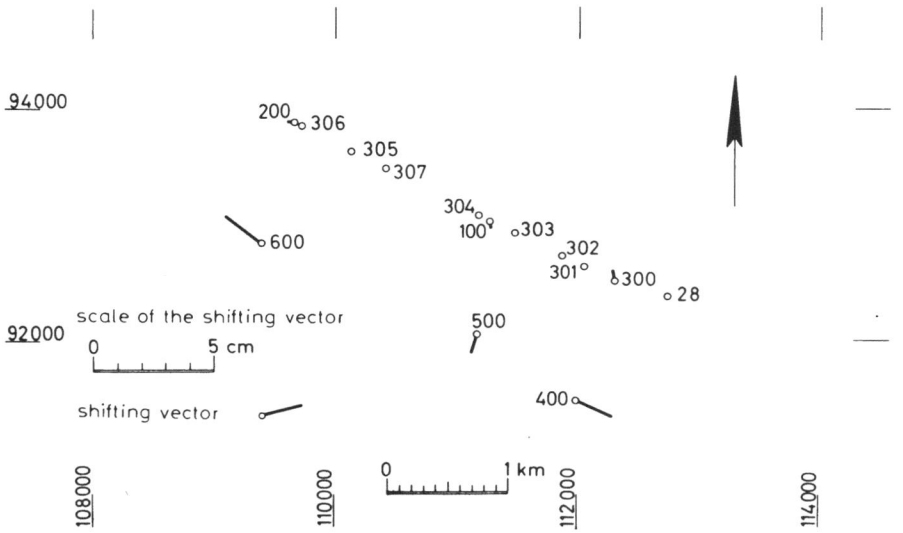

Fig. 2. Gjastikki deformation quadrilateral 1967-71.

In Fig. 2 the differences of 1971 to 1967 are shown, and there the expansion between the fix points 600 and 400 by approximately 2 cm is to be observed.

The height differences between the profile fix-points were determined in 1938 partly by simple levelling, partly by trigonometric heighting in 1965 and 1971 by precise levellings [5].

Here should be given again the available representation of the alteration of elevation numbers from 1938 to 1965 by Prof. Spickernagel with the remark about the differing accuracy of the surveys and the serious doubt about the given altitude differences occasioned by this. It shows negative height changes-subsidences?- in the fix-points lying in the fracture zone in the order of about 1 cm per year. The proof of these sizes is to be conducted by further precise levellings in the coming years (Fig. 3).

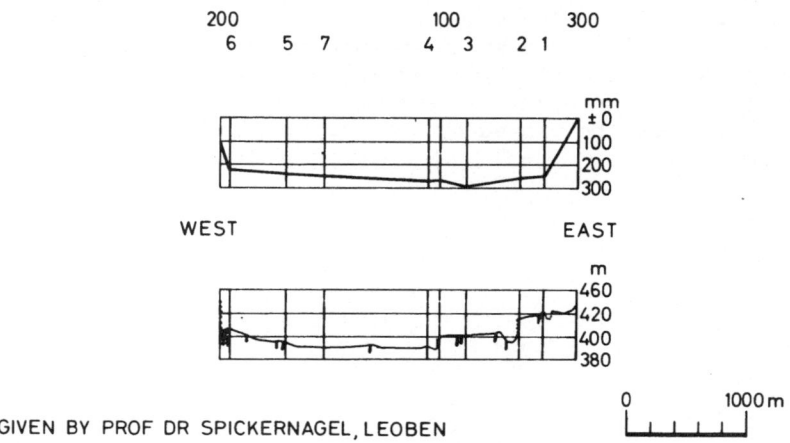

GIVEN BY PROF DR SPICKERNAGEL, LEOBEN

Fig. 3. Gjastikki profile height-changes 1938-65.

It is to be pointed out that the levelling team in 1967 made a topographical survey in a 150 m strip along the profile. In each of 5 chosen fractures were inserted, in 1967, 2 extensometers of recording accuracy ±0.2 mm and paper-feed of 4 cm per day. The recording conducted over 3 years showed no periodic or systematic difference in the width between the fracture walls.

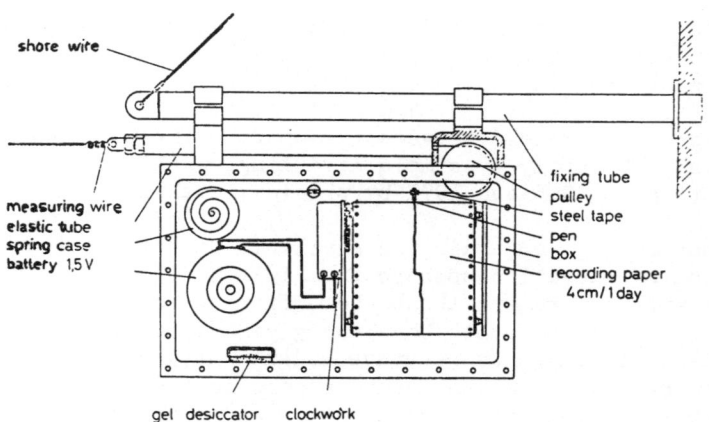

Fig. 4. Extensometer as used in Gjastikki 1967-70.

The supplementary 9 bench marks marked out in the vicinity of
each of these recording places showed no tilts up to 1971. Such
measurements of wall-distances are to be attempted again in 1975
with much greater accuracy (Fig. 4).

2.3.2 Klaustur

A further deformation profile was marked permanently in 1965 in
the Myvatn area, about 20 km eastwards from the lake, northwards
from the road leading to Grimsstadir - near the branching of the
road to Dettifoss - and in 1968 was increased to 2 quadrilaterals.
The co-ordinates of the fix-points were determined with precision
theodolite WILD T3 (1965), JENOPTIK THEO 010 (1968), ZEISS TH2
(1971) and distance measurers Geodimeter Model 4B (1965), Tellurometer MRA 4 (1968) and Tellurometer MA 100 (1971) both in the
geodetic control network and relative to one another.

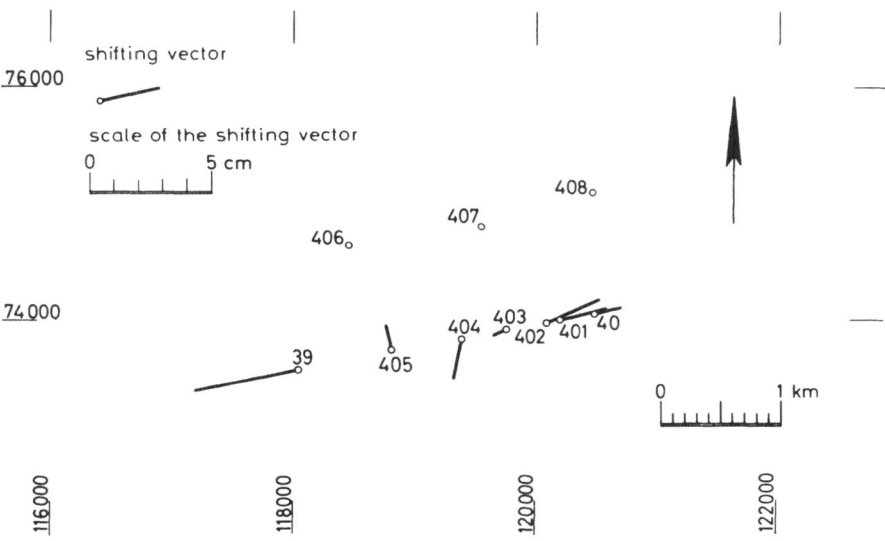

Fig. 5. Klaustur deformation profile 1965-71.

     The results of the analysis of difference for the measurements
of 1965 and 1971 are presented in Fig. 5. According to this the
significant changes of position of fix-points are proved, that is
for the whole of the profile length of 2.5 km, to an extent of
+ 6.5 cm, which means an extension of approximately + 1 cm per
year.

     A comparison of the elevation numbers of the bench marks deter-

mined by precise levellings in 1965, 1968 and 1971 allows minimal
changes in the height of some points to be recognized, but as the
sizes lie within the range of the normal error of observation they
must be determined more reliably by further measurements.

3. THINGVELLIR AREA

3.1 Geodetic control network

The geodetic control network crosses the Thingvallavatn and the
bordering fracture zone; it was reconnoitred and marked out in
1967. Stations are situated in the West and East on the shield
volcanoes Mosfellsheidi and Lyngdalsheidi, in the South on the
Hengill massif.

The direction observation of 1967 was carried out with the
precision theodolite WILD T3, the distance measurement with
Tellurometer MRA 3.

The whole control network was measured for a second time in
1971, for this the precision theodolite WILD T3 and Tellurometer
MRA 3 were again used as well as additional Tellurometer MRA 4 and
Geodimeter Model 8.

The two independently measured networks were adjusted with
directions and distances. This produced the following mean square
errors of unit weight.

1967	1971
direction distance	direction distance
$\pm 9.1^{cc}$ ± 12 cm	$\pm 6.2^{cc}$ ± 3 cm

The deformation analysis according to Pelzer for the geodetic
network with direction and distance measurements of 1967 and 1971
produces the rejection of the hypothesis that no alterations
happened - therefore the probability of changes of position of
individual fix-points. The greatest differences are however only
minimally significant. They are apparent changes, geologically
not maintainable, which allow one to suppose systematic errors,
although the frequency-controls of the distance-measurers undertaken before and after the field periods show only essentially
smaller changes than would be caused by the order of the changes
(Fig. 6).

It must be said, that with modern geodetic measurers and
analysis of the records no horizontal change of position of fix-
points in this geodetic control network could as yet be established.

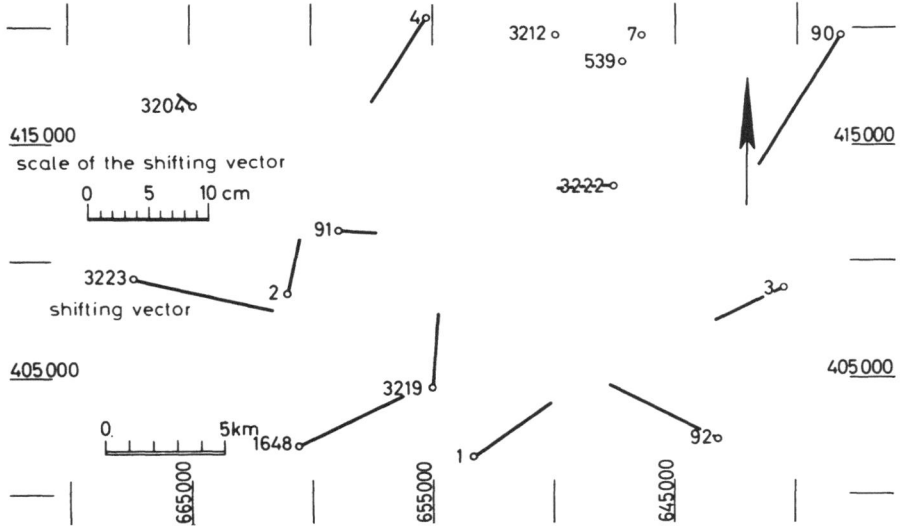

Fig. 6. Triangulation network SW-Iceland 1967-71.

Further repeat-measurings, if possible with still greater accuracy, are to follow.

3.2 Levelling loop

In 1967 in the Thingvellir area a levelling loop 62 km long with 50 bench marks was marked out around the lake. The measurements were carried out in both directions with precise levels ZEISS NI2 with plane-parallel plate and JENA KONI 007 and invar rods. The standard deviation was ±0.5 mm/km. The first repeat measurement took place in 1971 with the standard deviation ±0.3 mm/km.

The results of both measurements have not yet been published. The alteration of elevation numbers shows however, apart from individual local point changes, a clear tendency to height changes of the fix-points and thereby to vertical movements. The sizes are to be finally determined by repeat measurements in the coming years.

3.3 Deformation quadrilaterals

3.3.1 Position measuring

In 1967 two deformation quadrilaterals were marked out with 8 fix-

points to the north of the Thingvallavatn. In the first survey the
theodolite JENOPTIK THEO 010 was used for direction observations,
and the AGA-Geodimeter Model 4B was used for the distance measurements. The first repeat survey took place in 1971 with ZEISS TH2
and the distance-measurers Tellurometer MA 100 and AGA-Laser-
Geodimeter Model 8.

The mean square errors of unit weight from the direction-
distance adjustments of coordinates proved to be:

1967		1971	
directions	distances	directions	distances
$\pm5.1^{cc}$	±1.3 cm	$\pm4.8^{cc}$	±0.8 cm

The alteration analysis of the adjusted results by an error
probability of 5% showed significant changes for the co-ordinates
of stations 7,3212 and 5001. The distances between fix-points are
set out in the following table:

from - to	reduced distances		differences	
	1967 Geodim.4B	1971 Tellurom.100 " Geodim. 8	1971-1967	Geod.8-MA 100
5- 6	2001.668	-		
		2001.696	+0.028	
6- 7	2899.200	-		
		2899.230	+0.030	
7- 539	1431.455	-		
		1431.462	+0.007	
3212- 539	2901.466	-		
		2901.488	+0.022	
3212- 7	3533.458	-		
		3533.506	+0.048	
6- 539	3129.397	-		
		3129.438	+0.041	
3212- 6	1906.873	1906.881		
		-	+0.008	
5001-3212	1988.783	1988.786	+0.003	+0.018
		1988.804	+0.021	
5001- 5	2000.171	2000.200	+0.029	-0.001
		2000.199	+0.028	
5001- 6	2865.559	2865.594	+0.035	+0.004
		2865.598	+0.039	

The consideration of the adjusted distances - without taking
the weight of the instruments into account - shows almost exclusively positive differences from 1967 to 1971, which could, in spite
of the frequency controls before and after the measuring periods
point to remaining systematic influences, which act as scale
alterations, (Fig. 7).

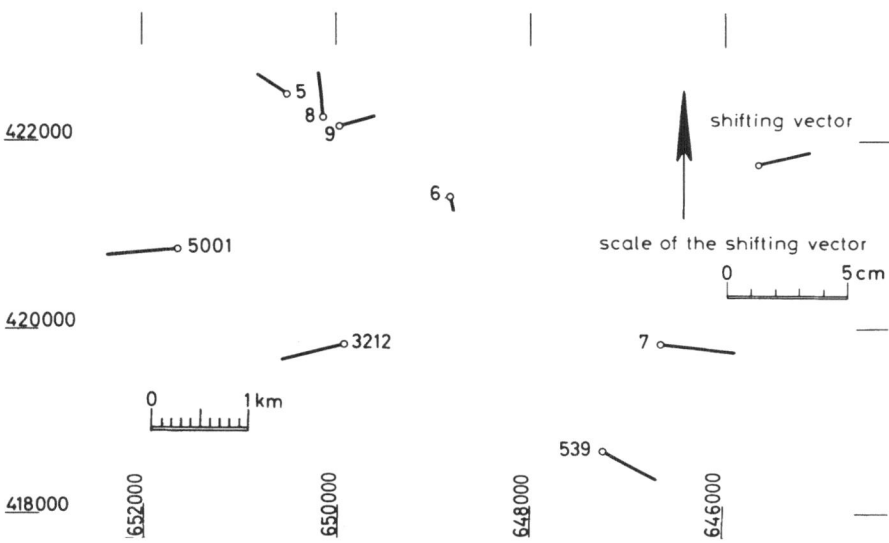

Fig. 7. Thingvellir deformation quadrilateral 1967-71.

As magnitude of the changes of position one can nevertheless ascertain from these results a lengthening of the distances normal to the fracture system of about + 1 cm/year. Further repeat measurements have to confirm these values.

It must be noted in this connection, that some of our stations in the deformation quadrilaterals are identical to or in relation to the fix-points of Prof. Tryggvason, Tulsa, and Prof. Mason, London. Although I do not know the final reduced distances, a comparison would be desirable.

In 1973 Prof. Bjerhammar, Stockholm, and his team have measured 8 distances in the quadrilaterals with a completely new prototype of the AGA-Geodimeter 700, for which no frequency corrections were yet known. The results with 7 positive signs are mostly about + 1 to + 3 cm longer than our distances in 1971.

3.3.2 Height measurements

The southern distances lines of the deformation quadrilaterals lie very close to the levelling loop, therefore till now special precise height measurements have not been attempted.

4. CONSEQUENCES AND CONCLUSIONS

4.1 Positional changes - horizontal movements

In the extended geodetic control networks there could not, as yet, from the measurements so far carried out with modern instruments according to considered procedures, neither in the Myvatn area nor in the Thingvellir area, be proved any significant horizontal positional changes of fix-points. The secular movements can indeed even by geodetic measurements of high accuracy and analyses which have been ascertained by mathematical statistics only be obtained from repeat measurements quantitively ascertained over decades.

On the other hand, from the test areas of smaller extension, laid out as quadrilaterals, horizontal movements could be determined as significantly secured. It is possible at present to indicate these, by various signs, perhaps as expansions and contractions with a magnitude of 1 cm/year.

4.2 Changes of height - vertical movements

The long levelling lines and loops which have been laid must still be measured several times with great accuracy, before conclusions about regional vertical movements can be made.

However, in the restricted test networks, some of the changes of height of fix-points which surpass the error of observation have again been determined.

4.3 Summary

However, the causes of the movements were not explicable by geodetic measurements alone, so that it is necessary for their significance and explanation to take into consideration the gravimetric, magnetic and seismic surveys. Results will be reported elsewhere; the complete analysis will follow after the next period of measurements.

The preparation of the future projects can, with the experience so far acquired, be carried out better than in the past. For this, particular value will be laid upon the use of improved and more accurate methods of measurement, as well as the evaluation and adjustment of the measurement data for the analysis of causes and effects of the errors of observation. Last but not least, however, the joint consideration of the results to date of all disciplines is important for the continuation of successful research work.

REFERENCES

1. K. Gerke, Ein Beitrag zur Bestimmung rezenter Erdkrusten-
bewegungen. Festschrift zum 70. Geburtstag von Prof. Dr.
Ing. Walter Grossmann, Stuttgart, 1967.
2. K. Gerke, Measurements of recent crustal movements in Iceland. Problems of recent crustal movements. Third International Symposium Leningrad U.S.S.R., Moskau, 1969.
3. F. Heumann, Untersuchungen im geodätischen Sondernetz in
Nordost-Island zu Messungen von 1938 und 1965. Dissertation
T.U. Braunschweig, 1972
4. H. Pelzer, Zur Analyse geodätischer Deformationsmessungen,
Veröffentl. d. Deutschen Geodätischen Kommission, Reihe C,
Heft 164, München, 1971.
5. H. Spickernagel, Höhenmessungen in Nord-Island, Mitteilun-
gen aus dem Markscheidewesen, Heft 4, Herne in Westfalen, 1966.

MARINE HEAT FLOW MEASUREMENTS IN THE NORWEGIAN-GREENLAND SEA AND
IN THE VICINITY OF ICELAND

Marcus G. Langseth and Gary W. Zielinski

Lamont-Doherty Geological Observatory, Palisades,
New York 10964 and Department of Geological Sciences,
Columbia University, New York, New York 10027, U.S.A.

ABSTRACT. We report 55 new measurements of heat flow in the
Norwegian-Greenland Sea and 29 new measurements on the flanks of
the Reykjanes Ridge. Interpretations of the data are based only
on the most reliable measurements in areas with a low probability
of environment disturbance to the heat flow. The results in the
Norwegian-Greenland Sea show the expected decrease in heat flow
with distance from the presently active spreading center. Based
on a simple thermal model of sea-floor spreading, the average
heat flow in the Norwegian-Greenland Sea over areas of the sea
floor that can be dated is 2.7 HFU, which is anomalously high
compared to other oceans of comparable age. Measurements at
distances greater than 100 km from the Reykjanes Ridge axis show
a steady decrease with age, and when age is taken into account
the heat flow is also anomalously high relative to other spreading centers.

1. INTRODUCTION

Iceland appears to lie within a broad oceanic region of relatively
high heat flow. This result implies the occurrence of relatively
high temperatures at shallower depths. To explain this phenomenon
it is not necessary to assume that comparatively more heat is being
brought to the base of the lithosphere in the vicinity of Iceland.
The fundamental requirement is that the ratio of the heat being
brought to the base of the lithosphere to that being dissipated via
the sea-floor spreading mechanism be greater for this region than
for other oceanic spreading centers.

Kristjansson (ed.), Geodynamics of Iceland and the North Atlantic Area. 277-295. *All Rights Reserved.*
Copyright © 1974 *by D. Reidel Publishing Company, Dordrecht-Holland.*

Much of the discussion of this conference has focused on the
fact that the spreading center in Iceland is anomalous compared
to most spreading centers in the world. Recently Iceland has come
to be viewed as a "hot spot" and considerable significance has
been placed on such hot spots as a key to processes in the mantle.
Evidence is increasing that Iceland is but the emerged part of a
much larger anomalous oceanic region. The ridge to the south of
Iceland, the Reykjanes Ridge, is characterized by relatively
shallow depth and positive free-air gravity anomalies. Similarly,
abnormally shallow basin depths and positive free-air anomalies
are found in the Norwegian-Greenland Sea to the north. If the
shallow ocean floor in these regions is due to thermal expansion
as a result of anomalously high temperatures in the lithosphere,
this should be reflected in the heat flow at the surface.

In this paper we report a large number of new measurements
made in the area during the past eight years principally from the
R.V.VEMA and a few measurements made aboard the Russian research
vessel, the AKADEMIK KURCHATOV (Tables 1 and 2). By comparison
with heat-flow results in other oceans, we will show that the
whole region north of the Charlie-Gibbs fracture zone at 53°N and
the Norwegian-Greenland Sea has anomalously high heat flow. The
direct implication of this result is that high temperatures exist
at shallower depths below this region as compared with other
oceanic spreading centers. These high temperatures could result
from an imbalance between spreading rate in the area and heat flow
from the mantle compared to spreading centers in more normal areas
of the ocean.

The observations reported here were all made with the Ewing
thermograd apparatus which is used on a piston corer [1]. By
means of thermistor probes attached to the core barrel, temperature measurements at up to five depths in the sediment are made.
The thermal conductivity of the sediment is determined by making
measurements by the needle-probe method [2] on the sample returned
in the corer. In Fig. 1 we show a typical record from the thermograd apparatus. After the corer penetrates the sediment, all of
the traces corresponding to sediment probes are displaced upward,
indicating higher temperatures. These temperatures when plotted
versus depth show a linear increase (see Fig. 2) of about
1.0°C/10 m with depth. This gradient results from heat flow from
depth. Fig. 2 also shows the results of conductivity measurements
on the core sediments sample plotted versus depth. The gradient
and conductivity determined over each interval are used to
calculate the heat flow, which is shown in the right-hand plot.
The heat flow calculated between the topmost and bottom-most
point is shown as the dashed line.

TABLE 1

GEOTHERMAL DATA IN THE NORWEGIAN-GREENLAND SEA

Cruise	T'Grad Station #	Latitude	Longitude	Depth m	Grad. °C/10m	Cond. Cal/°C cm sec	Heat Flow ucal/cm² sec	Eval.	Topography	Sediment		Water
V-23	47	74°02'N	7°24'E	1928	1.90	2.35A*	4.47	8	rough ridge crest	VP** thin covering	VP	NP
	49	77°57'	0°12'	3052	0.71	2.89	2.05	8	abyssal plain	thick, flat lying	NP	NP
	51	76°59'	7°05'	2941	1.94	2.42A	4.69	8	side of mountain	thin sediment pocket	VP	P
	53	72°04'	1°24'	2360	2.86	2.24A	6.41	9	axial mountain base	small sediment pond	VP	NP
	54	70°59'	6°41'	3043	1.11	2.44A	2.74	8	abyssal plain	thick, flat lying	NP	NP
	55	64°48'	1°19'W	2930	0.32	3.86	1.24	10	gently sloping basin	thick, layered	NP	NP
	56	63°39'	1°22'E	1743	0.13	3.80	0.49	5	continental margin, flat	thick sediment wedge	P	P
V-27	4	68°27'	13°32'W	1719	1.37	2.33	3.19	6	tilted sediment pocket	.4 sec over sloping subbot.	NP	NP
	5	70°15'	13°05'	1392	1.08	2.13	2.30	6	irregular bottom	thin sediment pocket	P	NP
	7	74°28'	1°41'	3607	1.14	2.70	3.08	7	small, rolling hills	thick, stratified	NP	NP
	9	73°04'	4°49'E	2297	N/L	2.39	-	7	axial mountains	U-shaped sediment pocket	VP	NP
	10	72°11'	8°35'	2537	0.85	2.34	2.00	7	abyssal plain	thick, flat lying	NP	NP
	11	71°19'	12°04'	2299	0.60	2.58	1.54	6	abyssal plain	thick, flat lying	NP	NP
	12	70°24'	14°47'	2429	0.78	2.48	1.94	5	gently sloping	thick, slightly tilted	NP	NP
	21	78°04'	9°25'	2919	1.06	2.95	3.13	5	large V-shaped valley	thick pocket	VP	P
	23	74°59'	10°50'W	2897	0.87	2.85	2.48	4	shelf edge	thick, stratified	VP	NP
V-29	16	68°47'	20°46'	1326	1.70	1.81	3.08	9	flat portion of trough	thick, stratified	NP	NP
	18	70°48'	18°19'	1677	1.00	2.50	2.50	9	flat, continental margin	thick, flat lying	NP	NP
	19	72°26'	13°39'	1283	0.54	3.19	1.72	7	flank of sharp ridge	thin wedge	VP	P
	21	75°10'	10°52'	2462	0.61	2.53	1.54	8	hilly, continental slope	small pocket	P	P
	22	76°49'	1°20'	3137	0.68	2.70	1.84	8	flat basin	thick, slightly sloping	NP	NP
	28	62°54'	0°35'E	1171	0.21	3.00	0.62	5	continental slope	thick sediment wedge	P	NP
	29	68°43'	12°43'W	1822	N/L	2.47	-	8	flat, large basement relief	1/3 sec. sediment covering	P	NP
	30	72°04'	9°04'	2398	0.88	2.35	2.07	8	rough bottom over irreg. subbot.	thick but irregular	VP	NP
	31	69°23'	4°24'	3449	1.11	2.26	2.51	8	base of mountain	small wedge at base	VP	P
	32	67°53'	1°56'	3376	1.09	2.48	2.70	8	rough	small sediment pocket	VP	NP
	37	74°00'	4°14'E	3148	0.81	2.28	1.85	5	very rough	small sediment pocket	VP	NP
	38	70°56'	9°23'	2716	0.66	2.00	1.32	7	flat	thick, flat lying	NP	NP

TABLE 1 (Continued)

STATION ENVIRONMENT

Cruise	T'Grad Station #	Latitude	Longitude	Depth m	Grad. °C/Dm	Cond. Cal/°C cm sec	Heat Flow ucal/cm² sec	Eval.	Topography	Sediment		Water
V-29	138	66°44'	6°44'W	2459	0.71	2.41	1.71	8	gentle slope	very thick	NP	NP
	139	67°48'	6°41'	2549	0.77	2.22	1.71	7	gentle slope	very thick, stratified	NP	NP
	140	70°09'	7°20'	1443	0.81	2.27	1.84	9	broad, flat terrace	thick, stratified	NP	NP
	142	72°58'	6°59'	2609	0.92	2.48	2.28	8	smooth, low relief peaks	thick conformable	NP	NP
	143	75°56'	5°07'	3129	0.70	2.35	1.65	7	smooth, gentle slope	thick stratified	NP	NP
	144	73°48'	0°06'E	2964	0.48	2.42	1.16	6	near scarp, undulating bottom	rapidly thickening pocket	P	NP
	145	70°54'	4°55'W	1939	1.23	2.79	3.43	7	steep slope	thin sediment pocket	VP	VP
	146	68°23'	5°26'	3411	0.93	2.18A	2.03	7	gentle slope	thick, uniform	NP	NP
	147	65°11'	0°05'	2856	0.30	3.77	1.13	6	gentle slope	thick, with unconformities	NP	NP
V-30	98	67°31'	15°06'	860	1.81	2.03	3.68	9	flat plateau	uniform 1/3 sec.	NP	P
	99	66°51'	9°02'	1595	0.90	2.02	1.82	9	flat	thick, uniform	NP	NP
	100	65°05'	7°08'	2027	0.67	2.26A*	1.51	8	near scarp, rugged basement topa	thick V-shaped pocket.	VP	NP
	101	65°08'	5°18'	3812	0.81	2.26	1.84	9	flat floor of graben	thick, stratified.	NP	NP
	102	64°28'	4°57'	3316	0.71	2.26A*	1.61	8	narrow trough	pond	P	NP
	103	70°19'	9°32'	1308	1.08	1.82	1.97	10	steep peaks	thin	VP	P
	104	70°56'	16°56'	1472	1.26	2.14	2.70	10	low relief	1/3 sec. draped sediments	NP	P
	105	71°29'	14°38'	1247	1.29	2.71	3.50	9	shallow broad ridge	thick	NP	VP
	108	76°25'	2°15'E	3183	0.97	2.72	2.34	8	near steep peak	thick layer abutting peak.	P	NP
	109	77°52'	4°08'	3075	0.89	2.67	2.36	7	smooth, floored, trough	thick	NP	NP
	110	78°19'	2°00'	2006	0.63	3.10	1.94	7	top of broad ridge	thick, stratified	VP	P
	111	76°17'	9°07'	2243	1.10	2.57	2.81	9	relatively flat	thick, stratified	NP	P
	112	76°41'	6°52'	2413	>4.7	2.35	>11	6	near rift valley edge	thin	VP	NP
	113	70°11'	1°49'W	2905	0.96	2.67	2.57	8	gentle slope	thick, stratified	NP	NP
	114	69°39'	2°30'E	3210	0.78	2.32	1.82	10	flat floored basin	thick, stratified	NP	P
	116	67°41'	8°24'	1918	0.75	2.76A	2.07	6	continental slope foot	thick, stratified	P	P
	129	68°10'	5°45'	1911	0.52	2.70	1.40	10	gentle slope	thick	NP	NP
	131	67°10'	6°05'	1236	0.38	2.58	0.98	10	fairly flat	thick	NP	uncertain

* (A) indicates that the conductivity is assumed based on nearby values
** The letters NP, P or VP indicate our assessment of the probability of a
significant disturbance to near surface heat flow by topography, sediment
distribution and/or bottom water. NP = not probable; P = probable; and
VP = very probable.

TABLE 2

NEW GEOTHERMAL DATA IN THE REYKJANES RIDGE AREA

Cruise	T'Grad Station #	Latitude	Longitude	Depth m	Grad. °C/ 10m	Cond. Cal/°C cm sec	Heat Flow ucal/cm² sec	Eval.	STATION ENVIRONMENT Topography	Sediment		Water
V-23	16	56°04'N	44°33'W	3259	0.73	2.03	1.48	10	flat near peak	P** fairly thick	NP	NP
	19	59°46'	39°24'	2776	0.76	1.98	1.50	10	sloping	NP thick	NP	P
V-28	9	60°25'	35°50'	3009	N/L	-	-	8	flat	NP thick	NP	P
V-29	133	60°28'	20°58'	2603	1.27	1.73	2.19	7	gentle slope	NP thick, undulating layers	NP	uncertain
	134	60°48'	22°26'	2087	1.17	1.87	2.20	10	rugged basement topography	P fairly thick	NP	NP
	135	61°11'	23°50'	1851	1.26	1.80	2.28	10	irregular basement topography	NP thick	NP	NP
	136	61°33'	25°08'	1441	0.54	1.79	0.96	10	atop plateau in rugged basement	NP thick, stratified	NP	NP
	149	60°43'	26°38'	1699	0.86	1.91	1.64	10	large sediment pond	P 1/3 sec., fairly flat	NP	NP
	150	60°41'	27°15'	1304	2.83	1.98	5.60	8	top of flat topped Mount	P 1/4 sec. fairly flat	NP	NP
	151	60°41'	27°50'	1527	4.28	1.91	8.18	8	rugged	VP small pocket	VP	VP
	152	59°22'	34°42'	3098	0.32	1.77	0.57	9	flat, broad basement depression	NP thick	NP	NP
	153	57°50'	40°08'	3210	N/L	1.94	-	8	flat, irregular subbottom	NP 1/4 sec., stratified	NP	NP
	154	58°02'	42°48'	3208	0.98	2.04	2.02	8	flat	NP fairly thick	NP	NP
	155	56°11'	44°42'	3300	0.82	2.01	1.65	9	flat corrugated	NP thick	NP	NP
V-30	85	53°38'	44°16'	3686	0.81	2.02	1.65	10	hilly	NP thick	NP	NP
	86	56°56'	44°53'	3504	0.79	2.03	1.61	10	flat	NP 1/3 sec., stratified	NP	NP
	88	56°50'	43°43'	3428	0.98	1.88	1.84	10	rolling	NP thick, stratified	NP	P
	89	55°01'	43°53'	3392	0.83	1.92	1.54	10	low relief hills	NP thick	NP	NP
	91	56°47'	38°34'	3306	0.88	2.28	2.01	10	broad topographical high	NP thick	NP	NP
	93	57°52'	35°30'	2636	1.07	1.83	1.96	9	irregular	P edge of sediment pocket	VP	VP
	94	58°34'	35°30'	2457	N/L	2.20	-	8	near sharp peak	VP sediment pond	P	P
	95	63°09'	35°32'	2542	0.82	2.22	1.83	10	gently sloping	NP thick	NP	P
	96	64°04'	30°13'	2311	0.68	1.83	1.24	9	irregular	VP thick, conformable	VP	NP
	132	54°03'	24°11'	3434	0.44	3.01	1.33	9	rolling	NP thick	NP	NP

** The letters NP, P or VP indicate our assessment of the probability of a significant disturbance to near surface heat flow by topography, sediment distribution and/or bottom water. NP = not probable; P = probable; and VP = very probable.

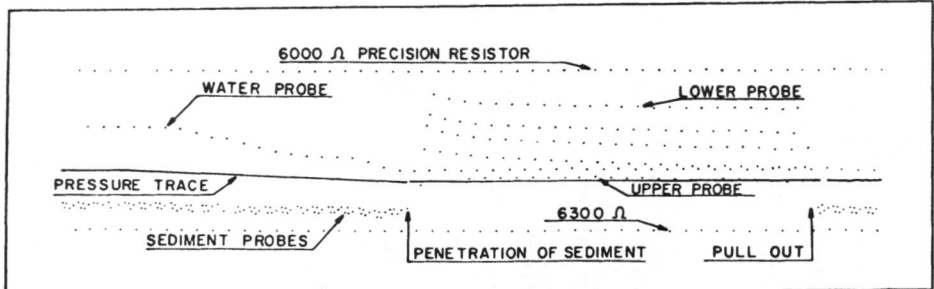

Fig. 1. A reproduction of a Ewing thermograd record showing the period during which the thermistor probes are in the sediment.

AKADEMIK KURCHATOV   STATION K1319

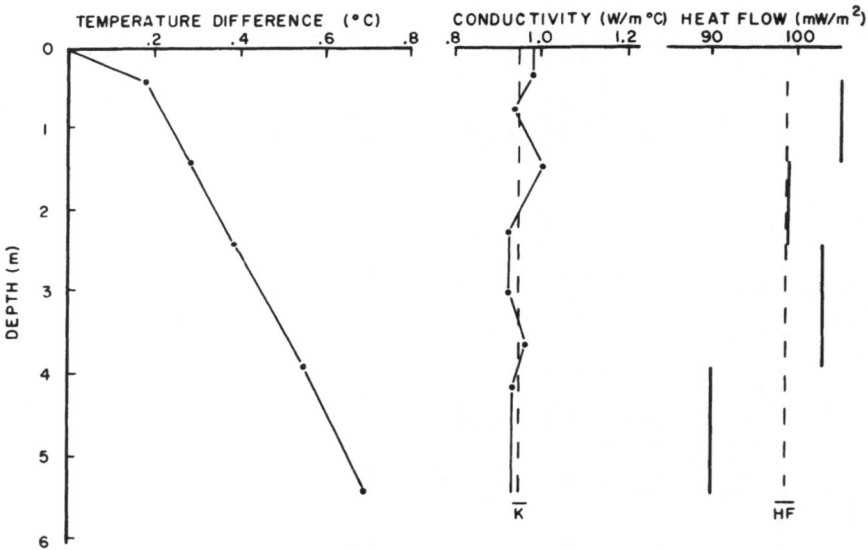

Fig. 2. Data from a thermograd station are shown plotted versus depth in the sediment. 41.8 mW/m^2 equals 1 μcal/cm^2-sec.

2. THE EVALUATION OF HEAT FLOW STATIONS

A most important part of the analysis is a critical evaluation of each heat-flow station. There are two parts to this evaluation: 1) an assessment of experimental procedure and instrument accuracy, and 2) an appraisal of the station's local environment and the probability of local disturbance to the near-surface heat flow.

The first part of the evaluation entails answering such questions as: How many probes penetrated the sediment and how deeply? Was the corer disturbed during the observational period? Was the recorder functioning properly? How accurate and representative are the conductivity measurements? Answers to these questions provide an evaluation of how reliable and accurate the heat-flow measurement at each station is. The evaluation for each station is given in Tables 1 and 2 and is based on a scale from 0-10. Generally, evaluations of 9-10 represent the most reliable and accurate stations; 7-8, good accuracy but reliability not firmly established; 4-6, stations of low accuracy and reliability; 0-3, unreliable measurements.

Regardless of how reliable a measurement is, it is equally important to know how representative it is of heat flow from the crust and mantle below the station. The "representativeness" of a station is assessed by examining the environment of each station to estimate the probability of local disturbance to the near-surface heat flow. In our evaluation we examine two major aspects of the environment: 1) the temperature stability of the bottom water, and 2) local topography and sediment distribution.

An appraisal of the temperature stability of the bottom water can be made from sediment and near-bottom water temperature measurements at the station. Measurements that show the heat flow is not uniform with depth in the sediment are probably made in areas of variable bottom water temperature. Large temperature gradients in the near-bottom water can make possible large excursions in bottom water temperature and consequently transient disturbances to the near-surface heat flow.

It is well known that the surface heat flow is disturbed in areas of high-relief topography and patchy sediment distribution. See, for example, Von Herzen and Uyeda [3]. Seismic profiler records and echo-sounder records can be used to determine the proximity of a station to a large topographic feature or a sediment-rock boundary. Stations in close proximity to such features have a high probability of local disturbance.

Our evaluation of the probability of local disturbance to the measured heat flow is given in Tables 1 and 2 together with a brief description of the topographic and sedimentary environment. In the interpretation of the heat-flow results to be presented in the following sections, we have considered only data which have reliability and accuracy evaluations greater than 5 and a low probability of being locally disturbed. In general we find removing the unrepresentative and unreliable measurements greatly decreases the scatter of values. As a result regional averages

and trends determined from the cleansed group of data should have
greater significance.

3. RESULTS FROM THE NORWEGIAN-GREENLAND SEA

Since 1966 Lamont-Doherty Geological Observatory ships have made
55 successful measurements of heat flow in the floor of the
Norwegian-Greenland Sea, which are shown on the map of Fig. 3.
These stations are geographically well distributed over the area
encompassed by the sea. At each of the stations the temperature
and conductivity data have been carefully evaluated to determine
the reliability of each heat-flow determination. The local
topography, sediment distribution and bottom water temperature
structure have been examined to assess the probability of local
disturbance to the heat flow measured at each station. These
assessments are presented in Table 1 together with the geothermal
data.

     Looking at the map of Fig. 3, we see that stations very near
the axis of the Mohns and Knipovich Ridges yield very high values
of heat flow. A heat flow greater than 11 HFU (1HFU $=10^{-6}$cal/cm^2
-sec) was observed at station V30-112. No measurements were made
close to the Iceland-Jan Mayen Ridge axis; however, our measurements on the nearby flanks have relatively high values, i.e. 2.3-
3.7 HFU. It is perhaps worth noting that, unlike other mid-ocean
ridges, no station near the axis yielded very low heat flow
although measurements at one station, V27-9, indicated heat flow
that was not uniform with depth.

     A line of heat-flow measurements on the western flank of the
Iceland-Jan Mayen Ridge was reported by Lachenbruch and Marshall
[4]. The locations of these stations are shown in Fig. 3. One
of their stations, which is very near to our station V28-16,
yielded a value of 2.67 HFU, whereas our station gave a value of
3.08 HFU. The principal difference was due to the gradient
determination. Lachenbruch and Marshall gave strong evidence for
appreciable variations in bottom water temperature in this area,
and since their measurements were relatively shallow, <3 m, there
is a possibility that the observed difference results from
transient heat flow in the upper sediment induced by bottom water
variations, particularly if the variations are aperiodic.

     We have examined the heat-flow distribution relative to
prominent morphological features and crustal age in three provinces
separated by the two prominent fracture zones - the Jan Mayen
fracture zone and the Greenland fracture zone. The values used
in the following discussion are only the most reliable measurements
that we have judged to have a low probability of local environmental
disturbance. The distribution of values in the complex of basins,

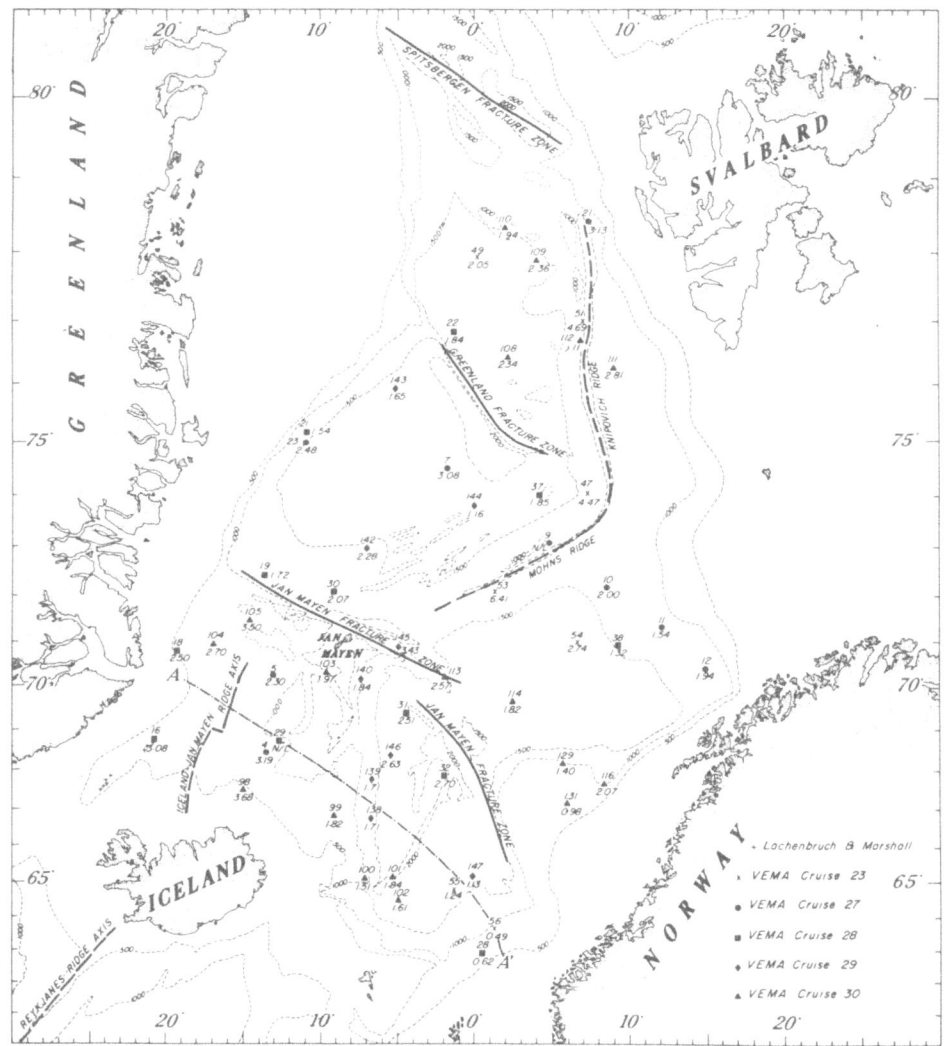

Fig. 3. Map showing the distribution of heat-flow stations in the Norwegian-Greenland Sea. The number above the symbol is the T'Grad station number. The number below the symbol is the heat-flow value for that station. N/L indicates a heat flow which is not uniform with respect to depth. The depth contours are in fathoms and are based on a map compiled by Talwani and Eldholm [7].

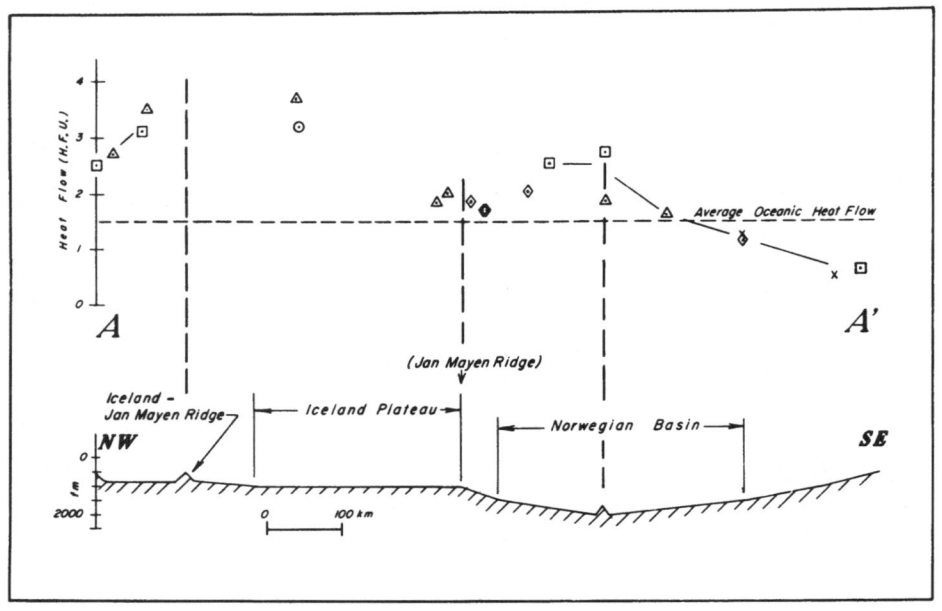

Fig. 4. Schematic section along A-A' in Fig. 3 across the
Norwegian-Greenland Sea south of the Jan Mayen fracture zone
showing the relation of heat flow to the various morphologic
features in that area.

plateaus and ridges south of the Jan Mayen fracture zone are shown
in profile by projecting the values on to Section A-A' in Fig. 3;
the profile is shown in Fig. 4.

   First we note a general decrease of heat flow from the northwest toward the southeast, reflecting the fact that the present
active center of spreading lies near the western end of section
A-A'. Four values on the western flank of the Iceland-Jan Mayen
Ridge show a uniform decrease of heat flow from high values
(3.5 HFU) to lower values (2.5 HFU) with increasing distance from
the ridge axis.

   Two values about twice the oceanic average (3.2-3.7 HFU) were
measured over the eastern flank of the Iceland-Jan Mayen ridge
complex. The values are relatively high considering their distance
from the axis of the ridge. It is possible that these measurements
are not associated with the present spreading center of the Iceland-
Jan Mayen Ridge since they lie east of the well-defined magnetic
pattern over this ridge [5].

A group of 5 measurements over the Jan Mayen Ridge and the region extending south from the ridge all indicate a heat flow of about 1.8 HFU. These values are relatively low compared to adjacent measurements and suggest that the Jan Mayen Ridge is thermally inactive relative to the Iceland–Jan Mayen Ridge and Iceland Plateau. In the Norwegian Basin relatively high values are observed over the area where Johnson and Heezen [6] have postulated a spreading center that became extinct at anomaly 7 time [7]. Values of 2.0-2.5 HFU would be expected over a ridge that became extinct 20-25 m.y.b.p. since, as we will show later, values in this range are found over crust of similar age in other parts of the Norwegian-Greenland Sea.

On the southeastern side of the Norwegian Basin there is a striking decrease of heat flow as the Norwegian continental margin is approached. Two values of 1.24 and 1.13 HFU were measured at the foot of the continental rise. These values are typical of older oceanic crust (>60 m.y.b.p.) in other areas of the world. The two easternmost values of 0.49 and 0.62 HFU are exceptionally low, but we believe these measurements are reliable. The two stations with very low values lie near the Faeroe-Shetland escarpment defined by Talwani and Eldholm [8]. The values are lower than the heat flow typically measured over the oldest oceanic and continental regions, 0.9-1.1 HFU; therefore, it is difficult to explain these very low heat flows in terms of a static crustal model. The conductivities measured at all four stations were abnormally high (3.0-4.0 mcal/cm-°C-sec). The sediments in the core samples contain a high percentage of sand and volcanic glass. Difficulties in conductivity measurement of sandy sediments are commonly encountered because of their tendency to lose water when opened. However, if we assume the conductivity measurements are unreliable and assume lower values typical of the western part of the Norwegian Basin, even lower heat flows would result.

Mohns Ridge appears to have been the most persistent spreading center during the evolution of the Norwegian-Greenland Sea Basin. Magnetic anomalies in the adjacent basins are well defined [7] so that we have plotted the heat flow on either side of Mohns Ridge against crustal age as shown in Fig. 5. On the same figure we show by solid dots the mean heat flow for various age provinces in the North Pacific reported by Sclater and Francheteau [9]. Comparison indicates that generally the heat flow over Mohns Ridge is greater than that of the North Pacific.

A similar plot is shown for data on the flanks of the Knipovich and Iceland-Jan Mayen Ridges (Fig. 6). A steady decrease of heat flow with age is seen and, as in the case of Mohns Ridge, these values all exceed the means of the North Pacific for regions of comparable age.

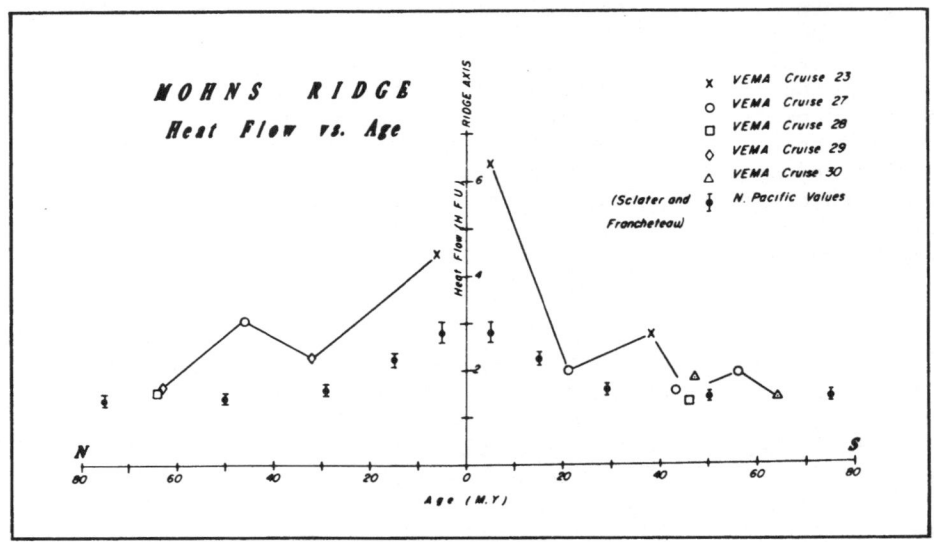

Fig. 5. Heat flow as a function of age of the oceanic crust for
the Mohns Ridge area. The North Pacific average heat-flow values
of Sclater and Francheteau [9] are plotted on both sides of the
ordinate for comparison. The ages are based on the interpretations of Talwani and Eldholm [7].

4. A THERMAL MODEL OF THE LITHOSPHERE BELOW THE NORWEGIAN-GREEN-LAND SEA

All of the heat-flow measurements from the Norwegian-Greenland
Sea where the magnetic anomalies permit an estimate of crustal age
shown in Fig. 7. This combined data clearly show that heat flow
in the Norwegian-Greenland Sea is higher than in the North Pacific
for ocean crust of the same age. Over crust about 60 m.y. old,
several stations in good agreement indicate heat flow is about
1.6 HFU.

We have attempted to fit the combined data of the Norwegian
Sea, using the cooling plate model of McKenzie [10]. In this
model the lithosphere is assumed to be a slab of uniform thermal
properties moving away from the axis at a constant velocity. The
accreting edge of the slab and the base of the slab are assumed
to be an isotherm and the surface temperature is assumed to be a
uniform colder temperature usually taken as zero. The lithosphere
cools by conduction as it moves away from the axis.

To obtain a reasonable fit to the combined Norwegian-Green-

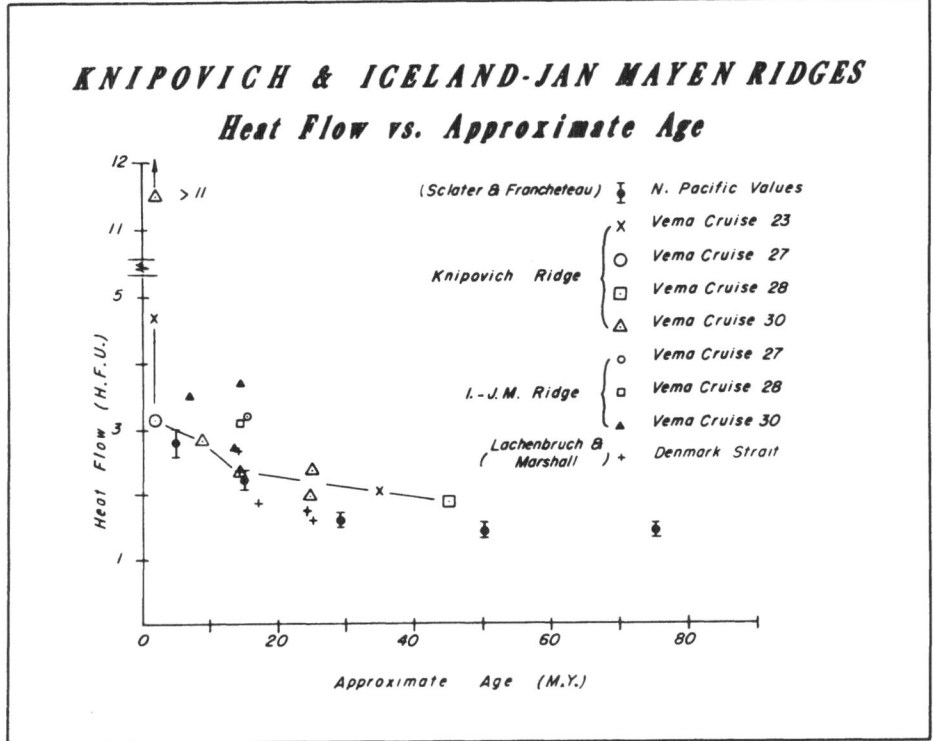

Fig. 6. Heat flow plotted as a function of approximate age of oceanic crust for the Knipovich and Iceland–Jan Mayen Ridge areas. The crustal age is based on the interpretation of magnetic anomalies by Talwani and Eldholm [7].

land Sea data we assumed a lithosphere 60 km thick with a temperature at the axis and base of the slab of 1475°C and a velocity of 1 cm/yr based on the magnetic lineations. The thermal conductivity of the lithosphere is assumed to be 0.007 cal/cm-°C-sec. The results of this model are shown by the solid line in Fig. 7. Such a fit is far from unique. An adequate fit could also be obtained with a basal temperature of 1250°C if the lithosphere were assumed to be 50 km thick rather than 60 km.

If we use the theoretical curve as a method to interpolate between the data and assume that heat flow within 10 km of the axis is 7.0 HFU, we find by integrating this curve that the average heat flow in the Norwegian-Greenland Sea is about 2.7 HFU – an average considerably higher than that over other oceanic areas of comparable age. It is well known that the elevation of the sea

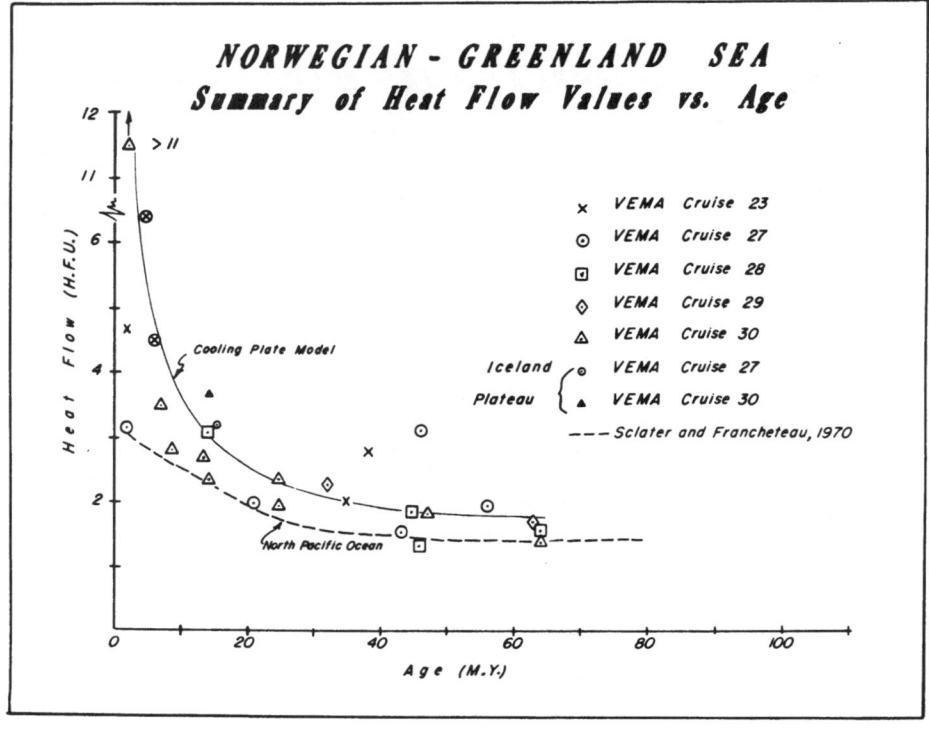

Fig. 7. Summary plot of heat flow as a function of age of the oceanic crust for the Norwegian-Greenland Sea where the age of oceanic crust is known from magnetic chronology. The dashed line is an interpolation of the mean heat flow-versus-age data of Sclater and Francheteau [9] for the North Pacific. The solid line represents a fit of the data to the cooling plate model of McKenzie [10] assuming a spreading velocity of 1 cm/yr, a thickness of 60 km and basal temperature of 1475°C.

floor in this area is also anomalously high. There is much evidence that the large wavelength topography of the sea floor results from thermal expansion of the upper mantle [9,11]. The correlation of high heat flow and shallow-floored basins strengthens the evidence that anomalously high temperatures are found in the lithosphere below the Norwegian-Greenland Sea.

5.   RESULTS FROM THE REYKJANES RIDGE

As part of a comprehensive geophysical survey of the axial zone of the Reykjanes Ridge in 1966, numerous heat-flow measurements

were made within 100 km of the axis [12]. The area of the survey
is shown by the rectangle in Fig. 8. Since that survey was made
we have had an opportunity to make 29 successful measurements over
the broad region between Iceland and the Charlie-Gibbs fracture
zone. Four of these stations over the axis of the ridge were made
aboard the AKADEMIK KURCHATOV last summer. The geothermal data
for these new stations except for the AKADEMIK KURCHATOV stations
are given in Table 2. The locations of these stations and their
heat-flow values are shown in Fig. 8. Most of the new observations
lie on the western side of the ridge in the relatively shallow
basin of the Irminger Sea between Greenland and the Reykjanes
Ridge.

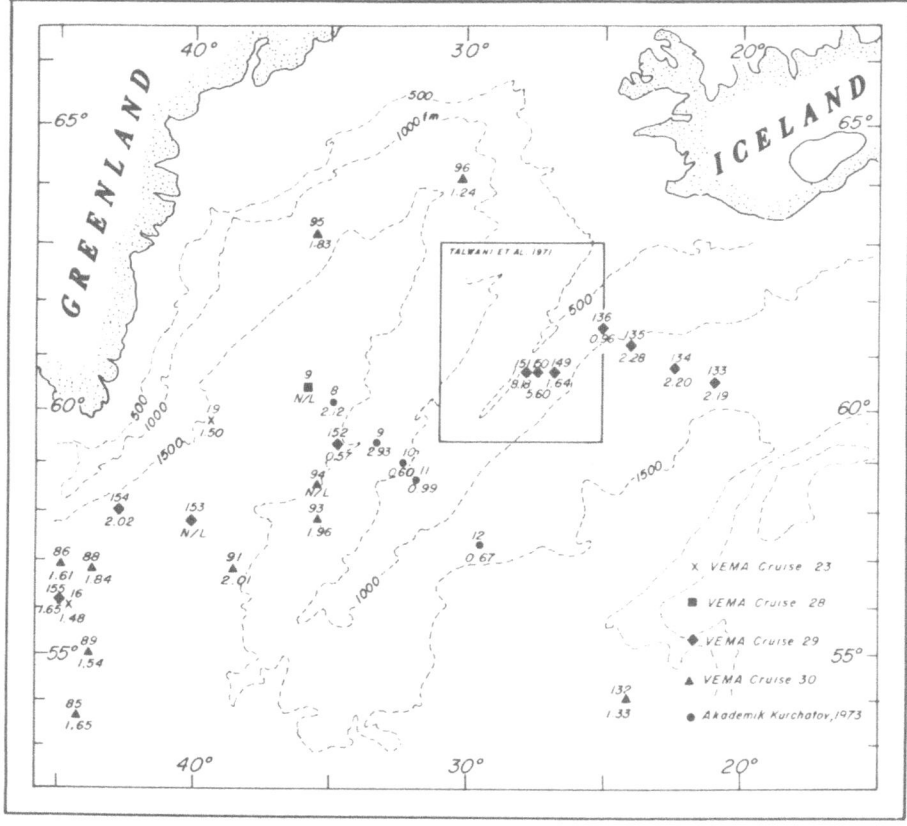

Fig. 8. Map showing the distribution of new heat-flow stations on
the Reykjanes Ridge. The numbers above the symbols are T'Grad
station numbers. The numbers below the symbols give heat-flow
values. Depth is in fathoms. The rectangle outlines the area of
an earlier comprehensive geophysical survey [12].

The earlier survey, during which numerous closely spaced
observations were made, showed a rather complex pattern of heat
flow relative to the ridge axis. One serious difficulty with observations over the shallower part of this ridge is the likelihood of variable bottom water temperatures. Temperature profiles
in the sediment down to 10 m indicated that sea-floor heat flow
in the shallower parts of the ridge (i.e. <1000 fm) was strongly
affected by recent changes in the near-bottom water temperature.
Consequently only gradient measurements deeper than 3-5 m would
result mainly from heat flow coming from depth. Even after the
effects of variable water temperature were eliminated to the
extent possible, a confusing pattern of heat flow relative to the
ridge axis resulted. In particular, a band of extraordinarily
low values was observed at a distance of 70 km from the axis, and
some stations within 30 km of the ridge axis yielded values from
1-2.3 HFU, which is near to or only slightly above the average of
ocean basins.

Evidence is now strong that in regions near the axis of mid-
ocean ridges where the sediment cover is relatively sparse, water
circulation in the fractured upper crust may severely disturb the
surficial heat flow [12,13]. If significant heat is being transferred by the convection of water, then the measurements of conductive heat flow in the sediment should underestimate the true
heat flow. Substantial water circulation in the upper crust would
also explain the large scatter of heat-flow values which has
always characterized measurements in the axial zone.

We note that some of our new observations quite near the axis
are below normal (three KURCHATOV stations and a VEMA 29 station).
On the other hand, a line of three stations on the eastern side of
the axis at about 60.5°N measures very high heat flows within a
few tens of kms of the axis. In general, the values within 100 km
should not be used to assess the heat flow from the Reykjanes
Ridge until the thermal processes that transfer heat from depth
to the sea floor are better understood. This is not true for
measurements at greater distances from the axis. In these areas
the sediment cover is generally thicker and more uniform which
masks the thermal effects of water circulation in the crust
beneath. The thicker sediment also diminishes possible effects
of basement topography. Far from the axis, water depths are
greater so that the probability of bottom water variations affect-
ing the heat flow are minimized. The uniformity of values in
these regions exemplified by the group of measurements south of
Greenland and on the eastern flank of the ridge (see Fig. 8)
further indicates the absence of large surficial disturbances in
those areas.

When we examine the distribution of heat-flow values relative
to the ridge axis, combining the new data with previous results,

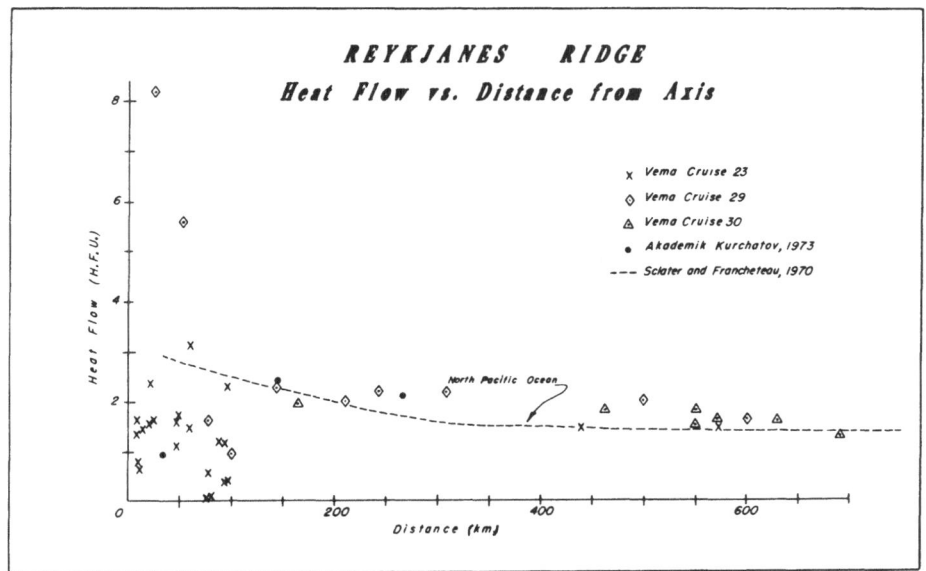

Fig. 9. Heat flow as a function of distance from the axis of the
Reykjanes Ridge. The dashed line is the same as in Fig. 7.

we obtain a quite different conclusion than previously reported
based on values close to the ridge axis. In Fig. 9 the values
are plotted versus distance from the axis. The dashed line represents the North Pacific heat flow-versus-age data assuming a
spreading rate of 1 cm/yr for the Reykjanes Ridge. The great
scatter and predominantly low values within 100 km of the axis
are very apparent, but it is clear that when data farther than
100 km from the axis are considered, the heat-flow values lie
above the North Pacific curve. The relatively high heat flow in
particular is supported by the very reliable and quite uniform
values between 400 and 600 km. These values average about 1.6 HFU,
which is significantly higher than the mean heat flow through the
floor of the North Atlantic basins south of 43°N [14].

The result that anomalously high heat flow characterizes this
region retlative to the North Atlantic supports the conclusion of
Haigh [15] that high temperatures lie at relatively shallow depths
below the basins adjacent to the Reykjanes Ridge compared to the
more southern parts of the North Atlantic. Haigh's conclusions
were based on thermal and petrochemical modelling of the gravity
and topography of the area.

6. CONCLUSIONS

Our new results show that Iceland sits at the mid point of a vast region of anomalously high heat flow which extends from 53°N to as far north in the Norwegian Sea as we have measurements (∿80°N). The anomalous heat flow is accompanied by relatively shallow depths and positive free-air gravity anomalies. These observations are consistent with abnormally high temperatures at relatively shallow depth compared to other oceanic areas of the same age.

The sea-floor spreading mechanism is an efficient way for the mantle to lose heat. Heat is absorbed in the partial fusion of mantle material to make the basaltic crust. Large quantities of heat are brought upward near the surface by the formation of the crust and injection of material at the ridge axis, and once near the surface more effective cooling is possible by conduction. The general observation that heat flow and elevation of the ridges as a function of age over most oceanic spreading centers are very similar suggests that there is balance between spreading rate and heat flow from the deeper mantle to the base of the lithosphere, the faster spreading rates being associated with greater heat flow from the mantle.

One way of defining the anomalous conditions in the Iceland area is to state that the ratio of spreading rate to heat being brought to the base of the lithosphere is less there than is found over most spreading centers of the deep oceans. The main features of the oceanic region surrounding Iceland may result from the fact that the plates bounding the spreading center are not moving apart at a rate fast enough to dissipate heat flowing up from below to produce the lower heat flow, deeper ocean floor and thinner crust that we observe in most other so-called normal oceanic regions.

This point of view suggests that the cause of a hot spot of the Icelandic type may lie in the lithosphere's ability to respond to asthenospheric motions below it. If for some reason the plates bounding the region are prevented from rifting apart at a rate that forms a more normal ocean, then a hot spot will result. This would be a sufficient condition by this reasoning to form all the features we associate with Iceland and surrounding seas. It would not be necessary to call on heat coming up from the lower mantle, as is proposed in the mantle-plume hypothesis although such a mechanism is not ruled out.

The point we wish to emphasize is that the first-order diagnosis of the "Iceland hot spot" is a maladjustment between spreading rate and heat flow from the mantle compared to other oceanic spreading regimes. This diagnosis should be the starting point of conjecture about what "hot spots" tell us about the nature and mechanics of the asthenosphere beneath.

ACKNOWLEDGEMENTS

The data reported here were collected during a continuing geophysical survey of the Norwegian-Greenland Sea carried out over
five summers during the past eight years. Much of the work was
done as a cooperative program with the Norwegian Geotraverse
Committee. The Lamont-Doherty Geological Observatory program
has been led by M. Talwani, and we are grateful to him and
O. Eldholm for use of their age interpretations before publication. We wish to acknowledge the indispensable help of Capt.
H. C. Kohler, the scientific staff and crew of the VEMA. Annette
Trefzer drafted the figures and Sivia Brodsky prepared the manuscript. G. Pálmason and G. Bryan read the manuscript and offered
helpful criticisms. The data acquisition and research were
supported by National Science Foundation grant GA27281 and Office
of Naval Research contract N00014-67-A-0108-0004.

Lamont-Doherty Geological Observatory Contribution No. 2125.

REFERENCES

1. R. Gerard, M. G. Langseth Jr. and M. Ewing, J. Geophys. Res.,
    67, 785, 1962.
2. R. Von Herzen and A. E. Maxwell, J. Geophys. Res., 64, 1557,
    1959.
3. R. Von Herzen and S. Uyeda, J. Geophys. Res., 68, 4219, 1963.
4. A. H. Lachenbruch and B. V. Marshall, J. Geophys. Res., 73,
    5829, 1968.
5. P. R. Vogt and O. E. Avery, J. Geophys. Res., 79, 363, 1974.
6. G. L. Johnson and B. C. Heezen, Deep-Sea Res., 14, 755, 1967.
7. M. Talwani and O. Eldholm, in prep., 1974.
8. M. Talwani and O. Eldholm, Geol. Soc. Am. Bull., 83, 3575,
    1972.
9. J. G. Sclater and J. Francheteau, Geophys. J. Astr. Soc.,
    20, 509, 1970.
10. D. P. McKenzie, J. Geophys. Res., 72, 6261, 1967.
11. X. LePichon and M. G. Langseth Jr., Tectonophysics, 8,
    319, 1969.
12. M. Talwani, C. C. Windisch and M. G. Langseth Jr., J. Geophys.
    Res., 76, 473, 1971.
13. C. R. B. Lister, Geophys. J. R. Astr. Soc., 26, 515, 1972.
14. M. G. Langseth Jr., X. LePichon and M. Ewing, J. Geophys.
    Res., 71, 5321, 1966.
15. B. I. R. Haigh, Geophys. J. R. Astr. Soc., 33, 405, 1973.

HEAT FLOW AND HYDROTHERMAL ACTIVITY IN ICELAND

G. Pálmason

Lamont-Doherty Geological Observatory, Palisades, New
York 10964, U.S.A. and Department of Geological Sciences,
Columbia University. (On leave from the National Energy
Authority, Reykjavik, Iceland.)

ABSTRACT. The physical characteristics of the hydrothermal activity
in Iceland are discussed in the light of recent exploration results.
A review is given of regional heat flow and its interpretation in
terms of plate tectonics. Finally, the problem of hydrothermal
activity at mid-ocean ridge axes is discussed in relation to
present knowledge of such phenomena in Iceland.

1. INTRODUCTION

Hydrothermal activity is a very common phenomenon in Iceland. It
bears a conspicuous relationship to the zone of rifting and
volcanism, where the most intensive hydrothermal areas occur.
Because of their economic importance as energy sources for space
heating and various other applications, the hydrothermal resources
have been studied on an increasing scale in the last few decades,
both by surface geological, geophysical and geochemical surveys,
and by exploration drilling. In the areas that have been exploited
to date, production drilling has been carried out, yielding
detailed information on lithology, aquifers, subsurface temperature,
rock alteration, and so on. A program of regional heat-flow
studies has been in progress for the last 10-20 years in order to
provide a better understanding of the background heat-flow pattern
and its relationship to the hydrothermal activity.

It is not possible to review here but a part of the work that
has been done to study the hydrothermal areas in Iceland. Emphasis
will be placed on those aspects which appear to be of importance
for current ideas of ocean-floor spreading. The chemical characteristics of the hydrothermal systems are discussed in another paper

Kristjansson (ed.), Geodynamics of Iceland and the North Atlantic Area. 297-306. *All Rights Reserved.*
Copyright © 1974 by D. Reidel Publishing Company, Dordrecht-Holland.

of this Institute [1].

2. DISTRIBUTION AND NATURAL HEAT DISCHARGE OF HYDROTHERMAL AREAS

Bödvarsson [2] has divided the thermal areas into two main groups,
the high-temperature and the low-temperature areas, on the basis
of subsurface temperatures, either measured or indirectly inferred
from surface chemical characteristics. The about 17 high-temperature areas which are estimated to account for some 90% of the
total heat discharge by hydrothermal activity are all located in
the active zone of rifting and volcanism (Fig. 1). Where drill holes

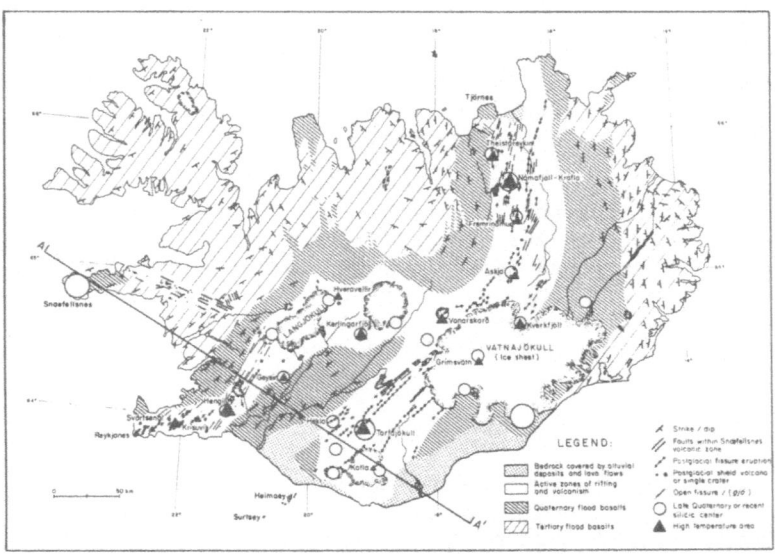

Fig. 1. Geological map of Iceland showing the location of the high-
temperature hydrothermal areas [23].

have been sunk in these areas a temperature of over 200°C at a depth
of a few hundred meters has always been found (cf. Fig. 2). The low-
temperature areas are widely scattered over the older regions with a
tendency to a somewhat greater concentration at the flanks of the
active zones. The eastern Iceland Tertiary areas are practically
without thermal manifestations.

    The natural heat discharge of the high-temperature areas is
rather poorly known. It varies greatly from one area to another.
The Reykjanes area at the tip of the Reykjanes Peninsula, which
covers an area of only about 1 km^2, has been estimated to have a
natural heat output of about 20 MW [3]. The partly subglacial
Kverkfjöll thermal area at the northern margin of the Vatnajökull
ice sheet has been estimated to have a natural heat output of
over 1000 MW [4]. A similar value is considered likely for the

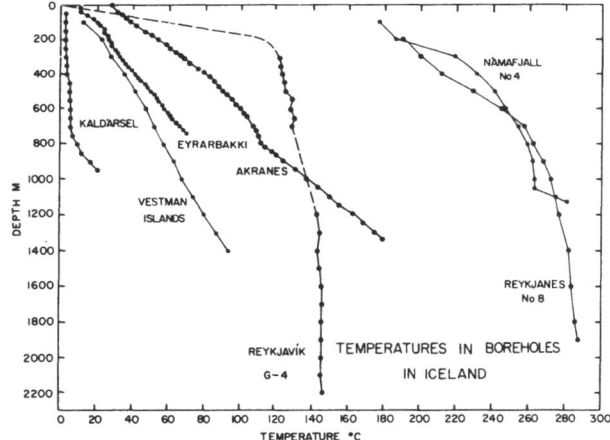

Fig. 2. Examples of temperature profiles from boreholes in Iceland illustrating disturbances by movement of water.

relatively large Torfajökull thermal area in central southern Iceland, which extends over an area of approximately 100 km^2. Other high-temperature areas are probably intermediate in heat discharge. For the total natural heat discharge of the high-temperature areas it appears reasonable to use an estimate of about 4000 MW [2]. There appears to be a variation along the volcanic zone in the intensity of hydrothermal activity, the intensity being greater near central Iceland than towards the north and southwest. A similar variation in the rate of extrusive volcanism was pointed out by Jakobsson [5].

3. SOME EXPLORATION RESULTS

At the present time exploration of the high-temperature areas has mainly been confined to the areas on the Reykjanes Peninsula and in the northern part of the volcanic zone in northeast Iceland. Of these areas the Reykjanes area at the tip of the peninsula is unique in that the geothermal fluid is sea water. The exploration of this area has been described by Björnsson et al. [3]. Drill holes to a maximum depth of 1750 m have revealed temperatures of over 290°C (cf. Fig. 2). Aquifers appear to be more abundant in the basalt formation below about 1000 meters depth than in the upper hyaloclastite formation. From a study of the rocks penetrated it was inferred that a minimum average subsidence rate of 0.5 mm/yr was taking place as a result of the piling up of new volcanic products at the surface. In other explored high-temperature areas borehole temperatures in the range 200–300°C have been found also. Maximum borehole depths are usually in the range

1000-2000 m. None of the holes drilled so far reaches the seismic layer 3 (P-wave velocity about 6.5 km/sec), but several of them penetrate well into basalts with P-wave velocity in the range 4-5 km/sec.

The hydrothermal areas are considered to be the outlets of extensive water circulation systems where the water is heated by contact with rock. The depth of circulation is not known. It has been suggested that the crustal seismic layer 3 might be relatively impermeable and constitute a bottom for the circulation systems [2]. This may be true in the case of the low-temperature areas where water temperatures are usually not much above 100°C. But it is likely that in the active zone of rifting and volcanism water circulation reaches well into layer 3, which is commonly found there at depths of 2-4 km [6]. Microearthquake surveys in the volcanic zone in southwest Iceland [7,8,9] have shown that the most active microearthquake zones often coincide with high-temperature areas. Focal depths of the microearthquakes appear to fall mostly in the upper part of seismic layer 3 [8] in a depth range of approximately 2-6 km. Although the exact relationship between the microearthquakes and the hydrothermal activity is not known, it must be considered very likely that water convection extends into layer 3 in the active zone of rifting and volcanism. This is also indicated by the high water temperatures found in the high-temperature areas.

It should be remarked here that the Icelandic high-temperature systems are all as far as is known of the so-called hot-water convective type. Vapor-dominated systems [10], which occur e.g. in Lardarello, Italy, and at the Geysers, California, have not been found in Iceland. This is probably related to the relatively large permeability of the upper crust in the active zone of rifting and volcanism and to an abundant supply of meteoric water.

The problem of the heat source for the hydrothermal systems has often been discussed in the literature [e.g. 11,12,2]. Two viewpoints have been advanced: One is that the heat source is the general outflow of heat from the Earth's interior; the other is that intrusions associated with volcanic activity supply the heat to the convecting water. The modern hypothesis of plate tectonics and crustal accretion at diverging plate boundaries unifies these two viewpoints into a single concept. In the zone of crustal accretion, dykes and other irregular intrusions are injected into the zone, and to a large extent the intensity of dyke injection controls the crustal temperature in the zone, and thus also the heat flow to the surface of the adjacent lithospheric plates. One may say that the general outward flow of heat is the heat source for the hydrothermal areas, but this relatively high heat flow is in turn brought about by the injection of numerous dykes and other intrusions in the zone of crustal accretion. The distribution of

high- and low-temperature areas in Iceland is in good general
agreement with this concept.

4. REGIONAL HEAT FLOW

A program of regional heat flow studies has been in progress for
the last 10-20 years. Holes drilled specifically for obtaining
subsurface temperatures are usually 100 m deep. Several deeper
holes, drilled for other purposes, have also been found to yield
useful temperature data. The distance of the holes from major
hydrothermal areas and the non-disturbance of the temperature profiles by flow of water have been used as criteria for the "representativeness" of the temperature profiles. In the case of single
boreholes these criteria may not be sufficient as disturbance by
water movement may occur beneath the bottom of the holes. If,
however, a systematic pattern of gradient values is observed over
an area several tens of kilometers in extent, one may have some
confidence that the pattern is indicative of temperature conditions
deep in the crust. The presently available data on regional heat
flow are given in Fig. 3 and in ref. [13], Table 2. Only gradients

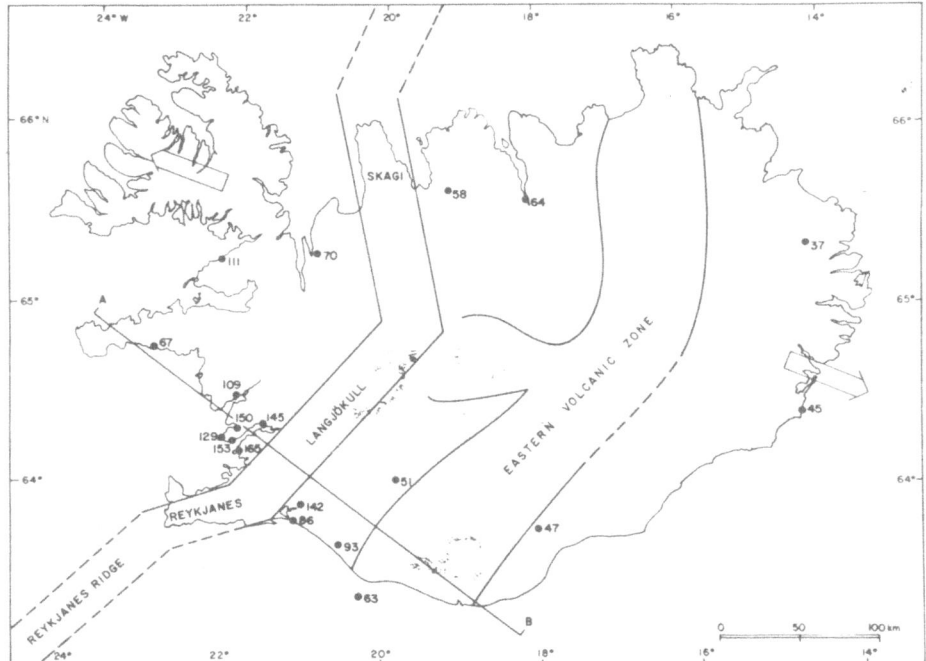

Fig. 3. Regional surface thermal gradients in boreholes in
Iceland [13].

have been measured, and a conductivity of 1.9 $Wm^{-1}°C^{-1}$ has been used to estimate heat flows. Conductivity measurements are in progress and they indicate that a somewhat lower value may be appropriate (B.J. Polyak, pers. communication). It is not likely, however, that the pattern of heat flow will change much from the gradient pattern as a result of new conductivity measurements.

The thermal gradients in southwest Iceland show a rather pronounced relationship to the Reykjanes–Langjökull zone which is the continuation landward of the Reykjanes Ridge axial zone. Fig. 4 shows the values projected on a line perpendicular to the zone. The trend towards higher values near the zone is rather regular, especially on the west side. Another conspicuous feature is the absence of a conductive heat flow anomaly associated with

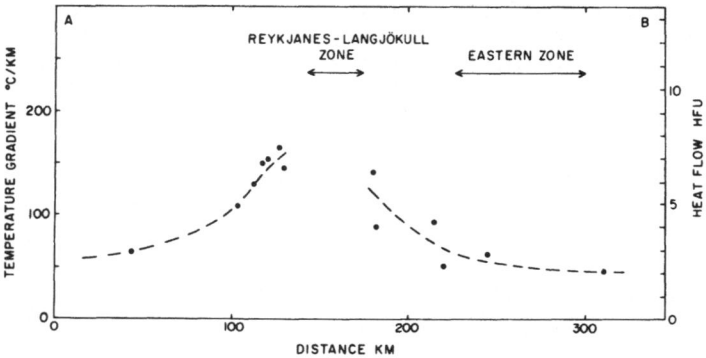

Fig. 4. Thermal gradients in southwest Iceland projected on the line A–B in Fig. 3 [13].

the eastern volcanic zone, despite the fact that high-temperature hydrothermal activity is abundant in that zone. These low values near the eastern zone are also confirmed by a recent unpublished gradient value at Thórshöfn close to the zone in northeast Iceland (gradient about 50°C/km) and by the relatively low values from eastern Iceland.

The absence of a flanking conductive heat flow anomaly associated with the eastern volcanic zone is in agreement with other evidence suggesting that the eastern zone became active only a few million years ago [14,15,16].

It thus appears likely that the main bulk of the Icelandic basalt pile was accreted in a western Iceland zone, possibly the present Reykjanes–Langjökull zone. In Fig. 5 the regional gradient data are plotted versus distance from the Reykjanes–Langjökull zone

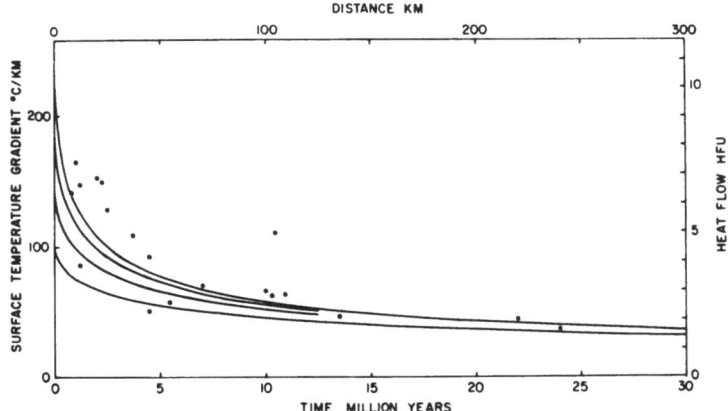

Fig. 5. Comparison of measured gradients in Iceland with
theoretical gradients for a cooling lithospheric plate [13].

as schematically shown in Fig. 3. Small corrections have been made
to account for the assumed shifting of the volcanic zones. The
solid lines show surface gradients from model calculations for
the Icelandic zone of rifting and volcanism [13]. The measured
values are slightly higher than the calculated ones. This can at
least in part be explained by the effect of erosion in the coastal
areas of Iceland from where most of the measured values come.
Otherwise the agreement between measured and calculated values
appears to be satisfactory. The regional heat flow pattern in
Iceland thus seems in a general way compatible with model calculations based on conductive cooling of lithospheric plates.

5. TOTAL HEAT TRANSPORT IN THE AXIAL ZONE

Heat is transported to the surface in the axial zone by the three
processes of extrusive volcanism, heat conduction and hydrothermal
activity. The first part can be estimated from average volume rates
of extrusive volcanism [e.g. 17,5]. The hydrothermal component can
also be estimated, albeit rather roughly [2]. The conductive
component cannot be measured directly in relatively shallow holes
in the axial zone because of thermal disturbances from water movement. It is possible, however, by model calculations of crustal
generation, using parameters from the Icelandic zone, to estimate
the sum of the conductive and the hydrothermal components [13].
The three components thus estimated are as follows, calculated
per unit length of the axial zone, which is taken as 300 km long:

Extrusive volcanism: 21 megawatts/km
Heat conduction: 21   "
Hydrothermal activity: 14   "

Within the uncertainty of these estimates the three components
can be considered approximately equal.

6. HYDROTHERMAL ACTIVITY IN MID-OCEAN RIDGE AXIAL ZONES

The possible occurrence of hydrothermal activity in submarine
ridge axial zones has been suggested several times [18,19]. It
is of some interest to discuss this possibility in the light of
known conditions in the Icelandic zone to try to predict how likely
such occurrences may be.

On the basis of theoretical studies of water convection in
permeable layers heated from below [e.g. 20,21,18,22] it may be
assumed that the significant properties controlling the onset of
convection are the thickness and permeability of the layer and the
thermal gradient before the onset of convection. The higher the
values of these three parameters, the more likely it is that
convection takes place. It is difficult to predict the permeability
and its variation with depth at mid-ocean ridge axes, but there
appears a priori no reason why it should be significantly different
from what it is in the Icelandic zone.

Some predictions about the undisturbed thermal gradient in the
axial zone of mid-ocean ridges can be made by comparison with
Iceland. The undisturbed gradient in the axial zone appears to be
controlled mainly by the two processes of dyke injection, which
heats up the crust, and subsidence resulting from lava flows piled
up at the surface. The subsidence has a cooling effect. The heat
source strength due to dyke injection depends on the spreading
rate and on the width of the dyke injection zone. Model calculations for the Icelandic axial zone [13] indicate an undisturbed
surface gradient of 2-300°C/km. At most mid-ocean ridge axes the
spreading rates are higher than in Iceland, the width of the dyke
injection zone narrower, and the lava production rate smaller than
in the Icelandic zone. All these circumstances tend to produce a
higher undisturbed thermal gradient in the upper crust of the
axial zone than in Iceland.

Thus, if permeability conditions are similar on submarine
axial segments to what they are in Iceland, conditions for water
convection in the upper crust are more favorable on the submarine
segments than in Iceland. The abundant hydrothermal activity in
Iceland then indicates rather strongly that similar phenomena
must be common on the submarine axial zones also. The Reykjanes
thermal area at the southwestern tip of Iceland is probably

transitional between the land- and marine-type hydrothermal
activity.

ACKNOWLEDGEMENTS

The borehole temperature data discussed in this paper are from
the files of the Icelandic National Energy Authority (Orku-
stofnun). The paper was written during the author's stay at
Lamont-Doherty Geological Observatory as Vetlesen Visiting
Professor.

Lamont-Doherty Geological Observatory Contribution No. 2124.

REFERENCES

1. S. Arnorsson, Thermal fluids in Iceland as regards composition
    and geological structure (this volume).
2. G. Bödvarsson, Jökull, 11, 29, 1961.
3. S. Björnsson, S. Arnórsson and J. Tómasson, Bull. Am. Ass.
    Petrol. Geol., 56, 2380, 1972.
4. J.D. Friedman, R.S. Williams, Jr., S. Thórarinsson and
    G. Pálmason, Jökull, 22, 27, 1972.
5. S.P. Jakobsson, Lithos, 5, 365, 1972.
6. G. Pálmason, Crustal structure of Iceland from explosion
    seismology. Rit 40, Soc. Sci. Islandica, 187 pp., 1971.
7. P.L. Ward, G. Pálmason and C. Drake, J. Geophys. Res., 74,
    665, 1969.
8. P.L. Ward and S. Björnsson, J. Geophys. Res., 76, 3953, 1971.
9. F.W. Klein, P. Einarsson and M. Wyss, J. Geophys. Res., 78,
    5084, 1973.
10. D.E. White, L.J.P. Muffler and A.H. Truesdell, Econ. Geol.,
    66, 75, 1971.
11. T. Einarsson, Uber das Wesen der Heissen Quellen Islands.
    Rit 26, Soc. Sci. Islandica, 91 pp., 1942.
12. G. Bödvarsson, Timarit V.F.I. (Reykjavik), 39, 69, 1954.
13. G. Pálmason, Geophys. J. R. Astr. Soc., 33, 451, 1973.
14. K. Saemundsson, Geol. Soc. Am. Bull., 85, 495, 1974.
15. R.L. Wilson and M.W. McElhinny, Geophys. J. R. Astr. Soc.,
    in press, 1974.
16. G. Pálmason, The insular margin of Iceland, in: The Geology
    of Continental Margins (Ed. C.A. Burk and C.L. Drake), in
    press, 1974.
17. S. Thórarinsson, Hekla and Katla, in: Iceland and Mid-Ocean
    Ridges (Ed. S. Björnsson). Rit 38, Soc. Sci. Islandica, 190,
    1967.
18. J.W. Elder, Physical processes in geothermal areas, in:
    Terrestrial Heat Flow (Ed. W.H.K. Lee). AGU Geophys.
    Monograph 8, 211, 1965.

19. G. Pálmason, On heat flow in Iceland in relation to the Mid-
     Atlantic Ridge, in: Iceland and Mid-Ocean Ridges (Ed. S.
     Björnsson). Rit 38, Soc. Sci. Islandica, 111, 1967.
20. E.R. Lapwood, Proc. Cambridge Phil. Soc., 44, 508, 1948.
21. R.A. Wooding, J. Fluid Mech., 2, 273, 1957.
22. I.G. Donaldson, Geothermics, Spec. Issue 2, 649, 1970.
23. G. Pálmason and K. Saemundsson, Ann. Rev. Earth Planet. Sci.,
     2, 25, 1974.

THE COMPOSITION OF THERMAL FLUIDS IN ICELAND AND GEOLOGICAL
FEATURES RELATED TO THE THERMAL ACTIVITY

Stefan Arnorsson

National Energy Authority,
Reykjavik, Iceland.

ABSTRACT. Hydrothermal areas in Iceland have been divided into
high and low temperature areas. In the drilled high temperature
areas underground temperatures are above 200°C in the uppermost
1000 meters but in the low temperature areas underground temperatures are below 150°C at similar depths. Underground temperatures
in the hydrothermal areas are rather strongly expressed by their
surface thermal manifestations, and undrilled areas are classified as high or low temperature on the basis of the type of these
manifestations. It is believed that there is a genetic difference
bewteen high and low temperature areas, the former deriving their
thermal energy from shallow level intrusions, but the latter by
contact with hot rock during a deep convection cycle of the water.
The high temperature areas tend to occur within volcanic complexes
of basaltic as well as differentiated rocks where intrusive activity
is intense. It is postulated that the bulk composition of thermal
fluids in Iceland is dominantly governed by three variables. They
are temperature, rock type, and influx of sea water into the heat
source areas of hydrothermal systems. Juvenile source of sulphur
and carbon may also contribute to the bulk composition of the fluids
in some of the high temperature areas.

1. GEOLOGICAL FEATURES AND THERMAL ACTIVITY

Thermal areas in Iceland have been divided into two types, high
temperature areas and low temperature areas [1]. The high temperature areas of which 17 have been identified, are all located within
the zones of post-glacial and late-quaternary volcanism (Fig. 1).
The low temperature areas are on the other hand located in tertiary
and early quaternary rocks. Drillhole data indicate that tempera-

Kristjansson (ed.), Geodynamics of Iceland and the North Atlantic Area. 307-323. *All Rights Reserved.*

tures are above 200°C in the uppermost 1000 meters of the high
temperature areas but less than 150°C in the uppermost 1000 meters
of the low temperature areas.

It is believed that there is a genetic difference between the
high and low temperature areas which is reflected in the mentioned
underground temperatures.

The high temperature areas are closely associated with central
volcanic complexes. These complexes are characterised with intense
volcanic activity and differentiated rocks [2,3]. Comparison with
their eroded tertiary analogues [see 4,5] strongly suggests that
intrusive activity is also intense. The intrusions are mostly sills
and dykes which are most commonly a few meters thick. In eroded
tertiary central complexes the intrusions may amount to as much as
50% of the rock by volume [4]. The life time of these complexes is
probably of the order of one million years and it seems reasonable
to assume that the intrusive activity covers similar span of time.
The intense intrusive activity which characterises the central
complexes where the high temperature areas tend to occur strongly
suggests that the thermal energy of these areas is derived from
shallow level intrusions. The heat in the low temperature areas

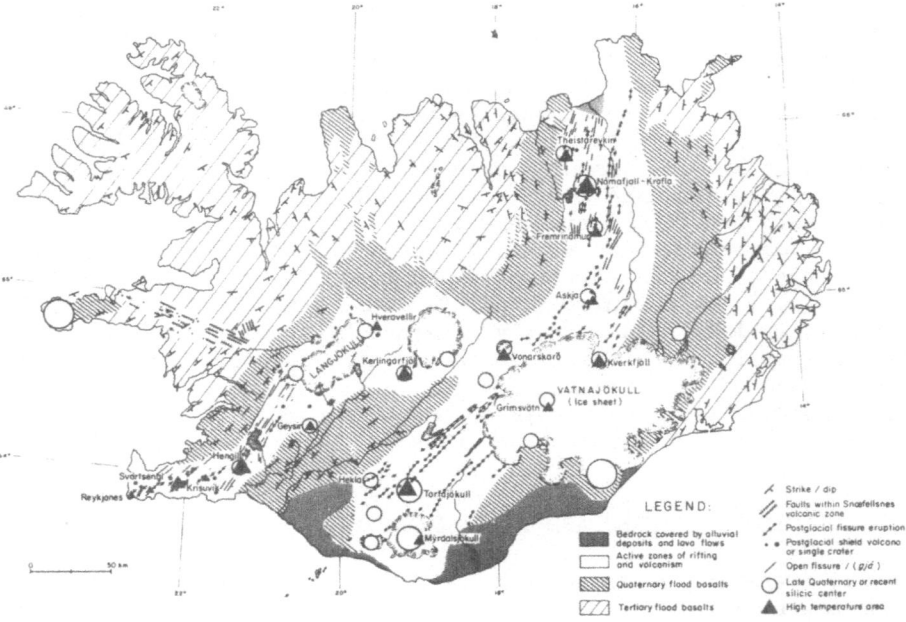

Fig. 1. The major geological elements of Iceland and the distribution of high temperature hydrothermal activity. Compiled by K.
Saemundsson.

is on the other hand considered to be extracted from the rocks with which the water of these areas comes into contact during its circulation to depths of few kilometers. This has been the prevailing view in Iceland for some 30 years since this theory was first put forward [6]. By contrast opinion has been somewhat divided regarding the high temperature areas and the view has also been held that the higher temperatures in the high.temperature areas are due to higher thermal gradients in the volcanic belts as compared with the older rocks and that heat provided from shallow level intrusions may not be so significant. In this respect it is noteworthy to point out that it is relatively easy to transfer heat from the small intrusions which are typical for the central volcanic complexes into hydrothermal systems.

Underground temperature conditions are rather strongly expressed in the type of surface thermal manifestations in the thermal areas. Thus alkaline water springs characterise the low temperature areas where little or no flashing of the water takes place below the surface and rock alteration at the surface is insignificant or non-existing. Flashing of ascending water in the high temperature areas results in drastic local alteration of the rock at the surface and produces completely different type of surface thermal manifestations. The flashing may reach depths of as much as 1000 meters depending on the underground temperatures. When the water flashes, the dissolved gases in this water are transferred into the steam. These gases are mostly carbon dioxide, hydrogen sulphide, and hydrogen. The ascending steam, containing these gases, mixes locally with downward percolatin rain water and the oxygen in this water oxidises the hydrogen sulphid to sulphur or sulphate. The resulting water becomes acid and it is this acid water that is responsible for the intense rock alteration at the surface in the high temperature areas. Steam vents, mud pools and native sulphur deposits are typical thermal manifestations in the hot altered ground of the high temperature areas. Alkaline thermal springs tend to be lacking.

Thermal springs in the low temperature areas are often linearily distributed along dykes and faults and it is evident that these structures facilitate movement of the thermal water to the surface, the faults probably because of their high permeability. Columnarily jointed dykes may also be permeable but it is evident that massive dykes may be a blockage to thermal ground water movement at depth and thermal springs will therefore emerge on the "upstream" side of these dykes. Scoriaceous tops and bases of lava flows provide the significant permeable horizontal channelways. Analysis of the mentioned structures which control water movement plays an important role in the location of drillholes in the low temperature areas.

Caldera complexes seem to be a characteristic feature of the tertiary silicic centres. Such a structure is prominent in one high temperature area, Askja. Recently caldera structures have been revealed in two other high temperature areas, Krafla and Torfajokull. These structures are not apparent in the surface topography due to accumulation of volcanic material within the cauldron subsidence [3,7]. All surface thermal activity is located within these calderas. Since the subsidence is considered to result from the emptying of a magma chamber in the roots of the complex, it follows that there is a close correlation between magmatic activity and hydrothermal activity as seen at the surface. In the Torfajokull high temperature area, the caldera is mostly filled with acid volcanics. Eruptive northeast striking tectonic fissures which run through the caldera complex have produced basaltic magma outside the complex but dacitic and rhyolitic magma within it. At the borders of the caldera composite lava has been erupted. This is considered to indicate the existence of acidic magma at depth under the caldera complex [3]. Xenoliths of gabbro and granophyre are rather common in some of the volcanics, particularly ignimbrites, which are located within the caldera in Krafla.

The view is commonly held now that acid magma under Iceland originates by partial melting of basalt in the lower part of the crust. The relatively high percentage of acidic rocks in Iceland would accordingly be attributed to anomalously thick crust in Iceland and acidic rocks would be expected to be scarcer on the Mid-Atlantic ridge where the crust is thinner. Bearing in mind the suggested genetic association between high temperature hydrothermal activity and silicic centers in Iceland, it might be expected that such activity would be not so widely distributed on the Mid-Atlantic ridge as it is in Iceland. It ought, however, to be pointed out here that there are four high temperature areas in the Reykjanes peninsula in southwest Iceland which are not known to be connected with any acid rocks although such rocks have been identified in all the other high temperature areas.

2. THE THERMAL FLUIDS

It is considered that the bulk composition of thermal fluids in Iceland is dominantly governed by three variables. They are temperature, rock type, and influx of sea water into the heat source areas of hydrothermal systems. Juvenile source of sulphur and carbon may also contribute to the bulk composition of the fluids in some high temperature areas. Isotopic studies indicate that the thermal water is of meteoric origin [8,9] excluding, of course, those waters which are substantially saline and are, therefore, partly or solely of sea water origin. Representative analyses of the various types of thermal waters are given in Tables 1 and 2. As can be seen from these analyses, the content of dissolved solids is typically of the order of 200-400 ppm in the low temperature areas where leaching alone accounts for the dissolved material. Comparable values for the waters of the high temperature areas are 700-1400 ppm. Of course, dissolved solids may be much higher in thermal water of mixed sea water-meteoric water origin whether high or low temperature.

Influx of sea water into hydrothermal systems is particularly noteworthy where the permeable late quaternary and pleistocene volcanics are located near the coast. Through reaction with hydrothermal minerals the ratio of the major cations in sea water (Na, K, Ca and Mg) are drastically changed in these thermal waters. Thus K^+ and Ca^{++} increase but Mg^{++} decreases much. Sulphate which is derived from the sea water may be largely removed from the water to form anhydrite, if its temperature becomes sufficiently high [10].

For the same temperature, waters issuing from acid rocks tend to be higher in dissolved solids than waters issuing from basaltic rocks. Fluoride is typically much higher in waters associated with acid rocks (see Table 1, analyses from Torfajokull and Geysir).

The influence of temperature upon the bulk composition of the water results from temperature dependent chemical equilibria between many of the dissolved components and hydrothermal minerals. Other dissolved components are not influenced by mineral solubilities or ionic exchange equilibria but by their rate of leaching from the rocks. There is a distinct tendency for leaching rates to increase with increasing temperatures so the concentrations of these latter mentioned components tend also to bear positive relation with underground temperatures.

Dissolved components whose concentration is dependent on mineral solubilities include silica, calcium, carbonate and sulphate. When water temperatures are above about 180°C, the dissolved silica concentrations are governed by quartz solubility but at temperatures below about 110°C chalcedony solubility governs the dissolved silica concentrations in solution. Insufficient data are available for the temperature interval 110-180°C to reveal whether chalcedony or quartz solubilities limit the concentration of the dissolved silica in the water. Equilibrium with both these minerals has been observed in drillholes but equilibrium with chalcedony appears to be more common (See Fig. 3). The thermal waters are invariably saturated with respect to calcite solubility so the formation of this mineral limits the concentration of calcium and carbonate in solution. Sulphate is limited by anhydrite solubility only in the high temperature waters. Waters from the low temperature areas are, as a rule, undersaturated with respect to anhydrite solubility and their sulphate content is considered to be governed by the rate of leaching from the rocks. More leaching occurs at higher water temperatures so there is a positive relation between underground water temperatures and the sulphate content of the low temperature waters. Fluorite saturation is also found only in the high temperature waters. When saturation is not reached, it is expected that the fluorite concentrations in the waters are governed by the rate of leaching from the rock.

TABLE 1

Chemical composition of selected waters from the high temperature areas.

Concentrations in ppm.

	Reykjanes well 8	Svartsengi well 3	Hveragerdi well 4	Krísuvík well 6	Námafjall well 4	Hveravellir Braedrahver	Torfajökull Eyrarhver	Geysir
Temp.°C	270	236	198	254	258	85	95	84
pH/°C	6.27/270	7.63	7.59/198	8.35/20	7.68/258	9.1/85	9.2/95	8.7/84
SiO_2	592	429	265	514	537	610	300	509
B	8.02	1.27	0.72	0.79	0.62	0.71	1.94	1.27
Na^+	9854	6322	142.0	700	129.9	161.0	372.0	209.0
K^+	1391	1012	11.4	119	22.5	18.2	23.9	22.4
Ca^{++}	1531	906	1.8	42.4	1.2	2.7	0.9	0.8
Mg^{++}	1.15	1.27	0.10	0.38	0.04	0.08	0.05	0.03
CO_2 total[1]	1437	174	210.1	62.2	82.8	34.1	60.4	134.6
SO_4^{--}	28.7	30.8	50.2	49.6	58.0	133.9	71.7	114.5
H_2S total[2]	31.5	4.3	30.1	6.6	158.1	4.1	24.9	0.7
Cl^-	18827	12635	97.7	1098	20.7	67.5	385.5	122.0
F^-	0.1	<0.1	1.8	0.5	1.4	3.6	27.0	11.5
Diss.solids	33653	22460	681	2605	956	1160	1351	1133

1) $H_2CO_3 + HCO_3^- + CO_3^{--}$

2) $H_2S + HS^- + S^{--}$

The composition of waters from wet steam wells with temperatures above 100°C are those of the water before flashing. The pH for these waters is computed.

TABLE 2

Chemical composition of typical waters from the low temperature areas in Iceland.
Concentrations in ppm.

	Laugarás	Deildartunga	Kristnes	Lýsuhóll	Selfoss	Seltjarnarnes
Temp.°C	98	99	75	40	79	83
pH/°C	8.8/98	9.70/13	9.89/24	6.5/40	8.2/79	8.9/83
SiO_2	127	125	90	187	81	110
B	0.32	0.29	0.26	1.06	0.63	0.47
Na^+	78.8	75.3	64.5	414.0	172.0	271.5
K^+	2.1	2.2	1.0	31.1	5.1	5.9
Ca^{++}	4.1	3.4	3.4	46.0	31.4	60.8
Mg^{++}	0.01	0.03	0.03	26.6	0.10	0.04
CO_2 total[1]	13.5	13.5	14.0	1518.3	19.3	4.5
SO_4^{--}	62.1	71.0	41.5	27.4	56.4	105.6
H_2S total[2]	0.7	1.7	<0.1	<0.1	0.2	<0.1
Cl^-	56.3	32.7	10.6	101.0	272.8	426.0
F^-	2.1	2.5	0.6	4.0	0.8	0.8
Diss.solids	372	358	250	1670	667	1111

1) $H_2CO_3 + HCO_3^- + CO_3^{--}$
2) $H_2S + HS^- + S^{--}$

Sodium, potassium, and magnesium concentrations are considered to be influenced by ionic exchange equilibria as has been demonstrated by experimental work and studies of thermal waters from hydrothermal fields outside Iceland [11,12,13]. The ratio of Na/K in Icelandic thermal waters corresponds well with equilibrium between alkali feldspars and sodium and potassium ions according to data of Helgeson [14].

The pH of the thermal waters is typically very high, or 9-10 when measured at 20°C. The saline waters have a pH of about 7-8. It appears that the pH is influenced by ionic exchange equilibria with alkali-bearing silicates and the pH or rather the hydrogen ion activity is proportional to the combined sodium-potassium activities.

It is clear that the atomic ratios of ions which combine to form hydrothermal mineral salts such as calcite or fluorite are dependent on temperature as well as rock type. For example calcium concentrations in the waters tend to decrease with increasing temperatures, but carbonate ion concentrations increase on the other hand. The difference between the influence of basaltic rocks on one hand and acidic rocks on the other is comparable with that of increasing temperatures.

The concentrations of hydrogen sulphide tend to increase with increasing water temperatures. In warm waters (temperatures less than about 50°C) the hydrogen sulphide is below the detection limit (0.1 ppm H_2S) but in waters feeding drillholes having temperatures of 250-270°C the hydrogen sulphide concentrations are as high as 250 ppm H_2S. The saline high temperature waters at Reykjanes and Svartsengi are, however, low in hydrogen sulphide relative to other waters with similar temperatures. This is considered to result from presumed higher oxidation potential of the saline waters.

In basaltic, as well as acid rocks, the sulphur concentrations are normally few hundreds of ppm and studies [15] indicate this sulphur is partly present as sulphate which replaces SiO_4^{-4} -tetrahedra in the frame structure of silicates, particularly feldspars. Some of the sulphur is also present as metallic sulphide, particularly in gabbroic rocks [15]. The sulphur content of subaerially erupted oceanic basalt is about 6 times lower than that of submarine erupted basalt according to Moore and Fabbi [16]. These authors consider that the difference is the result of retention of sulphur in basalt quenched on the sea floor and loss of sulphur by degassing at the surface. It seems likely that shallow level intrusions may expel sulphur during their relatively slow cooling and thus introduce this element into hydrothermal systems. It appears difficult to explain concentrations of as much as 250 ppm of sulphur mostly as hydrogen sulphide in some of the high temperature waters by leaching alone from rocks containing comparable concentrations of sulphur, and magmatic source for the major part of this sulphur is favoured.

It is evident that a considerable amount of sulphur has cumulated in the upper part of upflow zones of these areas but insufficient data are available to demonstrate quantitatively enrichment and/or depletion of sulphur in the rocks altered hydro- thermally by the sulphur rich waters. When magmatic source of some of the sulphur is postulated, the fact seems to be overlooked that the sulphur content of the water can be correlated with the rock type in the upflow zone and the water temperature in that zone. This correlation is believed to reflect an equilibrium condition that is not related to the source of the sulphur. Thus higher water tempera- tures increase the mobility of sulphur and basaltic rocks permit high sulphur mobility for high temperature waters. Other rock types that do not allow equally high mobility would cause cumulation of sulphur in the lower part of the hydrothermal systems if juvenile sulphur was brought into that system.

The problem with the geochemical behaviour of carbon dioxide in the thermal waters is similar to that of sulphur. Low temperature waters contain total carbonate which is of similar concentration as the carbon dioxide of ordinary ground water. Meteoric origin for the carbonate in these waters is therefore an obvious explanation. The tendency for increased concentration of total carbonate with in- creased water temperatures is explained by leaching of carbonate from the wall rock but it is natural to assume more effective leaching at higher water temperatures since the degree of rock alteration is distinctly related to these water temperatures. Some thermal waters are notably high in total carbonate. These include the sodium car- bonate waters of the Snaefellsnes peninsula in western Iceland and the high temperature brine at Reykjanes in southwest Iceland. The total carbonate content in the brine is about 1450 ppm and in the tepid waters at Snaefellsnes often between 1000 and 1500 ppm. To account for these high concentrations a magmatic or juvenile source for the carbonate is favoured.

Frequently, particularly for the hotter water, a separate gas phase is issued with the water, whether springs or drillholes. Accor- ding to [17], these gases are mostly carbon dioxide, hydrogen sul- phide, and hydrogen in the case of high temperature areas. Nitrogen, methane and argon are also present in lower concentrations (Table 3). In the low temperature areas nitrogen is practically the only gas. Carbon dioxide may though be present in small concentrations, parti- cularly in the hotter low temperature areas. The variation in the gas composition within and between high temperature areas can be re- lated to underground temperatures and flashing. Thus the hydrogen content of the gas tends to increase with increasing underground tem- peratures, probably as a result of equilibrium between sulphate, sulphide and water. Extensive steam formation by flashing under- ground is reflected in lowering of the carbon dioxide content of the

TABLE 3

Composition of typical gases from high and low temperature areas. Volume per cent.

	Námafjall well 4	Námafjall steam vent	Hveragerdi well 4	Hengill Middalur	Reykholt Skrifla	Laugaland Hörg.	Reykjanes steam vent
CO_2	30.3	34.7	82.2	89.6	13.1		94.3
H_2S	16.7	18.3	3.2	1.6	0.0		1.7
H_2	42.1	45.4	3.5	4.9	0.0		0.1
O_2	0.0	0.0	0.0	0.0	1.7		0.5
N_2	9.4	1.7		3.0	85.0	98.0	4.0
CH_4	0.5		0.3	0.7	0.3	0.02	0.01
Ar	0.3				2.3	2.0	
liters gas / kg steam	0.84/11°C		0.75/12°C				
steam fract.	0.15		0.14				

water relative to its hydrogen sulphide content. Extensive steam separation also reduces the nitrogen and hydrogen content of the gas.

This is so since these gases are only sparingly soluble in water and will therefore be transferred practically quantitatively to the steam that forms first. The total concentrations of gases in the steam of steam vents and drillholes depend, of course, on the degree of steam formation from the deep water. In wet steam drillholes, where the amount of steam separation is of the order of 20%, the total gas content is usually in the range of 0.5 to 5 liters of gas per kg of steam. In steam vents this content tends to be somewhat higher. There is a distinct tendency for the total gas content to increase with increasing feeding water temperatures of individual drillholes.

The low concentrations of hydrogen sulphide and hydrogen but high carbon dioxide in the high temperature fields at Svartsengi and Reykjanes in southwest Iceland are considered to be due to the salinity of the water. The saline water is presumed to be more oxidising than the typical low salinity high temperature water and it is this higher oxidation potential that is responsible for the low hydrogen and hydrogen sulphide.

Experimental work on interaction between sea water and basaltic rocks at elevated temperatures being undertaken at Harvard University indicates higher mobility of several transition metals in the heated sea water than is actually measured in the thermal brine at Reykjanes [18]. This difference is thought to be due to lack of chemical equilibrium in the experiments. The concentrations of the transition metals would be controlled by their rate of leaching from the rock. In the hydrothermal system at Reykjanes, the situation will be different as reflected by the hydrothermal mineral assemblage. It seems likely that the mobility of many transition metals in the Reykjanes thermal brine is maintained at low concentrations through equilibrium with the hydrothermal minerals.

It appears likely that heated sea water will not evolve into metalliferous fluid unless the salinity is greatly increased. In the highly permeable rock formations in Iceland and presumably also on the Mid-Atlantic ridge such increase in salinity of heated sea water is believed to be unlikely. Hydrothermal systems in these permeable rocks would be characterised by through-flow that does not permit any significant increase in the salinity of the water. If an impermeable cap rock sealed the hydrothermal field, for example sediments on the sea floor, an increase in salinity could occur through convection and successive evaporation of the rising water in the convection cell.

Sea water at 200-250°C that would be issued at the ocean bottom at some 1000 meters depth would be about 15% less dense than sea water at ordinary temperatures. It is, therefore, to be expected

that location of thermal sea water on the ocean bottom would be
difficult to discover, since this water would mix easily with the
sea water as a result of its lower density.

3. APPLICATION OF SILICA IN GEOTHERMAL STUDIES

It was pointed out previously that the concentration of dissolved
silica in thermal waters was either determined by the solubility

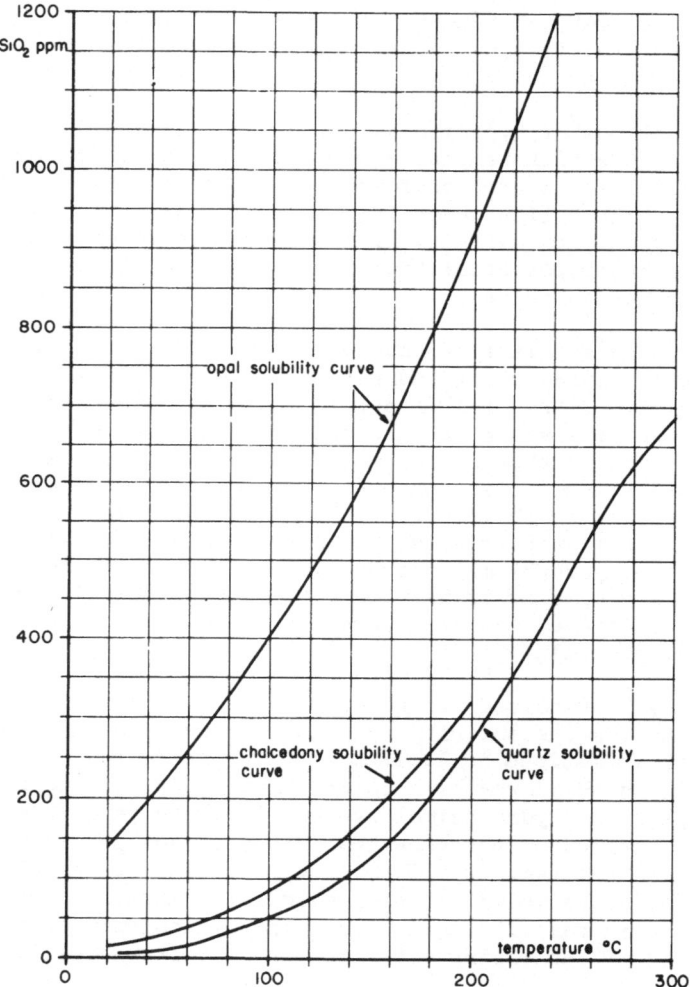

Fig. 2. The solubilities of quartz, chalcedony, and opal along the
three phase curve: Solid + water + steam. The data were derived
from [20-25].

of quartz or chalcedony. Silica in solution behaves like a weak acid, its first dissociation constant at 20°C being about 10^{-10}. It is only the undissociated fraction of the dissolved silica that participates in equilibrium with the solid phase [19]. The solubilities of quartz and chalcedony increase with increasing temperatures (Fig. 2). Therefore, the concentration of undissociated silica in the thermal waters increases with increasing water temperatures. The pH of the Icelandic thermal waters is frequently so high that a considerable fraction of the dissolved silica is dissociated. Reported analyses always give total silica, that is undissociated silica plus dissociated silica. In order to find out how much of the total silica in solution is involved in equilibrium with quartz or chalcedony, the dissociated fraction must be subtracted from the total silica concentration which is determined by analysis. The dissociated fraction is calculated from the analysed total silica content and the measured pH of the water. Temperatures derived from the concentration of undissociated silica in thermal waters assuming equilibrium with quartz or chalcedony are termed silica temperatures [26].

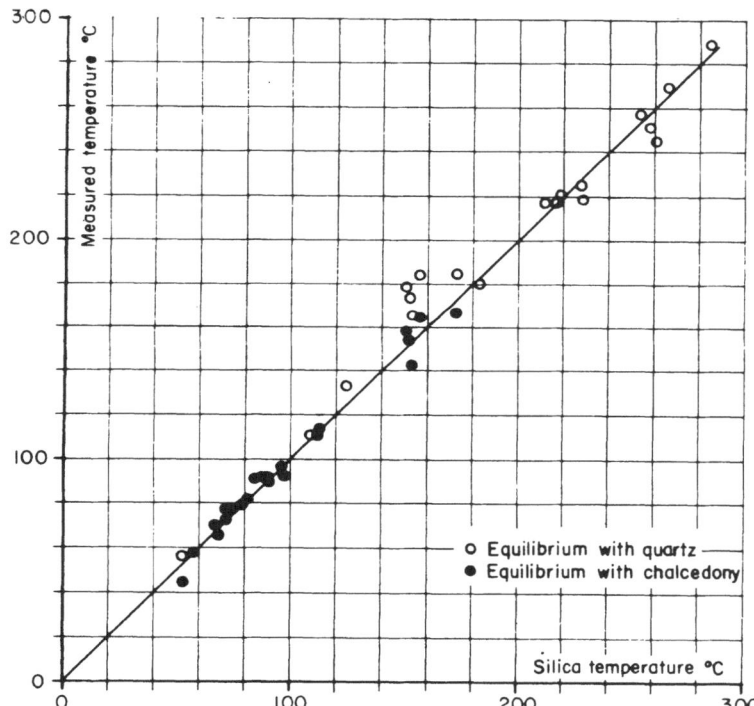

Fig. 3. Comparison between measured temperatures in deep drillholes and silica temperatures.

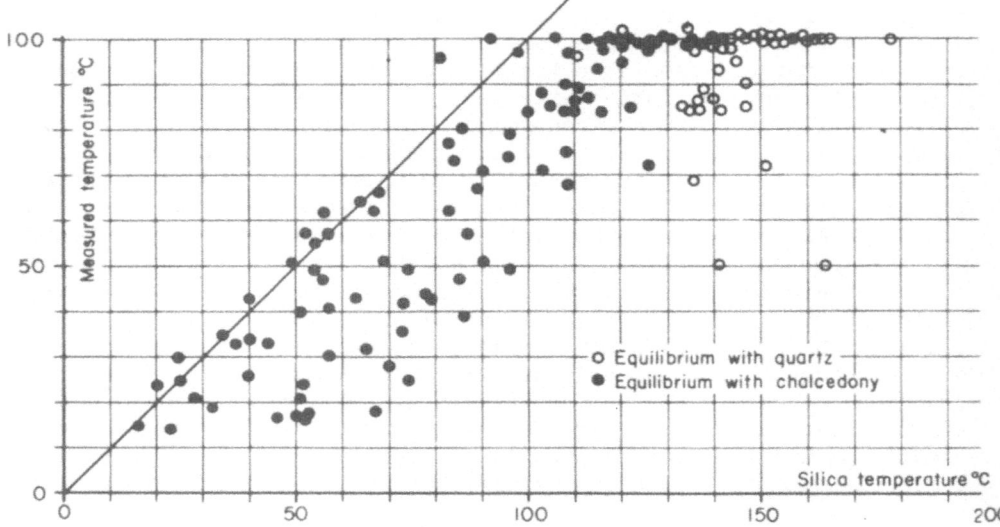

Fig. 4. Comparison between measured temperatures and silica temperatures in thermal springs.

Silica temperatures correspond well with measured temperatures in deep drillholes from high and low temperature areas in Iceland, if the silica temperatures indicate equilibrium with chalcedony when water temperatures are below about 110°C but equilibrium with quartz at temperatures above 180°C. (Fig. 3). Insufficient data are available for the temperature interval 110-180°C. Equilibrium with both quartz and chalcedony has been observed in drillholes in this interval although equilibrium with chalcedony appears to give a better fit. In contrast to the good comparison between measured temperatures and silica temperatures in deep drillholes, silica temperatures are generally higher than measured temperatures in thermal springs and shallow drillholes. In other words the waters in the thermal springs and the shallow drillholes are generally supersaturated with respect to both chalcedony and quartz solubility. In the low temperature areas, supersaturation by 10-40°C is common (Fig. 4) and may be as much as 150°C in the high temperature areas. This supersaturation results mainly from conductive cooling of the upflow water and very slow trend towards equilibrium between dissolved silica and chalcedony or quartz. As has been pointed out [26], supersaturation may also be caused by mixing with cold ground water in the upflow zone or lowering of pH, if the original pH of the water was sufficiently high for significant dissociation of the dissolved silica. Mixing with cold water and pH lowering does not, however, seem to be a common process, but if it occurs, the mixed waters can be identified by their lower content of dissolved solids, lower pH and lower temperatures as compared with other thermal springs in the neighbourhood.

The common silica supersaturation in thermal springs and shallow drillholes is taken to indicate that conductive cooling of upflowing water occurs frequently and that the silica can be applied to reveal the magnitude of this cooling and estimate what water temperatures would be encountered by drilling below this zone.

Estimated silica temperatures in those high temperature areas where thermal water springs occur lie in the range 174-274°C [27]. In the low temperature areas there is much variation in silica temperature within and between individual areas (Fig. 5). For the most active low temperature areas highest silica temperatures lie in the range of 126-178°C. There is no strong variation in highest silica temperatures of individual low temperature areas with distance from the active volcanic belts nor the age of the rocks from which the waters issue (Fig. 5). Yet the intensity of thermal

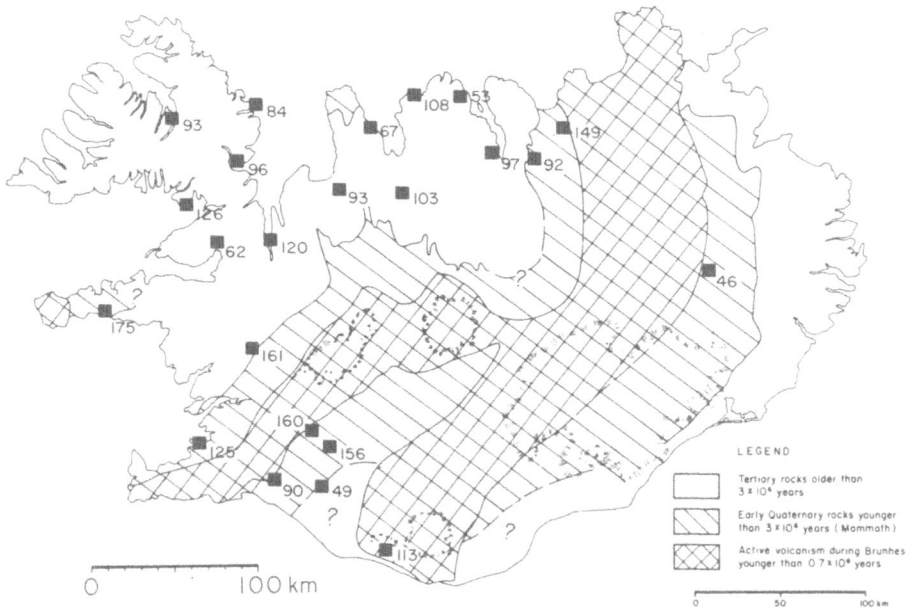

Fig. 5. Maximum silica temperatures in the various low temperature hydrothermal fields in Iceland.

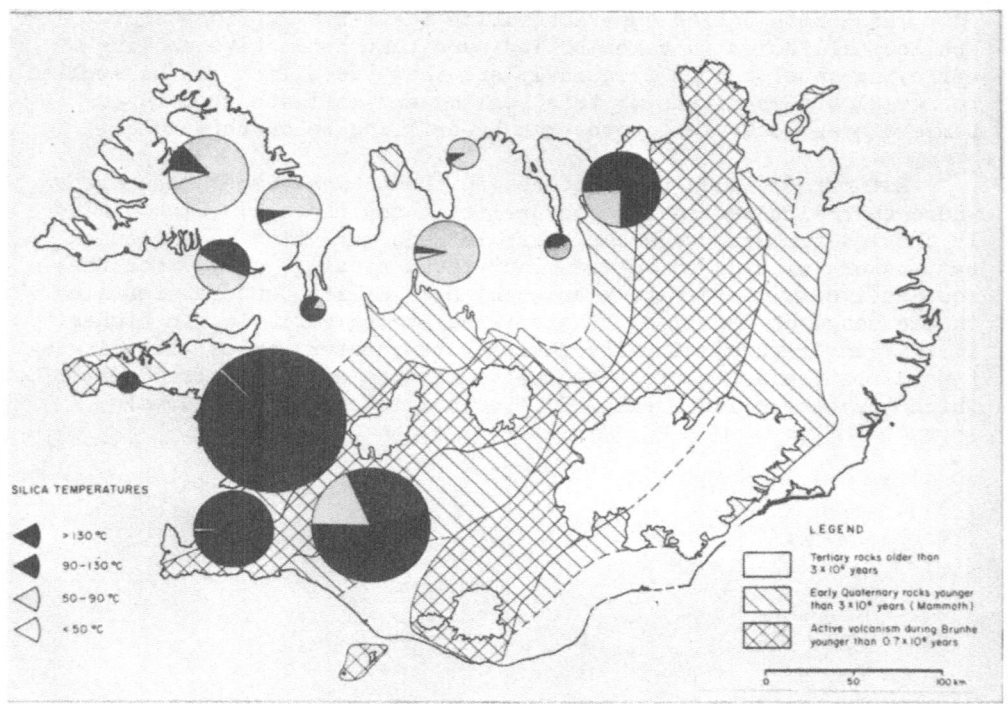

Fig. 6. Relation between silica temperatures, flow rates from springs, and age of rocks. The area of the circles is proportional to the flow rate.

activity, that is the combined magnitude of temperature and flow rate, in these areas shows a clear relation with the age of the rocks from which the waters issue (Fig. 6). This relation is considered to be a reflection of the regional thermal gradient and the permeabilities of the rocks. Both these variables increase towards the active volcanic belts and they constitute the prime factors necessary for the existence of surface hydrothermal activity.

REFERENCES

1. G. Bodvarsson, Physical Characteristics of Natural Heat Resources in Iceland, U. N. Conference on New Sources of Energy,
   Rome, 1961.
2. G. P. L. Walker, Quart. J. Geol. Soc. Lond., 119, 29, 1963.
3. K. Saemundsson, Natturufradingurinn, 42, 81, 1972.
4. G. P. L. Walker, Bull. Volcanologique, 29, 375, 1966.
5. H. Sigurdsson, Soc. Sci. Islandica, Greinar IV, 2, 1966.
6. T. Einarsson, Uber das Wesen der Heissen Quellen Islands, Soc.
   Sci. Islandica, 26, 1942.
7. K. Saemundsson, Namafjall-Krafla, Progress Report on the Exploration of the Geothermal Fields (In Icelandic), unpublished
   report of the National Energy Authority, 1972.
8. G. Bodvarsson, Jokull, 12, 49, 1962.
9. B. Arnason and Th. Sigurgeirsson, The Use of Hydrogen Isotopes
   in Hydrological Studies in Iceland, Paper submitted at the
   Hydrological Conference in Vienna, 1967.
10. S. Bjornsson, S. Arnorsson and J. Tomasson, Bull. Am. Ass.
    Petrol. Geologists, 56, 2380, 1972.
11. A. J. Ellis and W. A. J. Mahon, Geochim. Cosmochim. Acta, 28,
    1323, 1964.
12. A. J. Ellis and W. A. J. Mahon, Geochim. Cosmochim. Acta, 31,
    519, 1967.
13. R. O. Fournier and A. H. Truesdell, Geochim. Cosmochim. Acta,
    37, 1255, 1973.
14. H. C. Helgeson, Am. J. Sci., 267, 729, 1969.
15. W. Ricke, Geochim. Cosmochim. Acta, 21, 35, 1960.
16. J. G. Moore and B. P. Fabbi, Contr. Mineral. and Petrol., 33,
    118, 1971.
17. G. E. Sigvaldason, Bull. Volcanologique, 29, 589, 1966.
18. M. Mottl, Harvard University, U.S.A., written comm., 1974.
19. G. B. Alexander, W. M. Heston and H. K. Iler, J. Phys. Chem.,
    58, 453, 1954.
20. G. C. Kennedy, Econ. Geol., 45, 629, 1950.
21. G. W. Morey, R. O. Fournier and J. J. Rowe, Geochim. Cosmochim.
    Acta, 26, 1029, 1962.
22. R. O. Fournier, Proceedings of the International Symposium on
    Hydrochemistry and Biochemistry, Japan 1970, 1, 122, Clark,
    Washington D.C.
23. R. O. Fournier and J. J. Rowe, Am. Mineralogist, 47, 897, 1966.
24. S. Kitahara, Rev. Phys. Chemistry, Japan, 43, 131, 1960.
25. C. S. Hitchen, Inst. Mining and Metallurgy Trans., 44, 255, 1935.
26. S. Arnorsson, in press, 1974.
27. S. Arnorsson, Geothermics, - sp. issue 2, 2, 536, 1970.